DEVELOPING NEW FOOD PRODUCTS

for a

CHANGING MARKETPLACE

DEVELOPING NEW FOOD PRODUCTS
for a
CHANGING MARKETPLACE

Edited by
Aaron L. Brody, Ph.D., C.P.P.
John B. Lord, Ph.D.

CRC PRESS

Boca Raton   London   New York   Washington, D.C.

Library of Congress Cataloging-in-Publication Data

Main entry under title:
Developing New Food Products for a Changing Marketplace

Full catalog record is available from the Library of Congress

This book contains information obtained from authentic and highly regarded sources. Reprinted material is quoted with permission, and sources are indicated. A wide variety of references are listed. Reasonable efforts have been made to publish reliable data and information, but the author and the publisher cannot assume responsibility for the validity of all materials or for the consequences of their use.

Visit the CRC Press Web site at www.crcpress.com

© 2000 by CRC Press
Originally published by Technomic Publishing

No claim to original U.S. Government works
International Standard Book Number 1-56676-778-4
Library of Congress Card Number 99-66882
Printed in the United States of America   3 4 5 6 7 8 9 0
Printed on acid-free paper

Table of Contents

v

Foreword

Fᴏᴏᴅ product development is an excellent idea, and Developing New Food Products for a Changing Marketplace is an excellent book. The concept of gathering all of the pieces of product development together, integrating the portions in such a way that the silos that affect most development groups are eased carefully together is most welcome, and most unusual. The concept of discipline, applied to new food products, too often is translated to mean keeping "those guys" in line during development, marketing, etc.

The authors note that the emphasis for new products is on speed and that life cycles are shorter than ever. So the concern is that more new products will enter the marketplace, fail faster, and convince those operating in the new food product arena that the economic safety of mundane line extensions precludes real innovation. It is good to see prominent chapters on product policy and goals and on the role of business strategy, product portfolios, and product selection. The concept of a definition of a new product is refreshing: so is the statement that new product development is not a rationale to try to fool consumers.

A number of gurus participated in preparation of the book; including people from industry—notably Bob Smith, John Finley, Alvan Pyne, and others—and from academia and consulting—particularly John Lord and Aaron Brody. Although a certain amount of theory is included, it is exceptionally well backed up with real-life examples. It ought to be must reading for food science majors, as well as those pursuing any role in the food sector. The changes in this industry have been massive during the last several years, and have greatly affected the role of new products and of altered standard products as well.

Undergraduate and graduate food science students are often poorly prepared for the business of food science and often become disenchanted with the compromises that occur during the process of product development and improvement. Reading this volume should help prepare students and beginning scientists for the requirements of business and assist them in developing products that can be effectively marketed and processed. Anything that can help reduce the rate of failure in new products will be welcomed by everyone. This book just might do the job.

FRANCES KATZ, PH.D.
Director of Publications
The Institute of Food Technologists

Foreword

THERE'S plenty to say on the positive side of this new book, a true labor of love for Drs. Brody and Lord. In fact, the only negative thing I can think of is that packaging, as usual, is sort of at the end of the line. Of course, nearly one-quarter of the book is devoted to packaging—a great acknowledgment of the role of packaging in food product development. The first serious treatment of packaging doesn't occur until Chapter 8, and then packaging all but disappears again until Chapters 13 to 16. Such placement, perhaps done with some sort of acceptable logical organization in mind, nonetheless perpetuates a view of the world which is not productive: namely, that packaging and product are somehow distinctly different, and separate, considerations. First do the product. Then drop it in a box.

However, the real point should be that there is actually just one multifaceted personality on which to work in food product development. Ultimately, after all, the effort results in just one presentation to the consumer. It's often said that the package sells the product the first time; the product sells itself after that. Failure to understand and act on this simple premise may well account for a lot of product failures in the first place. Whether the package fails to catch the consumer's eye or provide an assurance of quality or prove its utility in the home after opening, the result is the same: the consumer makes another choice at the next opportunity.

The product and package is, at best, a coordinated, complementary system. Like other systems, it must start at the beginning, in the earliest planning stages, if it is to be successful at the end.

Whereas the definition of product is fairly straightforward, not so with packaging. Are we referring to the retail shelf package? Yes. To its graphic design? Yes. To the structural design engineering? Yes. To the distribution package? Yes. To the machinery line system which will produce, fill, and seal the product? Yes! Nothing in the world of packaging is simple, not even the apparently simple things. The trick is to provide an economically favorable packaging system which protects, preserves and communicates for the product. Material selection, machinery line design, graphic communication, and more all eventually have nearly equal status and, to the point, require informed marketers and technologists to make the final decisions.

Dr. Brody is one of the few individuals practicing today who truly understands the necessary melding of left and right brains, of creativity and engineering, of structural integrity and consumer preference in packaging decisions. Together with Dr. Lord, the two have assembled a ''who's who'' lineup of authorities to make this work broader in coverage, deeper in appropriate technology, more ambitious in scope, and fundamentally more useful than any previously recorded documentation of food product (and packaging) development. I would expect that of them. Whether pulling from the packaging design world with Roy Parcels, from the packaging regulatory world with Jean Storlie, or from the packaging engineering and technology world with Dr. Gordon Robertson and Dr. Steve Raper, Drs. Brody and Lord have gone to the top of the heap and pulled from the best. This book should be the centerpiece in the library of any food product development team, anywhere in the world.

WILLIAM C. PFLAUM
Executive Director
Institute of Packaging Professionals

Foreword

SAINT Joseph's University has had a long and very successful relationship with the food industry. Starting as a part-time program in 1960, the undergraduate food marketing major has become the University's largest and most well-known program, graduating hundreds of young men and women who have assumed positions of great responsibility and importance throughout the food industry. The Department of Food Marketing graduates approximately 120 students each year in our undergraduate and graduate food marketing programs.

The external recognition of our food marketing program comes in the form of the dozens of companies who each year recruit our graduates, the firms who sponsor their managers in our executive programs, and funding by the U.S. Department of Agriculture of the Center for Food Marketing which is housed, along with the Erivan K. Haub School of Business, in our newest campus building, Mandeville Hall. Both the Haub family and the Connelly/Mandeville families have vital stakes in the global food business. Their support is testimony to the value of our educational programs to that industry.

Given our commitment to the food industry, it comes as no surprise that John Lord and Aaron Brody have embarked upon this effort to pull together some of the leading food industry experts and create a truly cross-functional book describing the complex process of developing new food products. One of the hallmarks of Saint Joseph's University is our commitment to the liberal arts and sciences. This text brings together the sciences and business in a unique way and demonstrates the importance of a broad-based approach to education and to business strategy.

As a University president, teacher of executive students and author, I believe that *Developing New Food Products for a Changing Marketplace* represents an important advancement for the food industry and for education. The opportunity to bring food and packaging technology into a marketing classroom, and the equivalent opportunity to bring marketing into a food science classroom, should be greatly beneficial to students as well as to their teachers. Industry professionals who make up the cross-functional teams used to create and execute new products in this changing marketplace will likewise benefit from the combination of science and business represented herein.

NICHOLAS S. RASHFORD, S.J.
President
Saint Joseph's University

Preface

HARD to believe, but once upon a time, not so very long ago, new food products were the exclusive province of the white coat clad, bench technical people in the recesses of their laboratories. They brewed their liquids, mixed their ingredients, blended their flavors, heated and cooled, conducted their private taste panels and proclaimed, "This is our company's new food product! Go forth and sell it!"

Difficult to conceive, but ignoring both consumers and their marketing representatives within the corporate hierarchy was common practice, and in some companies new food products continue to be the technically driven.

At the opposite pole were companies that employed hired MBA-guns to "develop" new food products. To them, the transformation of their pet focus-group-tested concept into an edible reality was a simple overnight exercise by a lab "technician." That type of thinking still prevails in many places.

We cannot possibly forget the place of packaging in this scenario: after the product had been fully developed and the manufacturing folks had started engineering the line, someone remembered that this product must be thrown into a package ". . . and it has to be done by a week from Tuesday to meet the sales promotion schedule which coincides with the new TV ad campaign on the first pitch of the opening game of the World's Series . . ." "oh, and don't forget to call in the ad agency this afternoon to design the package . . ."

"Who did the shelf life testing?"

"Where are the market launch print insert materials?"

"Has anyone checked with the legal department to find out if this new plastic is FDA approved and if the package copy claims are O.K.?"

"... I don't care, the price has to be $x. You are going to have to find a way to trim costs ..."

Never happened ... can't possibly happen now ... not in our company. Don't you believe it!

When John Lord and Aaron Brody sat down at Saint Joseph's University Department of Food Marketing back in 1994, Professor Lord was teaching a senior course in Product Policy and Aaron Brody was consulting in food product and packaging technology. Product Policy was a marketing course covering what were then perceived to be indispensable food product development elements: identifying consumer needs, concept generation, concept evaluation, screening, targeting, product protocol, launch strategies, marketing research, advertising and promotion, post launch, and so on. No technical development, no packaging, nothing that smacked of nutrition or chemistry or shelf life.

Other universities were teaching food science and technology students about product formulation, pilot processing, flavor panel testing, instrumental analyses, and quality assurance. Hardly a nod toward consumers, marketing research and analysis, product positioning, or anything that resembled marketing.

At some Schools of Packaging students were learning package structural design, testing, materials, permeation, shelf life, migration protection, and all those other subjects that together summed to the physical package. Graduate Schools of Business Administration were expounding on theories to accelerate the new product development process. Schools of Art were mentioning package design as an afterthought, as if it were a wayward son, and ... but we imagine that you can imagine the situation back then ....

That day in Philadelphia, John and Aaron conceived a revolutionary paradigm to teach product policy to Food Marketing majors as the food product development integral of food marketing, food technology, and food packaging. Everyone in class might not become an expert in every facet, but each would know that many components together constitute food product development—and would be at least cognizant of the various inputs required.

We searched for a textbook for our students. We scoured the classics in product development, packaging, marketing, and food technology but could not a find a single book that captured the universe of coherent new food product development as we envisioned it.

Our solution was simple: for each class session, we prepared notes from our experience (and we certainly had lots, good and bad) and copied selected references from the literature (probably violating all sorts of copyright laws and rendering ourselves as criminals) and recited from the gospel according to the writers. We also invited a host of experts from industry to demonstrate and lecture on their specialties. Professionals conceived and formulated products before the students' eyes, showed slides and videos of their new food product development projects, and then asked the students to taste their outrageous foods. They ranted over distribution, went into rapture over the launch tactics and sampling, pontificated on nutritional values, and on and on it went.

A few students were bored; some were overwhelmed; some were underwhelmed; but a majority (most of whom are now practicing what they learned with food companies and consultancies) grasped the concept that new food product development was a systematic integration of many diverse disciplines. They realized that the too-high percentage of new food product failure may, in large measure, be attributed to a paucity of participatory elements, and not just a failure to identify consumer needs or to meet consumer expectations. This Product Policy course highlighted all the elements, and enhanced and enriched them with anecdotes and sea stories from the salts who had lived them. The course was capped by a term project in which student teams developed a new food product with its marketing plan, including a budget, packaging, formulae, consumer test results and all the rest of the elements that would make for a successful new food product.

The course was a success and continues to be taught, not only to seniors, but also to executives in a weekend education program, as a short course with the University of Georgia, and in conjunction with Pennsylvania State University.

The course has been taught without a text—until now. Aaron and John invested several years of our lives formalizing our notes and persuading our colleagues to contribute to this book. We have chapters from some of our original course lecturers, such as Stan Segall from Drexel University on laboratory procedures and Roy Parcels, a Packaging Hall-of-Famer on package design. We have some of the real pioneers, such as Bob Smith, the force behind Nabisco SnackWell's®, on new food product organization, Al Pyne from CPC International on laboratory testing, and Howard Moskowitz on sensory testing. We have some world class pros, such as University of Georgia's Romeo Toledo on food science, Eric Greenberg on legal issues and Tetra Pak's Gordon Robertson on shelf life testing. We have Jean Storlie, on labeling, Liz Robinson of New Product Sightings on the need for companies to introduce new products, Purdue University's John Connor on the framework of our food industry, and University of Missouri's Steve Raper and Arthur D. Little's Marv Rudolph on quality function deployment. No more knowledgeable and seasoned group could have been assembled to define new food product development.

We are proud of this accomplishment. Our contributors are proud of their participation. We all hope that we have interpreted our market accurately and have targeted this market with a product that meets its needs and fulfills its promises. We hope this will become the standard food product development reference for food industry marketing, new product, brand and product development executives and practitioners as well as the standard textbook in university business and food science and technology departments.

This book is a complete manual of theory and practice for food product development. It follows, therefore, that this preface also must be complete, with our gratitude expressed to all who made the book possible: our students, academic colleagues, and professors and the professionals with whom we debated and from whom we grew; Joe Eckenrode and Eleanor Riemer, our publishers who pushed us all the way, and Wendy Peeples Hill, who labored so marvelously with us on every detail.

However, as with all lifetime endeavors, our greatest thanks go to our colleagues and families who surrendered some of themselves to ensuring that we did indeed finish this work.

John wishes to acknowledge Ned Dunn, C. J. McNutt Professor of Food Marketing and former CEO of Harris Teeter, and Richard Kochersperger, Director of the Center for Food Marketing, both at Saint Joseph's University, for their valuable insight and comments. I want to thank food marketing students at Saint Joe's for the past 24 years (and the next 24 years) without whom there would have been little motivation to pursue this project. Mostly, I want to thank my wife Joan for her outstanding patience during the many weekends and evenings that this text consumed (the second time I have asked her to do this, because she also saw me through my doctoral education). I want to also thank my daughter Megan, who has decided to pursue a career in biology but who nonetheless helped to edit several chapters of the book, and to acknowledge my sons, Sean and Ryan, who had to endure some (but not really all that many) hockey, basketball, and other games without dad present because he was working on "that book."

Aaron's wife Carolyn has known so many of these 25-hour days from her husband but remains forever "my beloved and my beloved is thee." Aaron's eldest son Stephen now appreciates the body of work generated by his dad; and his wife Susan who is so devoted to excellence, their children and our wonderful grandchildren, Michelle Jennifer and Derek Jason. Our middle son Glen has matured into a master and his wife Sharon is one of the great mothers of history to their children and our so active grandchildren, Camryn Alexander and Skyler Alexis. Our baby son, Robyn is so dedicated to performance, his consummately professional wife Kellie, and their daughter and our granddaughter Natalia Sienna, and their son and our grandson Pierce Aaron, who had the good sense to appear in time to have his name included in this book. I love you all plus qu' hier et moins que demain—toujours. Thank you all for being my family.

AARON L. BRODY
JOHN B. LORD

Foreword Authors

Frances Katz, Ph.D.
Editor, *Food Technology*
Director of Publications
Institute of Food Technologists
221 N. LaSalle St.
Chicago, IL 60601

William Pflaum
Executive Director
Institute of Packaging Professionals
  Education Forum
481 Carlisle Dr.
Herndon, VA 22170

Nicholas S. Rashford, S.J.
President
Saint Joseph's University
5600 City Ave.
Philadelphia, PA 19131-1395

Bruce R. Stillings, Ph.D.
President, 1998-1999
Institute of Food Technologists
221 N. LaSalle St.
Chicago, IL 60601
President
Food & Agriculture Consultants,
  Inc.
482 Blackstrap Rd.
Falmouth, ME 04105

Contributing Authors

Aaron L. Brody, Ph.D., CPP
Managing Director
Rubbright·Brody, Inc.
P.O. Box 956187
Duluth, GA 30095-9504
Adjunct Professor
University of Georgia
Department of Food Science &
  Technology
Athens, GA 30602-7610
Adjunct Professor
Saint Joseph's University
5600 City Ave.
Philadelphia, PA 19131-1395

John M. Connor, Ph.D.
Professor
Agriculture Economics Department
Purdue University
1145 Krannert Bldg.
West Lafayette, IN 47907-1145

John W. Finley, Ph.D.
Senior Technology Principal
Kraft Foods
800 Waukegan Rd.
Glenview, IL 60025

Eric F. Greenberg
Partner
Ungaretti & Harris
3500 Three First National Plaza
Chicago, IL 60602-4283

John B. Lord, Ph.D.
Academic Associate Dean, Graduate
  Education
Haub School of Business
Saint Joseph's University
5600 City Ave.
Philadelphia, PA 19131-1395

Howard R. Moskowitz, Ph.D.
Principal
Moskowitz/Jacobs, Inc.
1025 Westchester Ave.
White Plains, NY 10604

J. Roy Parcels
Dixon and Parcels Associates, Inc.
19 W. 44th St.
Studio 1010
New York, NY 10036

Alvan W. Pyne, Ph.D.
A. W. Pyne & Associates
Technology Assessment/Brokering
215 Brookvale Rd.
Kinnelon, NJ 07405

Steven A. Raper, Ph.D., CPP
Associate Professor
Dept. of Engineering Management
University of Missouri-Rolla
204 Engineering Management Bldg.
1870 Miner Cir.
Rolla, MO 65409-0249

Gordon L. Robertson, Ph.D.
Tetra Pak Asia
P.O. Box 145 Jurong Point
Singapore 916405

Liz Robinson
President
Platinum Communications
7900 S. Cass Ave.
Dariel, IL 60561

Marvin J. Rudolph, Ph.D.
Director of Food Product and
  Process Development
Arthur D. Little, Inc.
15 Acorn Pk.
Cambridge, MA 02140

Stanley Segall, Ph.D.
Professor of Nutrition and Food
  Science
Department of Bioscience and
  Biotechnology
Drexel University
Philadelphia, PA 19104

Robert E. Smith, Ph.D.
955 Eagle Point Rd.
Newport, VT 05855

Jean Storlie, MS, RD
15001 64th Ave. N.
Maple Grove, MN 55311

Romeo T. Toledo, Ph.D.
Professor
Department of Food Science &
  Technology
University of Georgia
Athens, GA 30602-7610

CHAPTER 1

The United States' Food Industry and Its Imperative for New Products

AARON L. BRODY
JOHN M. CONNOR
JOHN B. LORD

America's food industry is a large industry that grows in total volume only with the population. Within that slowly growing volume, however, is a powerful dynamic of change. As population segments change in age, lifestyles and ethnic make-up, their eating habits change. To meet the ever-changing needs and desires of consumer groups, the food industry's delivery systems change—and are faced with challenges of quality, safety, consumer satisfaction, economics and simply the ability to perform. Agriculture, food processing and distribution have undergone profound changes, not the least of which is the imperative to develop and offer more and different products for their changing marketplace—a formidable opportunity for many.

THE United States' food processing industries often appear to be taken for granted—a feature of the economic landscape so unremarkable as to be nearly invisible. Our food industries may be viewed as commonplace because the methods of production are in some cases quite ancient, or because food factories appear to be organized on a small scale. It may be that processed food products seem so familiar as to be humdrum, or that processing is believed to be merely a modest extension of agriculture or the consumer's own kitchen. Or perhaps our product development may be chaotic or even archaic.

Some operations employed by food processors, as, for example, flour milling, are indeed prehistoric in their origins, but the methods and equipment in use today bear little resemblance to colonial grist mills. Most food processing technologies in place today are the result of modern scientific discoveries and decades, if not centuries, of technological refinement. And even if our food processes do not match the technological elegance of chemical or pharmaceutical processing, they are still far ahead of the remainder of the world—and thus represent a major planetary resource to feed its population.

1

The majority of our country's 15,000 food processing plants are quite small (fewer than twenty employees). Yet there are at the same time some food factories that rival the nation's largest in size. Small scale at the plant level belies a mode of business organization at the corporate level that is as modern as any in the manufacturing sector, if not more so. Scores of raw and semi-processed foodstuffs must be assembled to finish a given processed food, and each input most likely has a unique procurement system. Once finished, the typical food or beverage is distributed through multiple channels, each of which has special business practices, and, all the while, new products are being fashioned and introduced into the network to satisfy the almost-insatiable demands of an even-widening consumer population.

Food processing was traditionally closely linked to agriculture or domestic household activities, and in many university settings still is. Many processing industries were originally part of farm operations (such as butter or cheese making) or were skills found in domestic kitchens (pickling or baking). Contemporary food processing is today similar to all other manufacturing, if not as sophisticated. The rupture between farming and processing is not yet complete, but it has gone far. It is not impossible to find otherwise-educated consumers who believe that chicken breasts originate in the backrooms of supermarkets or that mushrooms are grown in cans—although sometimes, one wonders.

As we shall demonstrate, the demand for the food processing industry's output is derived from the changing needs and wants of consumers. Hence, the nature of household demand for food shapes the strategies, activities, and market channels used by food processors. As consumers have become more numerous and diverse, market channels have evolved to cater to emerging segments. Attention has been devoted to these changes, especially those in food retailing and hotel/restaurant/institutional (HRI) or food service. In addition to reaching consumers through the appropriate marketing channel, it is of paramount importance to have the right products and to communicate the availability and attributes of these products to consumers.

Designing strategies to meet consumer needs does not occur in a vacuum. Product development and manufacture is necessarily dependent on the availability of inputs and the means by which both the cost and quality food processing inputs can be managed. Because food is such a basic human need, its development, production, marketing, and now safety have long been considered areas of vital national interest. Consequently, a myriad of regulations has evolved that covers many aspects of food processing and marketing and especially safety.

DEFINITIONS: WHAT IS FOOD PROCESSING INDUSTRY?

Commercial food processing is the branch of manufacturing that starts with raw animal, vegetable, or marine materials and transforms them into intermediate foodstuffs or edible products through the application of labor, machinery, energy, and scientific knowledge. Various processes are used to convert relatively bulky, perishable, and typically inedible agricultural and some other materials into ultimately more useful, convenient, waste-free, relatively shelf-stable, and palatable foods or potable beverages. Heat, cold, water removal, chemical and biological reactions, and other preservation techniques are applied to enhance distribution. Packaging confers portability as well as extending shelf life by protecting against the environment. Changes in product forms often reduce preparation time for consumers, as will be shown. Increasing palatability, storability, portability, and convenience are all aspects of ''adding value.'' In other words, food processors utilize factory systems to add economic value by transforming raw materials grown on farms or fished from the sea into useful products. Steers become meat; wheat becomes flour; corn becomes fructose; and tuna becomes canned tuna.

The principal and unique economic function of food processing is to convert various food materials into finished, consumer-ready products, but the economic contributions do not end there. Food processors perform many value-added economic functions that are shared with other food marketing organizations: farm product assemblers, grocery wholesalers, food transporters, food retailers, and food service (HRI) operators. For example, food processors add value by transforming products through space and time. That is, most processors are willing to deliver or arrange for delivery of their finished products to grocery wholesalers or retailer warehouses or even outlets. Moreover, the procurement operations of food manufacturers arrange for the orderly flow of the many material inputs and supplies required to manufacture foods. Often, although much less than in the past, food processors will maintain a significant inventory of material inputs and finished goods; in some food processing industries, such as canned vegetables, processors typically have had a year's supply of canned goods on hand at the end of the canning season, although this is changing with global raw material sourcing, financial pressures, and just-in-time operations. Many food processors act as wholesalers by purchasing finished products from other processors for resale. In short, processors perform several distribution functions essential for ensuring a steady food supply throughout the year and in all parts of the country.

Perhaps more important than physical storage and movement is the information function of food processing firms. Because of their central position in the U.S. food system, food processors have abundant, and at times unique, access to sources of information on the quality, quantities, and prices of processed foods. Food processors, especially those specializing in making semiprocessed food ingredients, spend significant corporate resources collecting, studying, and forecasting information on agricultural supplies in their region or possibly worldwide. The collective judgment of purchasing agents in a given food processing industry can quickly drive farm prices up or down.

Other food processors, particularly those manufacturing consumer products, employ teams of analysts that are experts on business practices in the food distribution industries and that follow consumer expenditure trends in minute detail. What is not generated internally within food processing companies can be purchased from specialized grocery information consulting firms.

The information collected about consumer demand and agricultural supply conditions comes together at the processing nexus. Tight supplies for a given food product at the retail level eventually translate into higher processor prices, a greater willingness to pay for key inputs, and a price signal to farmers to expand production or sell off their stored crops. It works in the other direction, also, with an unexpectedly short crop impelling manufactures to encourage consumers to reduce their purchases by raising prices on those food products that incorporate the crop. In short, processors are in a strategic position to use price increases to signal lower-than-expected supplies to farmers, consumers, and all the other middlemen in the food system; unexpected abundance is indicated by holding prices steady, price cuts, or more generous deals to retailers.

In addition to price-quantity information, processors are repositories of knowledge about food and agricultural product quality: fine gradations in the flavor of coffee beans, the moisture content of a shipment of corn, the shelf life of products under various handling conditions, and nutritional characteristics, to name a few. Thus, food processors are typically in the sole position to formulate and design foods, taking into account consumer preferences, distributors' demands, ingredient availability, scientific knowledge of biological properties, technological feasibility, and profitability—a complex but critical task. Processors share some of their food quality information through ingredient labeling, nutrition labeling, and other programs that assist consumer choices. Through quality control and product testing, food processors assume much of the responsibility of protecting the safety of the nation's food supply.

FOOD EXPENDITURES—AN OVERVIEW

Total expenditures for food in 1995 climbed 4% to $669.4 billion. At-home food expenditures rose 3.4% to $360.1 billion representing 53.8% of total food expenditures, while food away-from-home spending rose 4.7% to $309.6 billion accounting for 46.2% of total expenditures, and rising (see Table 1.1).

These United States Department of Agriculture figures include expenditures for food at all retail outlets (food stores, restaurants, etc.) as well as "service" establishments (meals at lodging places, snacks at entertainment facilities for example) plus allowances, on a retail dollar equivalent basis, for food served in institutions (schools, hospitals, etc.), in the travel industry (airline meals for example), and in the military. It covers expenditures by individuals, families,

TABLE 1.1. Expenditures for Food: 1976–1995.

Year	Food at Home* (Millions)	Food Away from Home** (Millions)	Total	Percent of Total Home	Percent of Total Away
1976	$127,835	$76,833	$204,668	62.5%	37.5%
1977	137,332	84,835	222,167	61.8	38.2
1978	151,083	96,084	247,167	61.1	38.9
1979	168,203	109,171	277,374	60.6	39.4
1980	185,788	120,296	306,074	60.7	39.3
1981	198,283	130,914	329,197	60.2	39.8
1982	205,583	139,776	345,369	69.5	40.5
1983	216,390	150,883	367,273	58.9	41.1
1984	227,547	161,046	388,593	58.6	41.4
1985	235,687	168,831	404,518	58.3	41.7
1986	244,431	181,695	426,126	57.4	42.6
1987	254,612	200,226	454,838	45.0	44.0
1988	267,515	218,567	486,092	55.0	45.0
1989	286,557	232,708	519,265	55.2	44.8
1990	311,584	249,205	560,789	55.6	44.4
1991	324,599	256,531	581,130	55.9	44.1
1992	326,419	266,010	592,429	55.1	44.9
1993	332,217	281,965	614,182	54.1	45.9
1994	348,342	295,285	643,627	54.1	45.9
1995	360,053	309,306	669,358	53.8	46.2

* Food for off-premise use.
** Meals and snacks.
Source: USDA.

businesses, and government. It does not include expenditures for alcoholic beverages sold at stores or restaurants.

Food expenditures in 1996 indicate total spending advanced at a more modest 2.4% to $685.6 billion of which food at-home was $370.1 billion or 54% of the total (up 2.8% from 1995) and food away-from-home was $315.5 billion or 46% of total expenditures (up 2.0% from 1995). See Tables 1.2 and 1.3.

FOOD—A SMALL SHARE OF DISPOSABLE INCOME

Expenditures for food continue to account for a shrinking portion of disposable personal income as shown in Table 1.4. In 1995, total food spending accounted for only 11% of disposable income (the lowest percentage in world history). Of that 11%, food at-home accounts for 6.7% (down from 6.9% in

TABLE 1.2. Food-at-Home Expenditures, Food-at-Home Price Changes—
1970/71 to 1995/96.

Annual Percent Change	Expenditures	Prices*
1970/71	+5.1%	+2.5%
1975/76	+6.7%	+2.1%
1980/81	+6.7%	+7.2%
1985/86	+3.7%	+2.9%
1990/91	+4.2%	+2.6%
1995/96	+2.8%	+3.7%

* Based on Consumer Price Index for food-at-home.
Source: USDA.

1994 and 6.8% in 1993), whereas food away-from-home took 4.3% of disposable personal income—unchanged from 1994 and 1993.

It is interesting to note that in 1950 food consumed 20.6% of disposable
income, in 1960 it was 17.4%, 1970 13.8%, 1980 13.4%, and 1990 11.6%.

"REAL" CHANGES IN FOOD SPENDING

Increases in food spending are predominantly dependent on population
growth, which between 1993 and 1995 averaged 1% annually—just marginally
lower than the 1.06% annual average increase between 1980 and 1982. Changes
in prices and income also play a role.

Historically, the trend for food away-from-home spending, in general, has
been more volatile. From the early 1980s through 1996 the rate of change in
prices and the rate of change in expenditures for food has narrowed considerably (see Table 1.5).

TABLE 1.3. Away-from-Home Food Expenditures, Away-from-Home Food
Price Changes—1970/71 to 1995/96.

Annual Percent Change	Expenditures	Prices*
1970/71	+6.7%	+5.1%
1975/76	+12.8%	+6.8%
1980/81	+8.8%	+9.0%
1985/86	+7.6%	+3.9%
1990/91	+2.9%	+3.4%
1995/96	+2.0%	+2.5%

* Based on Consumer Price Index for food-at-home.
Source: USDA.

TABLE 1.4. Food Expenditure as a Share of Disposable Income: 1975 to 1995.

Year	Disposable Personal Income (Billions)	Expenditures for Food		
		At Home	Away from Home	Total
1975	$1,159.2	9.9%	4.0%	13.9%
1976	1,273.0	9.7	4.1	13.8
1977	1,401.4	9.4	4.2	13.6
1978	1,580.1	9.2	4.3	13.5
1979	1,769.5	9.2	4.3	13.5
1980	1,973.3	9.1	4.3	13.4
1981	2,200.2	8.7	4.4	13.0
1982	2,347.3	8.5	4.5	12.9
1983	2,622.4	8.3	4.5	12.8
1984	2,810.0	7.9	4.3	12.2
1985	3,002.0	7.7	4.3	12.0
1986	9,187.6	7.5	4.3	11.8
1987	3,363.1	7.4	4.4	11.8
1988	3,640.8	7.2	4.3	11.5
1989	3,894.5	7.2	4.3	11.6
1990	4,166.8	7.3	4.3	11.6
1991	4,343.7	7.4	4.2	11.8
1992	4,613.7	7.0	4.2	11.1
1993	4,789.3	6.8	4.3	11.2
1994	5,018.8	6.9	4.3	11.2
1995	5,306.4	6.7	4.3	11.0

Source: USDA.

CHANGES IN THE UNITED STATES FOOD INDUSTRIES

According to a recent, widely publicized report from McKinsey and Company consultancy, the HRI segment is expected to capture between 80 and 100% of food industry sales increment in the decade between 1995 and 2005, at the expense of traditional supermarkets. Actually, the HRI sector has been increasing its share of the food business for decades and now accounts for more than 46% of total dollars spent for food in the United States. Together with competition from participants relatively new to the industry such as club stores and supercenters, plus home meal replacement specialists, this trend has created a tremendous pressure on supermarkets which find themselves in a much more intense battle for "share of stomach."

CHANNELS OF DISTRIBUTION FOR THE FOOD INDUSTRY

The basic channels of distribution for the food industry can be summarized as follows:

TABLE 1.5. Per Capita Food Expenditures.

Year	U.S. Resident Population (Thousands)	1996 Prices		1988 Prices	
		At Home	Away from Home	At Home	Away from Home
1976	217,563	$588	$353	$1,011	$645
1977	219,760	625	386	992	676
1978	222,095	680	433	1,046	728
1979	224,567	749	486	1,063	742
1980	227,225	818	529	1,049	766
1981	229,466	864	571	1,046	779
1982	231,664	887	603	1,054	772
1983	233,792	926	645	1,040	764
1984	235,825	965	683	1,037	767
1985	237,924	991	710	1,073	786
1986	240,133	1,018	757	1,083	798
1987	242,289	1,051	826	1,100	797
1988	244,499	1,094	894	1,100	818
1989	246,819	1,161	943	1,102	859
1990	249,403	1,249	999	1,094	894
1991	252,138	1,287	1,017	1,089	901
1992	255,039	1,280	1,043	1,102	912
1993	257,800	1,289	1,094	1,105	898
1994	260,350	1,338	1,134	1,080	903
1995	262,755	1,370	1,177	1,034	930

Source: USDA.

- producers: agriculture, cattle/poultry/pork, and fishing
- intermediate processors: ingredients and supplies
- final processors: finished food product manufacturers
- wholesalers and distributors
- grocery retailers and hotel/restaurant/institutional outlets
- home or hotel/restaurant/institutional consumers

There are many variations and complexities in food channels, and many types of operators at each level in the channel. There is also a wide range of different types of what might be termed facilitating organizations. These organizations perform one or more vital activity to help food products move down the channel from raw material production to availability of final products to consumers. They include a wide variety of retailers, common and contract carriers which provide transportation services, warehouses that provide both storage and cross-docking services, plus advertising agencies, marketing research companies, information technology and communications providers, and financial institutions.

FOOD PROCESSING

Producers include all entities that produce products in an original or unprocessed state. Raw, unprocessed fruits and vegetables, freshly caught fish, beef "on-the-hoof," and unprocessed grains are examples. The food processing sector of the economy is highly complex and differentiated. Companies such as David Michael and Otten's Flavors produce ingredients and flavors used by other food processors. Companies such as Cultor Food Science and A. E. Staley are also primarily in the ingredients business, while McCormick, which specializes in spices, produces both intermediate and finished goods.

Many food processors operate at both the intermediate and final level. For instance, Knouse Foods (Peach Glen, Pennsylvania) manufactures processed fruits and juices under several brand names (e.g., Lucky Leaf Applesauce) and provides fruit fillings for bakers such as Tasty Baking in Philadelphia. Ocean Spray, like Knouse Foods, a grower's cooperative, manufactures cranberry products as ingredients for other processors and as final products for both the retail and food service sectors. Even Procter and Gamble, whose varied product lines include both nonfood and food items, produces products such as fats, oils, and Olean® (the brand name for Olestra, an artificial fat replacer) for use by food processing companies such as Frito-Lay.

Some processors produce only for the retail sector, some produce only for the foodservice sector, but many produce for both. Some manufacturers produce exclusively branded food products while others package exclusively private label products for wholesalers, retailers and even hotel/restaurant/institutional outlets. Some companies offer both their own branded products and produce for private label business.

Other companies that fit under the broad definition of processors include manufacturers whose primary business is contract manufacturing and packing (typically called contract or copackers) to whom food companies outsource some portion or even all of their production. Most food processing companies purchase packaging materials from package material suppliers. Other processors will take a partially finished or intermediate product, such as a filling, and then manufacture and produce the final product. A very few manufacturers outsource all packaging to a packaging specialist.

WHOLESALING

The wholesaling sector includes different types of food wholesalers and distributors. The retail side includes full-line wholesalers, specialty or limited assortment wholesalers (such as those who feature frozen foods or produce), and rack jobbers who distribute products directly to the retail store shelves. The leading U.S. wholesaler is Supervalu. Supervalu, like Fleming, the second place wholesalers, sponsors a voluntary group of retailers and also distributes to independent grocers. Some wholesalers, such as Sysco, serve the hotel/restaurant/institutional trade.

Brokers are also wholesale level organizations. Unlike wholesalers which are merchant middlemen (that is, they take title to the product they handle), brokers serve a purely facilitating function in that they bring sellers and buyers together, mainly serving in place of a manufacturer's sales organization.

Most of the large supermarket chain organizations have vertically integrated supply operations and act as their own wholesalers. For example, most food manufacturers selling to Acme Markets, the supermarket share leader in Philadelphia, Pennsylvania, sell direct or through brokers to Acme, which purchases in wholesale quantities and stocks warehoused items in Acme warehouses.

FOOD RETAILING

The two primary divisions of food retailing are the traditional grocery channel and the nontraditional grocery channel. The traditional grocery channel includes conventional supermarkets, superstores (30,000 square feet and larger), food and drug combination stores (at least 25% of the merchandise mix in nonfoods and drugs), warehouse stores, limited assortment stores, convenience stores and other. The nontraditional grocery channel includes hypermarkets, wholesale club stores, supercenters, and deep discounters. Operators of supermarkets and superstores can be either chains (e.g., Kroger, Ralph's, Harris Teeter) or independents. Independent supermarkets can be members of wholesaler-sponsored voluntary groups, such as IGA, or cooperative buying groups, such as Associated Wholesalers.

One of the major growth segments in food retailing is the supercenter. Industry predictions place annual supercenter sales at $33 billion by 2000. A supercenter has a minimum size of 150,000 square feet, a merchandise mix that includes complete discount store departments and a full supermarket including all perishables, and common checkout for all items in the store. The leading supercenters are Wal-Mart, Meijer, Fred Meyer, and Super K-Mart. Another growth area is gourmet and specialty stores, such as Zagara's in metropolitan Philadelphia. Zagara's sells natural and organic foods, gourmet foods and prepared foods. A new hybrid retail-restaurant format is represented by Eatzi's, which specializes in very upscale produce, baked goods, meat, and fish, with an emphasis on prepared foods, primarily for takeout. An increasing number of other retail formats also sell food, including drug stores such as Walgreens, Rite Aid, and Eckerd.

Home delivery is beginning to return to the food industry in a significant way via such operations as Peapod and Shoppers Express. These Internet-based services allow household customers to conduct virtual shopping using a personal computer at home or in the office, communicate and pay for an order electronically, and have the order delivered direct to the home. For the most part, these operations are on a small scale and have yet to turn a profit, but as a new generation of computer users matures into the major food buying group, more and more home shopping or one-on-one marketing will take place.

HOTEL/RESTAURANT/INSTITUTIONAL

The hotel/restaurant/institutional industry includes that which encompasses all meals and snacks prepared away from home, including all takeout food beverages. The foodservice sector is divided into three major components: commercial restaurants, noncommercial facilities, and the military. The commercial sector includes eating and drinking places (restaurants, ice cream stands, bars and taverns, delis and pizzerias, bagel shops, social caterers), managed services (commercial foodservice management companies such as ARAMARK which have operations in plants, offices, schools and colleges, health care facilities, transportation facilities, and sports and recreation centers), and lodging places.

Restaurants come in several varieties, including full service restaurants and quick service or fast-food restaurants. Full service restaurants can be classified as fine dining and family restaurants, with the determining factor being average check per person. Noncommercial restaurant services include business, educational, institutional, or governmental organizations that operate their own foodservice operations. Military includes officers' and NCO clubs plus military exchanges. An interesting development during the 1990's has been the take-home restaurant, best characterized by Boston Market, which is in the business of providing home-cooked meals to people who lack the time, energy, ability and inclination to cook. This development has recently extended to the creation of a new concept called Foodini's, owned by the Chevron Corporation, which is a prepared meals store and gas station at one location.

There are several significant drivers of product development for foodservice establishments. First, with fine restaurants typically an exception, is a focus on more convenient or "speed scratch" items that require less preparation. Products that require little or no preparation are more important today because of labor shortages and increasing labor expense. Second, with an increasingly ethnically diverse population, we have a greater interest in foods that are "authentic ethnic" or fuse together different ethnic profiles, plus we have a greater interest in products with a higher flavor profile. Third, the increasing interest in health and wellness is reflected in healthier menu choices. However, it is important to note that people dining away from home typically are not willing to sacrifice flavor and that even healthy items must taste and look good.

FOOD CONSUMERS IN THE UNITED STATES

Today's food consumers, driven by changing demographics and lifestyles, have adopted different patterns of food consumption, in turn leading to differences in menu planning, food acquisition and food preparation. Increasingly, consumers demand that food be available when and where, and in the quantities and varieties, they want to support their more mobile and more diversified lifestyles. The task of the food system is changing from "bringing the consumer to the food" to "bringing the food to the consumer."

As it has for at least two decades, the changing role of women in society, specifically the increase in the percentage of adult women working, has been the single most important factor impacting food preparation and consumption patterns. The percentage of women over the age of 16 who work outside the home is stabilizing at just over 60%. Nearly 60% of children under the age of eight have mothers who work outside the home as compared with 18% in 1960. When women, especially mothers of younger children, work outside the home, households in effect substitute money for time. There are more dual-income households but much less time in those households for its members to perform routine household production tasks, such as cooking and cleaning up.

Family composition is also changing with attendant effects on household activities. The "traditional family" is no more, as evidenced by the statistic that only about 8% of households today feature a working dad, a stay-at-home mom, and children (as compared to 43% of households which met this definition of "traditional" in 1960). Today, approximately 30% of all households have two working adults. About 30% of all households have a single parent, and about 25% of the population lives alone. Households are becoming smaller, with the average now slightly fewer than 2.6 persons. The effects of these changes are far-reaching for the food industry because households with two working adults, or led by a single parent, or inhabited by a single person, are households that in many cases are time-poor and focused on finding more convenient ways of accomplishing household tasks.

As a result, convenience has become a significant driving force in the food industry. Time has become the currency of the 1990s. In dual-income households especially, consumers are interested in convenience and willing to pay for it. Life is much more diversified, with many more activities for people today. Family members engage in many activities and are constantly on the go. Virtually any suburban mom or dad is familiar with the feeling that her or his primary parental role seems to be providing transportation to kids who always have to be somewhere to do something. Given that mom is working and that everyone in the family is doing so many things, it is not surprising that the result is that mom is no longer in the kitchen spending 1-1/2 to 2 hours cooking dinner everyday, and is no longer available to teach daughter (or son) to cook.

Consumers with busy and active lives do not sit down for three "square meals" with the family, as was the tradition. Meals mean different things to different people at different times. Sometimes, meals have much social significance and represent opportunities for family and/or friends to gather and share. At other times, meals are strictly fuel stops with emphasis on finding something quickly that tastes good and satisfies one's hunger. Because of our hectic schedules, we have less opportunity to sit down with our families, and, as a result, fewer "all-family" meals are eaten than in the past. It is not unusual to have three or four different dinner times in a five-person household because all of the people living in the household participate in different activities and have different obligations. Much of our eating now takes the form of "grazing," which means that we grab small amounts of food several times during the day. The new dining rooms of America increasingly are the car ("dashboard dining") and the office ("desktop dining").

Because of the family's changing composition and lifestyle, meal planning and preparation are changing as well. Household food preparation time is steadily decreasing from 30 minutes to 15 minutes daily. Very few consumers (10%) plan meals more than one day in advance. In fact, up to 79% of consumers do not know what they are going to have for dinner within four hours of the meal. Although Americans are preparing the same number of meals in the home as they were in 1990, the food preparation techniques are different. Namely, fewer items are being prepared from scratch. In 1986, 64% of meals included at least one "homemade" item. By 1997, this number decreased to 56%. The second difference in food preparation techniques has to do with stove use. According to the NPD, in 1985, 69% of meals were prepared using the conventional stove; by 1997, only 57% of meals were prepared this way.

To spend less time and effort preparing meals, consumers are increasingly "outsourcing" meal preparation. Consumers are taking greater advantage of restaurants which are increasingly becoming "prepared foods supermarkets." Consumers are not eating out more, but carrying out more, as restaurants are becoming the standard of quality for prepared foods.

Consumer behavior research indicates that the definition of "family" is no longer restricted to the traditional definition. A variety of relationships and numbers of people constitute families today. With a greater number of women in the workforce, and no substitute at home to prepare meals, family and individual eating habits are changing. In short, consumers are seeking products that do not require much time, effort or thought to prepare. Prepared foods are an essential resource for time-pressed consumers.

ETHNIC DIVERSITY

In addition to changes in consumer behavior affecting convenience, changes based on ethnic make-up of the consumer also are occurring. In 1980, the racial breakdown in America was Caucasian, 80%; African-American, 11%; Hispanic, 7%; and Asian and Native American, 2%. Population projections to the year 2010 forecast major increases in Hispanic (to 14% of total population) and Asian (to 5% of total population) peoples who have been termed the "new Americans." The African-American population is also expected to increase to 13% of the total population, whereas, the Caucasian population is expected to decrease to 67% of the total population.

Social scientists predict that by the year 2050, Asians, African-Americans, and Hispanics together will constitute 47% of the U.S. population. The analogy increasingly being used to describe America's population is not America, the great "melting pot," but America, the great "salad bowl" or "stir fry!" Clearly, these changing demographics will have implications for the food industry as well. Consumers are increasingly demanding different spices, seasonings, flavors, blends, vegetables, meats, breads, and other foods. Food processor/marketers have begun marketing to America's diverse population and must continue to take advantage of new markets within our own nation.

THE AGING OF THE AMERICAN POPULATION

In addition to changing consumer habits and ethnic diversification, a third consumer consideration is the aging of the baby-boomer generation. By the year 2014, approximately 78 million Americans will be over the age of fifty. The percentage of the population over 50 will grow (26 to 39% by 2020) because more people are living longer. In 1935, only 15% of the population lived past age 65. By 1990, almost 80% of the population survived age 65. In fact, it is predicted that the number of people over age 85 will increase fivefold by the year 2050.

Older consumers have special food and nutritional needs. The food industry must be ready to meet the needs of those with special diets and unique preferences. As consumers age and health problems pose new challenges, they will demand foods that help in maintaining health. Nutritional information and labeling will have to be clear, concise, and meaningful. Aging also often brings a dampening of the senses. Smell and taste decline. Therefore, food marketers should consider the importance of enhancing flavors in products developed for older individuals. Visual acuity diminishes as well, which has implications for label design and the need for larger print. Muscle mass and strength also deteriorate as one ages, thereby creating a need for innovative packaging structure. Furthermore, although enhanced food products with unique labeling and packaging may be more expensive to manufacture, marketers would do well to remember that older consumers will likely be willing to pay a slightly higher price as individuals over age 50 hold 77% of the nation's financial assets.

HEALTH AND FOODS

Although aging Americans pay the most attention to the types and amounts of foods they eat because of health issues, Americans in general are becoming more health conscious. There is increasing recognition that prepared foods impact quality of life. Many people are trying to take a proactive position regarding health. Instead of treating health conditions with drugs from the medicine cabinet, many advocate eating "healthy," natural or enriched foods from the kitchen cabinet as a means to prevent or combat health problems.

Despite the desire to eat "healthy," the consumer is often confused about how to eat "healthy." Consumers are bombarded with information from television, magazines, doctors, and the Internet about what constitutes a healthy diet. Often, the data change and can be contradictory. Recently the emphasis in what are currently termed "wellness foods" has switched from removing the "bad stuff," such as fat and sodium, to adding the "good stuff," such as folic acid, antioxidants, and other nutrients.

Consumers are becoming more interested in "medicinal foods" and nutraceuticals. The U.S. market for these products is being driven by the baby-boomers and is growing at an 11% annual rate. Research indicates that 52% of shoppers believe that foods can be used to reduce their use of some drugs and medical therapy. In 1994, only 42% of those surveyed believed this. Furthermore, 33% of shoppers regularly choose foods for specific "medicinal" purposes.

It is clear then that as consumers take a turn toward more active maintenance of wellness, health will become an even greater driving force in the types of foods that will be demanded by consumers in the 21st century. Attention to health along with an aging population will have attentive marketers segmenting mature consumers by health condition rather than age. Nutritional individualization with regard to health condition will be a marketing emphasis.

CONSUMER PREFERENCES

In addition to factors that are a function of society and lifestyle (health, age, and convenience), consumer preferences are also changing. Health is important, but not at the expense of flavor. Consumers want variety, information, and new eating experiences. Modern communications technology and trade agreements among nations have made the world smaller. Products from around the world are available as they have never been before. Consumers want access to novel and interesting foods that are fresh, convenient, and tasty. Consumers, however, also want and need a lot of information. Increased attention to health along with the availability of unfamiliar or unique foods plus a strong consumer demand for simplification drive consumers' demand for information about preparation, serving, accompaniments, nutrition, etc. This need for information has implications for labeling and packaging design.

Adapting to, and taking advantage of, technology is the second challenge to the food industry as we face the 21st century. Bioengineered foods are already being produced. For example, genetically engineered, corn-fed, virtually fat-free pigs, cancer-fighting bread, and vegetarian milk and meat are being developed. In addition to bioengineering, ingredient technology, production processes and equipment, and packaging are areas in which taking advantage of new technologies can significantly reduce costs and increase efficiency and productivity.

Like other business, the food processing/marketing industry can and should take advantage of opportunities to use technology to increase administrative and logistical efficiency as well. Distribution management systems are available to streamline supply chain logistics. Web pages and Internet advertising are technological tools that food marketers must utilize to remain competitive. Electronic marketing through relational databases, and electronic shopping and home delivery are innovations that food businesses should also consider.

FOOD SAFETY

The food safety issue of greatest concern is that of reducing the risk of foodborne illness. Each year, 9,000 deaths and up to 33 million or more outbreaks of sickness are attributed to food poisoning. This issue is attracting greater attention and assuming greater importance for a number of reasons. First, interest in home meal replacements (HMRs) and minimally processed foods is increasing because of better quality and convenience. Although the cases of food-borne poisoning may not change because of this fact, liability may become a greater issue as law suits against food processor/marketers, retail grocers, and HRI outlets increase. Second, the creation of a global food supply implies greater reliance on imported produce. Imported food is subject to different and many-times-less-stringent handling and processing regulations than in the U.S. thereby increasing the likelihood of foodborne pathogenic microorganisms. Finally, growth in markets for organic, minimally processed, and no- or less-preservative foods further increases the risk of food poisoning.

A coordinated approach to food safety systems is needed. A single uniform and consistent set of standards among federal and state agencies would be a great improvement over the current policy structure. In addition, consumer education about food handling and about where food problems are likely to occur must be improved. Adequate consumer education calls for partnerships among industry, government, and the media.

FOOD BUSINESS

Another issue we must acknowledge is mergers, acquisitions, and industry consolidation. We are at the beginning of a new economic era. Wealth is no longer generated through the construction of proprietary infrastructures, but rather through the elimination of redundant infrastructures. This thinking is based on the theory of maximizing transactional value while minimizing transactional cost.

In the first quarter of 1998, 204 mergers and acquisitions took place in the food industry, a pace higher than that of 1997. Statistics indicate that retail consolidation is increasing, thus suggesting that more retail consolidation is likely to take place. The consolidation is driven by opportunities for increased buying power, substantial cost savings, smoother transition and blending due to information technology, and bigger store brand programs.

Another important issue that cannot be overlooked has to do with the critical social changes taking place in regard to income distribution. Only a small percentage of the United States' population is gaining financially, leading to an even greater concentration of wealth among a very few. The middle class is shrinking and a bi-modal income distribution is emerging. As the rich become richer, the poor are becoming poorer, *and* hungrier. Many lower-income consumers do not have access to the full food retailing spectrum. Finding ways to meet the food and nutritional needs of the hungry is a challenge we in the food industry must accept.

For food businesses to thrive, economic development incentives must be initiated. Tax policies and developmental incentives could be a part of this effort, as well as labor force education and training programs. Infrastructure development is another area where government can play an important role in improving the viability of food businesses. Reasonably priced access to energy and adequate transportation and communication networks are essential to the food industry. Furthermore, streamlined and relevant laws and regulations that are enforced can make a real difference to the bottom line and operating efficiency of companies in the food industry.

This opening chapter has provided an overview of some of the most important issues facing the food industry today and into the 21st century. Perhaps the issue that the food processing/marketing industry itself is best designed to play the biggest role in is that of the changing consumer. Developing convenient, user-friendly, novel, and healthy food products that meet the needs and wants of a diverse population of 21st century consumers is a challenge we will confront. Using modern technology to implement changes in products as well as to improve food safety and retain quality will be essential to long-term industry sustainability. In the drive to meet the needs of the changing consumer, we must be aware of the challenges and benefits of food processing/marketing industry consolidation. Further, as suppliers of food, we have a responsibility to play a role in abating the national and world hunger problem. We must communicate our financial, transportation, communication, and energy needs to government so that decision makers better understand how to serve the food processing/marketing industry.

Finally, we must continue to deliver new, safer, better, more convenient foods to a dynamic population—which translates into the need for effective and efficient food product development systems.

BIBLIOGRAPHY

Anonymous. 1997. "Food Consumption, Prices and Expenditures, 1970–1995." Washington, DC: U.S. Department of Agriculture.

Anonymous. 1997. *Food and Beverage Marketplace.* Lakeville, CT: Grey House Publishing.

Anonymous. 1998. "The Food Institute's Food Industry Review 1998." Fairlawn, NJ: The Food Institute.

Connor, John M. and William Schick. 1997. *Food Processing: An Industrial Powerhouse in Transition,* 2nd Ed. New York: John Wiley and Sons.

CHAPTER 2

The Marketing Drive for New Food Products

LIZ ROBINSON

Historically new food product development was the prerogative of the technocrats who invented a product and gave it to the sales force. With the realization of the marketing concept, consumers were recognized as the decision makers. Food product developers had to be attuned to consumers and not solely to technologies. One result has been a ceaseless torrent of products, a few innovative, many interesting, and a lot imitative. Many who manage food product development appear to have adopted an erroneous philosophy that if you swing enough times, you will eventually hit a home run. However, you may strike out 27 times in a row in several consecutive games—before you even foul off one. New food product development should be a systematic effort founded in a strategic plan to please—and even delight—consumers.

THE global market for food, beverage, and tobacco is $4 trillion in U.S. dollars. Processors and marketers are scrambling to meet the ever-increasing consumer demand leading to exponential growth of new products. From our vantage point at the monthly reporter on new products, *New Product Sightings®*, we have seen product proliferation up close and personal. With literally thousands of new products launched globally every month, we find it begs many questions. How do you insure the successful development and launch of a new product? What was the difference between the successful launches of Diet Coke®, fresh-cut salad in a pouch or tray, and Dean's Milk Chugs® portion-size fluid milk in brightly designed bottles (Figure 2.1), and the failures of New Coke®, Simple Pleasures® Ice Cream, and Jakes Cola®? Was it the overall marketing, some manufacturing edge, a better understanding of the underlying consumer need, more appropriate timing, or plain luck? These are the questions that food and beverage product development groups ask everyday. Many think that they know the answer to the question of what

19

FIGURE 2.1 After decades of packaging fluid milk in transparent glass, translucent high-density polyethylene and line-printed gable-top paperboard cartons, a new philosophy emerged in the late 1990s. Fluid milk packages are increasingly fully decorated glass or plastic bottles using reverse-printed top-to-bottom shrink film labels conforming to the bottle shape. These samples are of a dairy-based beverage that has been thermally sterilized to achieve ambient-temperature shelf stability. Note the product name, "Smooth Moos."

makes a product a success or failure; however, given the failure rate of new products, it would seem that most don't.

According to a recent report from the management-consulting firm of Booz Allen Hamilton, only one product in 58 makes it through the product development process to be launched. The stakes are high! According to the survey firm IRI, in 1997 the top ten new food and beverage products generated sales revenue of $2.2 billion. For food and beverage companies, the access to this revenue stream is absolutely vital for company survival. Some companies realize half of the their sales and 40% of their revenue from products that are less than 5 years old. For these and other manufacturers that want to continue to thrive and survive they must figure this out.

Food processor/marketers have made and will continue to make significant investments to improve their new product development process. During the 1980s complex multistep processes were institutionalized. These systems detailed every step and companies had to march lockstep through the process. The process became an unwieldy behemoth. For these companies creativity and speed were removed from the process in favor of analysis and in cases, paralysis.

The world of food product development has changed. It has shifted into high gear. Based on our organization's audits of the major world markets, we estimate that globally a new product is launched every twenty minutes. Speed is the fuel that drives the process. Speed to market, speed versus the competition, speed of consumers' evolving changing needs, and speed necessary to produce new products to meet the needs of the company bottom line. Short-term gains and short-term losses versus long-term gains and losses and the balance that goes along with it are important to the product development team.

Where can companies and individuals go to learn about how to make successful new food products organization? In 1999, few universities, among them, Saint Joseph's University Department of Food Marketing, teach a fundamental integrated approach to new food product development. Individual components such as packaging, consumer insight, or manufacturing integration might be taught, but the intrinsic relationship of this and other key components are not explored and explained. Additionally business schools tend to utilize a disciplined structured approach to new product development. The marketing managers that are turned out from this traditional education track are thrust into situations where a creative approach to new products is required. This "outside the box" thinking, "changing the behavior" paradigm and conceptual process utilization is the antitheses of the MBA's training and leaves graduates hard pressed to proceed and even to succeed in this arena.

Corporations are forced to learn their food and beverage product development craft at the school of hard knocks. Some attempt to short cut this process by hiring management consultants. What corporations tend to forget is that on a basic level, the process should consist of essentially four key components:

- assessing senior management's commitment to new product development
- finding the right idea
- developing the business case for the product
- development and commercialization of the concept

This is not intended to oversimplify a very complex and multistepped process. However, all of the individual components of food and beverage product development fall under one of these headings. The important thing to remember is that all of these are important for success. Each must thoroughly and competently be completed. No matter how many times a new project is undertaken, a corporation must step back and review.

It is easier to be successful if an organization understands management's commitment to the new product process. There are several directional signposts that provide clues to management's commitment toward new product development. One clue is the quality or mix of the staffing that is provided to drive, direct, and manage the new product development process. A committed organization understands the fine line that must be walked between seasoned food and beverage product development staff members and the fresh perspective of new staff. On one hand, food and beverage product development requires the experienced analysis of complex, detailed consumer information that must be synthesized with business and financial information, as well as product development information, technological assessments, and manufacturing practicalities. On the other hand, new food and beverage product development requires a fresh perspective, one that might provide a new twist to an old idea, outside the box thinking, and studious avoidance of the "been there, done that syndrome." Creative ability and practical analytical ability must be equally valued, incented, and rewarded. Both have their necessary place within the process. Both skill types are critical.

New food and beverage product development requires an organizational structure that rewards the entire team for their ability to communicate, work together, make decisions, and creatively execute their jobs. A management that is committed to building and creating new products provides an organization that is not overly bureaucratic and layered but has sufficient points for senior management involvement to insure their buy-in. An overly bureaucratic organization will lose its competitive edge and ability to speed to market. With too many layers there will be a centering effect where a product may lose its edge or its consumer focus as it must work to appeal to the largest numbers within the organization.

From an organizational perspective, it has been consistently found that the team approach to new food and beverage product development offers the best hope for product commercialization. There are two schools of thought on the best design for product development teams. One believes that a single team should work on a product from the ideation stage through commercialization and the launch. The second believes that there should be two teams involved. The first team, which works on design and development, while the second team works on commercialization. The theory is that design and development requires a different skill set than commercialization. Although this is true, the downside risk is that in the transition from team to team, there is a risk that there will be a shift in the objectives or a change or modification of some critical element of the product. Last, there is a potential for conflict of interests between the groups.

In either case, the key is to create a team and an environment where all of the players feel involved and committed to the overall success of the product, whether they are from packaging, R&D, sensory, manufacturing, finance, marketing, or consumer research. Their involvement will keep the project grounded in reality and increase the product's chance for success.

Commitment to new food and beverage product development can also be observed by assessing management's commitment of money and time to the process. In order to optimize the product and maximize its potential, the new product development team needs time to create, refine, and test the product and money to provide the necessary resources to do their jobs correctly. Cutting short on either of these two things can seriously jeopardize the successful launch of the product by compromising the product's potential.

Along with providing an organizational structure to facilitate development, time, and money to adequately create the product, management must clearly articulate the contribution new products must make to the bottom line and the time frame that these contributions are expected. Articulating these will help the new food/beverage product group determine if they are looking for several line extensions that will provide quick short-term hits delivering sooner to the bottom line or longer-term new products designed to establish their own identity and live as a brand for a long time.

Once senior management's commitment to the new food/beverage product process has been determined, then the process of finding the right ideas and developing the food/beverage products begins (Figure 2.2).

The first step is to undertake a strategic assessment of what platforms should provide the basis for new food/beverage product development. It is absolutely critical to understand the core competencies and the unique strengths that the company has. A company's core competencies could be formulation ability, manufacturing and production capabilities, legal, distribution, sales, and/or marketing prowess. Leverageable unique strengths that can provide a competi-

FIGURE 2.2 Modified-atmosphere-packaged fresh-cut vegetables for salad were a major growth success of the 1990s. The logical extension was ready-to-eat convenience fruit. For the 1996 Centennial Olympics in Atlanta—and the doubtless other markets—the processor/marketer offered "Grape Escape" (note the name) packaged, ready-to-eat, stemless and seedless grapes.

tive edge or point of difference could be sourcing, brand name, trademark ownership, and/or patent ownership. Without an understanding of strengths and competencies, a tremendous consumer product can be undone or usurped in the market by a competitor with developed skills and competencies which can be leveraged to outperform the initial launched product (Figure 2.3).

While the business assessment is under way, the consumer assessment needs to occur. Despite what manufacturing may tell you or what R&D may tell you, the food/beverage product development begins and ends with the consumer. It is critical to understand the consumer mindset and motivation by studying their values, attitudes, lifestyle, and demographic trends. What are the consumers' needs? What lifestyle, attitudinal, or demographic trends exacerbate this need or provide an opportunity to fill a need? What are the consumer roadblocks to your success?

After identifying consumer trends and determining the company's strengths and competencies, a company is ready to develop products. First, establish clear benchmark introductory criteria for each phase of food/beverage product development. Criteria should be based on tangible measurable markers such as purchase intent, household penetration, revenue potential, or payout potential.

FIGURE 2.3 AriZona Tea has generated sales growth in part through the unique product positioning and, in part, through singular packaging. This oversize glass bottle with a wide-mouth opening is decorated overall with process pictorial reverse-printed shrink film. The appearance is that of high-quality fired glass, but at much lower cost due to the use of film that conforms to the glass contour after heat-shrinking.

Clear measurable criteria keep companies from hanging on to mediocre ideas far too long. If a food/beverage product cannot be re-engineered or modified to increase the consumer need and ultimately the demand for the product, thus meeting the benchmark goals, it will not succeed. It should be discarded before more resources are spent on it. This introductory criterion needs to be put into place before the process begins. Although many people believe they can spot a good idea, it is best to have set pre-established criteria. This removes the personal bias and attachment that can develop to the product. Once the criteria are set the search for new ideas may begin.

Brainstorming is a tremendous way to tap into the broad skills and talents of a fundamentally diverse set of people to create possible new product ideas. Brainstorming teams should be made up of a larger group than just the core product development team. Advertising agencies, product development partners such as flavor houses, ingredients suppliers, or package supply firms should be included to capture their insight and knowledge of cutting edge trends in the industry. This collective group should be challenged to stretch to come up with as many ideas as possible.

Once the initial ideas are collected, the group should cluster the ideas into possible platforms for development. Further brainstorming around these platforms should take place to round out other possible consumer ideas that should be there.

Following this brainstorming, broad consumer screening of the ideas should take place. This will help to eliminate weak ideas and help focus the development group on which product or groups of products should be taken on to the next step for prototyping. Multiple new food/beverage products should be in the development pipeline. As the product is developed it is important to involve the consumer all along the way. Ultimately, all of the components come together and the entire concept, package, and product should be tested in a simulated test market to better project and estimate volume potential.

Although focusing on only one product at a time does maximize the resources, it does tend to create a situation where bad ideas are held onto longer than they should be because there is not another new product waiting in the wings and teams will be back to square one. Having multiple new food/beverage products in development facilitates the removal of mediocre ideas sooner rather than trying to maximize the most of a not great idea. Additionally, multiple new products in development increase the probability of one of them being successful.

Running in parallel and as part of the product development process is the business development process. A complete business plan is necessary to sell the product to senior management and to assure that team members buy in on all aspects of the product development process that have occurred. This document should contain

- the business objective for this project, including the near term and longer term goals
- the concept itself in as much detail as possible
- the consumer proposition and how this product fulfills the consumer's unmet need
- a detailed timetable for development and commercialization including key dates, milestone markers, and the timing and financial implications for missing timing targets
- manufacturing information, including where the product is manufactured, sourcing, cost of goods, any implications for plant staffing, equipment, or line needs
- a detailed marketing and support plan including the awareness and trial assumption based on the advertising and promotional support provided
- sales and distribution plans for the product and its impact, if any, on facings for other products in the current line; other necessary information to be included are who will handle the sales of the product, estimates on slotting, and timing projections for achieving specific distribution levels
- regulatory and legal issues, if any, that must be dealt with
- full financial analysis of the project including estimated market size for the product, cost of goods analysis, a detailed profit and loss projection, and a 3- to 5-year revenue projection

This business case should be a working document that contains the most compelling and complete information available on the product. This valuable document serves as the project road map ensuring that no steps are missed and that teams are kept informed. It also serves as a tremendous transition document for those companies where there is a development team and a commercialization team. With this document, the new team understands the steps that have already been taken, the consumer insight that drove the product development work that occurred, and what the next steps are in the development and launch sequence.

One final step to maximize the learning from each new food/beverage product development program is a postmortem with all team members. The development process should be reviewed and the key learning, both positive and negative, should be highlighted recorded and analyzed. Unfortunately, this last step is seldom taken, but in the long run is a vital and instructive step for the future development of new products within a company.

The authors of this book perceive the objective of this book as delivering the fundamental principles and practices of new food/beverage product development, detailing and expanding upon the elements we have outlined here. This book and its authors will be remiss if they do not impress upon the readers the importance and absolute need for integration of technology, consumer understanding, and marketing in order to maximize a product's potential.

Good luck and good learning as you enter this exciting world of new product development!

BIBLIOGRAPHY

Cooper, Robert G. 1993. *Winning at New Products.* Reading, MA: Perseus Press.

Crawford, C. Merle. 1997. *New Products Management.* New York: Irwin/McGraw-Hill.

Gruenwald, George. 1992. *New Product Development: Responding to Market Demand.* Lincolnwood, IL: NTC Business Books.

McMath, Robert M. and Thom Forbes. 1998. *What Were They Thinking?* New York: Random House.

Urban, Glen L. and John R. Hauser. 1980. *Design and Marketing of New Products.* Englewood Cliffs, NJ: Prentice-Hall.

Product Policy and Goals

JOHN B. LORD

New food product development programs should be elements of marketing strategies that, in turn, should fit the organization's goals. Each product development program should have objectives that mesh with those of other products in the organization's portfolio—and that can compete effectively in the marketplace. New food products may be innovative, adaptive, imitative, line extensions, or new forms. Whatever else new food products are, they should not change for the sake of change or be developed in a vacuum. That so many new food products fail soon after introduction should signal the need for rational routes to avoid the pitfalls and improve the probabilities for success.

INTRODUCTION

Food processors seeking to create successful new products in the late 1990s have faced increasingly difficult challenges. These include increasing competition from both existing and new types of competitors, heightened pressure on both wholesale and retail shelf space, pressure to shorten development cycles and get to market more quickly, increasingly complex technical challenges (e.g., foods that provide both whole health and good taste), and increasing regulatory complexity in areas such as labeling, food preservation, and so on. After over a decade of increasing numbers of new product launches in the food industry, *New Product News* (1998) reports that from 1996 through 1998, the number of launches declined by almost 20% from the high in 1995 of almost 17,000 new food items.

Industry observers posit several reasons for this downturn in new product activity (although keep in mind that even with recent declines, new product activity is still significantly greater than in the mid 1990s, as measured by the number of launches as tracked by *New Product News*). Companies are increasingly focusing on core brands, recognizing that there is less risk and a potentially faster return from leveraging equity in popular brands than in investing in innovative new products. Efficient consumer response (ECR) (see Chapter 4) and category management have led to fewer "me-too" product launches. ECR is an industry-wide initiative by which food processors, wholesalers and retailers have combined forces in an attempt to reduce excess cost in the retail food channel, primarily through a reduction in excess inventories and excess promotional dollars. Category management is one component of ECR that changes the way categories in supermarkets are managed, so that the emphasis is on achieving a most-profitable mix of items in the category relative to consumer purchasing patterns, as opposed to the traditional emphasis on gaining as much shelf space as possible for each individual SKU (stock keeping unit or item).

It has become increasingly difficult for companies, particularly midsized ones, to get new products on the shelves due to market saturation in many product categories. Although population in the U.S. is growing at less than 2% per year, the average supermarket carries 30,000 items today, up from 6,000 just a generation ago. Companies have responded to a history of product proliferation and an increasingly strident retail sector by simplifying product lines and making SKU elimination a priority. Finally, there have been an increasing number of mergers and acquisitions, sales, and spin-offs leading to the question: who owns which brand?

According to Hill (1997), new products and line extensions typically represent 10 to 15% of category volume each year. Over a 3- to 5-year period, new products will thus represent a very meaningful percentage of total category volume. The real challenge is to introduce products that add a unique difference to the category from a consumer and retailer perspective. Suppliers that have taken ECR seriously are carefully reviewing their existing lines and their new product launches with a more stringent emphasis on adding category value. The most successful retailers today are assigning category management partners to work with them in developing optimum approaches to managing specific aspects of new product launches, including distribution, shelf space/location, promotion planning, and pricing.

The challenge for the retailer is to be able to separate those brands that add category value from those that do not and partner with manufacturers living according to that discipline. Manufacturers must eliminate unproductive SKUs and focus new item activity on a category building perspective that will likely lead to fewer but more significant (i.e., with greater potential for leading real category growth) launches. However, manufacturers, especially those who are publicly held, operate under volume and profit pressure, so they must accomplish this goal carefully to avoid short-term volume hits.

THE IMPORTANCE OF PRODUCT DEVELOPMENT

New products represent one of the few growth avenues left to packaged goods companies pressured by Wall St. to increase profits. Downsizing and ECR have cut costs, and increasing pressure from private brand competition precludes price hikes. Growth will typically not come from category growth (except in rare instances), economic/population growth, price increases, or cutting costs. New products and new SKUs are still the best way to drive top-line growth. Brand and line extensions are still a good way to leverage consumer awareness and reduce risk and entry cost, but innovative new products potentially reap more rewards.

In their excellent book entitled *Leading Product Development*, Wheelwright and Clark (1995) argue that "the consequences of product development have a direct impact on competitiveness. They mean the difference between falling behind a leading competitor in the marketplace and being the competitor who provides leadership, compelling others to meet similar standards." The major focus of a firm, and the bulk of its assets, are tied up in how it delivers value to its customers (Figure 3.1). When a company is saddled with old products,

FIGURE 3.1 Lunch kits have become among the more popular products targeted at children—and more recently at older children. Cited by environmental activists as overpackaging, by some health food commentators as having too many calories and too much fat and sodium, and by single adults as too expensive, these products are widely accepted by children and their mothers. Part of the desirability is the quality retention afforded by modified atmosphere packaging in multicavity thermoformed barrier plastic trays.

the wrong products, or even the right products at the wrong time, the value it provides to customers, and therefore the value of the company itself, is severely limited. Thus, Wheelwright and Clark conclude, "if the firm does product development badly, its assets—particularly its equity with its customers—will wither and erode."

Product development is fundamental to the success of the organization. Wheelwright and Clark (1995) point out that "the development of a new product is the development of every aspect of the business that the product needs to be successful. And consistently successful new products need every aspect of the business working in harmony." Moreover, product development is the means by which a company builds the competencies and capabilities that set the stage for its future. In order to provide the value its customers will seek in the future, the firm must possess the human, financial, and structural capabilities to meet future customer requirements. These capabilities are created, enhanced, and renewed via product development activities. Product development is therefore central to future success of the firm.

It is not sufficient to have development capabilities. These capabilities must be organized and activities carried out in such a way that firms minimize product development cycle time. During the 1990s, success in new products means achieving speed to market with products that solve a customer problem in a demonstrably superior way. Peterson (1993) stresses the importance of getting to market quickly. By bringing a product to market quickly, a company can reap several potential benefits. You begin to sell sooner, leading to fewer lost sales. The learning curve begins to take effect sooner, leading to lower costs. Firms can pre-empt competition and develop a reputation for market leadership. Being first in the market helps a firm to attract customers because there is less ad clutter. Quick-to-market helps to create customer loyalty and allows the firm to target most attractive market segments. First-mover advantages also include the opportunity to create barriers to entry by forging contracts or good relationships with customers. Finally, getting to market quickly can reduce risk both by incorporating the most recent technology and by gaining first access to scarce resources.

PRODUCT DEVELOPMENT AND BUSINESS STRATEGY

It is important to understand the role of new product development in the larger context of business unit strategy. New product development activity, first and foremost, must be consistent with and flow from the strategic direction of the firm or the business unit. The strategic plan will chart a course for the business, specify key targets and objectives, including the time frame in which these targets are to be met, and the direction or directions for achieving growth.

Strategic marketing decisions are made at four levels, although for smaller and very focused companies with only one "business" the first two levels are the same. The four levels are (1) the organizational level, (2) the business unit level, (3) the product line level, and (4) the brand (item, SKU) level. At the organizational level, the major issue is determining what business or businesses the organization wishes to compete in, and how organizational resources are allocated across these businesses. A company like Kraft or Procter & Gamble is in several businesses which may differ in terms of customers served, competitors, suppliers, and so on. A company like Tasty Baking, a snack food manufacturer located (and a tradition) in Philadelphia, really has only one business—snack cakes, pies, and donuts—and for Tasty Baking, the business and the organizational levels are identical.

At the business unit level, the major issues are choosing, first, the market or markets we will attempt to serve, and, second, the product lines will we employ to serve these markets. In addition, we need to decide how to allocate the resources of the business unit across those product lines and customers. Portfolio theory, including commonly cited approaches from the Boston Consulting Group and GE-McKinsey, suggests that the most attractive markets and the product lines with the greatest growth potential, which leverage our most significant business strengths, receive the lion's share of our resources. On the other hand, weaker product-market combinations receive that level of investment necessary to maintain position, or, in extreme cases, exit the industry.

At the product line level, the major decisions involve which specific items in the line should be carried, how broad or narrow and how deep or shallow the line should be, and how production, distribution and marketing dollars be should be distributed among the different items. Finally, the item or brand level involves tactical decisions in terms of positioning, advertising and promotion, pricing, packaging and distribution for that particular product, using the resources allocated to the item at the product line level of strategy.

The strategic business plan specifies the objectives we have set for the business, typically in financial terms such as return on invested assets, earnings before income taxes, etc. These financial objectives lead to marketing objectives, typically stated in 'terms of revenue and market share growth. Almost inevitably there will be a gap, "the growth gap," between where we want to be and where we will be at the end of the planning period without some changes in our strategy. Therefore, the strategic plan also specifies avenue(s) for growth over the planning horizon. Typically in food companies, a significant portion of this growth will be achieved by launching new products.

It is important to note here that not all performance improvement strategies necessarily involve growth. If you start with the most basic model for measuring performance, which is

$$(\text{Revenue less expenses})/\text{invested assets}$$

financial performance can be improved in one or more of three ways: (1) increasing revenues via one or more growth initiatives, (2) reducing expenses, or (3) reducing your asset base. In the 1990s, firms have looked to build shareholder value and please the investment community by reducing expenses, many times involving downsizing or "right sizing," and perhaps reductions in marketing support, and/or or reducing the asset base by selling off or dissolving divisions, brands, and manufacturing operations.

GROWTH STRATEGIES

New product development is part of business unit strategy. Traditional marketing management texts (see, for example, Aaker, 1998) typically discuss new product development as one of the strategic alternatives in the growth matrix. This model posits that the two major elements of strategy are products and markets and that all products and markets can be divided into those which are current and those that are new. Using a two by two matrix with four cells, we can identify four different growth strategies:

- *market penetration,* which involves achieving growth by selling more of our existing products in existing markets
- *market development* (or market extension), which involves achieving growth by taking existing products into new markets or market segments
- *product development,* the creation of new products for markets we currently serve
- *diversification,* which involves new products for new markets

Thus, firms have several options for growth, which, of course, are not mutually exclusive. They can grow using current products in current markets through a strategy of market penetration, which might involve changing and/or increasing advertising and promotion, temporary price reductions, promoting new uses (for example, demonstrating uses for soups as sauces and ingredients in casseroles), and so on. Alternatively, companies can grow by taking existing product to new markets using a market development or market extension strategy. Campbell's is a good example of a company who has been looking hard at expansion in overseas markets, while Tasty Baking, long a local and then regional company centered in Philadelphia, has now expanded distribution to 46 states plus Puerto Rico. After all, if you have close to an 80% share of the domestic prepared soup category, as is the case with Campbell, or a close to 70% of snack cakes, how much can you grow via additional penetration of that market?

Another market development strategy is frequently employed in the food industry. A foodservice supplier that packages its product for sale in supermarkets, or, conversely, a food manufacturer that develops a foodservice package exemplifies this. Of course, firms can grow by creating new products for existing, or at least related, markets. This is what we call new product development. Judging from the pace of new product activity over the past 15 years, product development is clearly the growth strategy of choice for many food companies.

Authors such as Aaker (1998) have expanded the original two-by-two growth matrix by adding another dimension, based on Michael Porter's value chain, and thereby defining expansion along the value added channel, otherwise known as *vertical integration,* as a growth strategy. A firm can vertically integrate upstream by acquiring or developing its own sources of supply or downstream by acquiring or developing its own distributors. The two major motives driving vertical integration are (1) control and (2) operational efficiencies. A food manufacturer who owns or otherwise controls farm production is vertically integrated, as is a supermarket that owns its own private label manufacturing operation or its own wholesale distribution operation.

Going outside the existing scope of our current business involves a strategy of diversification (new products aimed at new markets) (Figure 3.2). This diversification can be *related,* in terms of the type of customer, technology employed, production process, distribution channels, and so on, or *unrelated,* meaning we get into a business which is entirely new to the company. Diversifi-

FIGURE 3.2 Orbitz is a unique beverage product concept in which solid "flavor" spheres are suspended in the liquid to deliver a different appearance through a transparent bottle and a different mouth feel.

cation can be accomplished via internal development, through establishing partnerships or alliances, licensing, or via acquisition. An example of related diversification is Procter & Gamble's joint venture that moved P&G into OTC pharmaceuticals with Aleve® (a brand later sold). An example of unrelated diversification is Kodak buying Sterling Drug Company (which Kodak later divested).

Any interested observer of the food industry will readily conclude that food processing companies have and continue to use all of these strategies for growth. One of the most common avenues for growth employed by food companies in the 1990s has been expansion via acquisition and licensing of brands. In fact, it is not too far-fetched to say that one needs a "scorecard" to keep up with which company owns which brands, given the very large number of brands which have been bought and sold during the past decade.

The focal point of this book is new products, developed internally. We will not discuss mergers and acquisitions that bring companies whole new businesses and stables of brands, such as the purchase of the Chex® Cereals line by General Mills from Ralcorp. We do however, discuss development of new products which use brand names licensed from other companies either individually or as co-brands, such as Delicious Chiquita Banana Ramas Cookies or General Mills Reese's Peanut Butter Puffs cereal.

STRATEGIC NEW PRODUCT DEVELOPMENT

According to Gill et al. (1996), there are "Seven Steps to Strategic New Product Development." Step one spells out where we want to go while steps two and three indicate where we are now. Steps four and five narrow down the range of options for getting where we want to go, and steps six and seven specify how we will go about completing the tasks that need to be accomplished.

The first step involves setting new product development targets and creating a product development portfolio. This is the collection of new product concepts that are (1) within our ability to develop, (2) are most attractive to our customers, (3) deliver short-term and long-term corporate objectives, and (4) help to spread risk and diversify investments. The output at the end of step one is a clearly articulated target indicating which and how new products will contribute to overall business goals.

The second step is the situation analysis, involving both the identification of key strategic elements and environmental variables, and the sources of information needed to yield planning premises. The output at the end of step two is a compilation of current information that gives a picture of the customer, the competition, and our competencies and capacities.

The third step is opportunity analysis, by which we map the strategic geography, understand the structure of markets, and delineate market gaps that might be filled. The output of this process will be a map of the opportunities.

The fourth step involves identifying potential new product options that fit the strategic geography delineated in step three. The output of step four is a complete list of all new product options in easily comparable format.

The fifth step involves establishing threshold criteria that will provide minimum acceptable performance targets. Portfolio criteria allow a business to create balance and diversity in the product line. At the conclusion of step five, we have a set of both threshold and portfolio criteria that decision makers have agreed to use in selecting the new product portfolio.

The sixth step involves creating the portfolio. This involves operationalizing the strategy determined in steps one through five by making specific line item decisions. At the end of step six, the output is a portfolio of new product options that fulfills the new product target, addresses key concerns in the customer and competitive domains, maintains and grows corporate capabilities, and can be developed within set budget limits. The seventh step is management of the portfolio.

PRODUCT DEVELOPMENT OBJECTIVES

Fuller (1994) maintains that growth initiatives are only one of the forces driving new product activity by food companies. Companies create new products to revitalize product lines as older products reach the maturity or decline stage of their life cycle. This generally means that changing consumer tastes, changing technology, competitive dynamics, new regulations, or changes in public policy have had a negative impact on overall market position of current products and created potential opportunity for new items. Cooper (1993) identifies some additional driving forces for new product development. These include the desire to grow into a new geographic market, the goal of gaining greater local and regional market penetration, the desire to reduce dependence on what may have become commodity items and provide more value added, and the need to expand our product and/or business base.

Companies introduce new products as part of both offensive and defensive strategies. In an article entitled "Cutting Edge," Mehegan (1997) reported that Schick bet $13 million in media spending that the proprietary technology of a new razor called the Protector. Schick's objective was to reduce the potential impact of the much-anticipated launch of market leader Gillette's "next generation" shaving system, the Mach3, which was introduced in 1998. Schick used a preemptive strategy, and employed a niche marketing approach targeted to younger shavers and African Americans with skin problems who have concerns about safety and comfort more so than close shaves. Conversely, one could argue that Gillette's continuing research and development efforts, leading to new generations of shaving technology, not only create a better profit mix in the product line through higher margins, but also serve to preempt other shaver manufacturers from gaining an advantageous market position.

Food companies frequently look to expand their market position and the range of offerings available to customers. Tasty Baking, with a dominant share in snack cakes and pies in its core market areas, added donuts to its product line, allowing better positioning for the breakfast occasion. New products, or at least packaging variations, are created to meet the specific needs of distributors and retailers. Many manufacturers have created club-packs and variety packs to meet the needs of club stores. Tasty Baking rolled out a line of cupcakes, called Tropical Delights, in summer 1998 as a response to their Puerto Rican distributor who told the company he could sell snack cakes with tropical fruit flavors such as papaya and coconut. Some firms add new products to reduce costs, by more efficiently utilizing existing production facilities, research and development resources, and so on.

Ultimately, companies create new products for a multitude of reasons that boil down to two primary reasons, which are (1) to enhance short-term earnings and (2) to enhance the long-term value of the brand and of the organization. There are numerous motives for looking to increase short-term earnings. The first motive is the pressure managers typically have to meet quarterly (or monthly or annual) revenue and profit targets. To the extent that rational managers are being evaluated and rewarded on the basis of meeting these short-term goals, they will behave in such a manner to do so, and a line extension or a copycat or cloned product is sometimes a good way to proceed. Top managers of publicly held corporations are under obvious pressure from shareholders to enhance the value of the organization as measured by the stock price (and by dividends). These can also be improved by new product activity. Wall Street loves companies who are innovative and active in new products, which set off streams of revenue growth.

THE ROLE OF TOP MANAGEMENT IN SETTING NEW PRODUCT STRATEGY

Wheelwright and Clark (1995) identify three principal ways in which senior management should be involved in new product development:

(1) Set direction and get people in the organization aligned; establish and articulate a vision.
(2) Select, train, and develop people capable of realizing the vision.
(3) Create, shape, and influence how work gets done in order to ensure that it gets done in the best way possible.

Senior managers select the core team for the project; serve as source of energy to sustain the project; serve as commitment managers, influencing, guiding, facilitating, and reviewing commitments; and play the roles of sponsor, coach, and process improver. The roles of senior management are identified as

- direction setter, which involves envisioning the future of the business
- product line architect, which involves determining what should be developed and in what sequence, and what the connections should be between various products that effectively build brands and business identity
- project portfolio manager
- process owner/creator, which involves managing the development process
- team launcher, which involves developing and/or approving charters for individual projects.

These charters have been designed to realize specific objectives established by the product line architecture and the business strategy; they set out the project's business purpose and provide the framework within which the project team will operate.

THE PRODUCT INNOVATION CHARTER

According to Crawford (1997), successfully innovative companies are characterized by (1) having clearly defined missions, (2) seeking future customer needs, (3) building organizations dedicated to accomplishing sharply focused goals, and (4) partnering with their customers in the product development process. Successfully innovative companies focus their new product efforts with a product innovation charter, a strategy statement that guides new product development by establishing the agenda and direction for the organization's new product development activities.

Wheelwright and Clark (1995) note that senior management must "lay out the product line's evolution in terms of product types, their relationship to current and future offerings, and the timing of their introduction. Moreover, they must do so in a way that fits the marketplace, the competitive environment, and the firm's resource realities." The four issues that must be addressed as part of establishing the future of the product line, according to Wheelwright and Clark (1995), include:

- Position—deciding where in its lineup and price/performance set the business will need new products. Attention is given to opportunities in the market, possible competitive moves, gaps in the product mix, and the breadth of line needed to accomplish the business objectives.

- Type—deciding the types of products that will be most effective in meeting specific customer and consumer needs. How the firm chooses to meet these needs, whether through innovative new products that create technical platforms, products that build on existing technical or brand platforms, or competitive clones will have significant impact on the economics of the business and its ability to respond to contingencies.
- Timing—deciding when new products should be introduced. Timing is predicated on the pace of technological change, price/performance economics, and the evolution of the market and changes in customer tastes, plus the ability of customers to learn new habits. In addition, the ability of the firm to fund development at a sustainable pace must be considered in making decisions about how to time new product entries.
- Relationships—deciding how new products will be related, i.e., building on the same technical platform, using common equipment and/or processes, distribution channels, advertising and promotional vehicles; targeting the same customers.

The product innovation charter specifies direction for these issues and others.

According to Dimancescu and Dwenger (1996), the charter spells out the project or program constraints and empowers a team to act around explicit expectations. When a project team is assembled, one of the empowering documents is a charter prepared by senior management. The charter outlines objectives for the proposed project and specifies the constraints within which the new product team will operate.

Six elements constitute a full-blown charter (Dimancescu and Dwenger, 1996):

(1) Rationale for the project and its links to the firm's vision and strategy

(2) Specific project goals expressed strategically and in financial terms

(3) The context of the project by market segments being targeted, relevant competitors who need to be accounted for, technologies, and other external factors that will affect the project

(4) Project process ground rules for management review at milestone check-points or gates

(5) A precise definition of expected deliverables

(6) Constraints that may affect the project, such as management assumptions, resource availability, and budgetary limits (An effective charter is one that provides clearly articulated expectations and objectives for the project

team, reducing the chance of a project that is not consistent with the strategic direction of the business. Additionally, "the charter is a significant element of a team-based reward system because it spells out precise objectives against which actual performance can be measured.")

Similarly, Wheelwright and Clark (1995) identify six elements of a charter:

(1) Reasons for a project and its links to the firm's vision and strategy
(2) Specific project goals
(3) Context of the project: markets to be targeted, competitors to be attacked, technologies, external factors that will affect the project
(4) Ground rules for management review at checkpoints/gates
(5) Precise definition of expected deliverables
(6) Constraints—management assumptions, resource availability, budgetary limits

Charters are generally focused on market opportunities as opposed to specific products, and are designed to bring common focus to opportunities. Crawford (1997) identifies two major dimensions with which charters are concerned: technology and the market. Opportunities are defined in terms of these two dimensions, and any opportunity has both technology drivers and market drivers. Technology drivers for food companies would include research and development, for example, fat replacement or other ingredient technology, process technology, supply chain and logistics technology, plus distribution and order handling technology. Market drivers include customer groups or segments, functions performed by the product, customer benefits, and usage or consumption occasions. The charter should provide specifics regarding all of these dimensions.

In this context, Cooper (1996) pointed out the critical role of synergy that is vital to new product success. "Step-out projects" have much higher rates of failure than projects that build on synergy, of which two types are relevant:

• Technological synergy—the degree to which the project builds on in-house development technology, utilizes inside engineering skills, and can use existing manufacturing resources and skills.
• Marketing synergy—a strong project/company fit in terms of sales force, distribution channels, customer service resources, advertising and promotion, and marketing intelligence skills and resources (Figure 3.3).

The product innovation charter also specifies the company's posture with respect to both innovativeness and timing. Firms can choose to be pioneers, that is, first to market with innovative products. This approach promises great potential return, but is fraught with significant risk. Alternatively, companies can adopt an approach of adaptation, which means early entry into the market

FIGURE 3.3 Reconstitution of dry foods may be performed in their own packages that holistically provide protection during distribution, a cooking vessel, and a serving dish. Dry hot cereal is reconstituted by adding water to the tapered paperboard cup and heating the package in a microwave oven.

with the intention of improving on a competitive product. Finally, a firm may choose to take the least risky approach of imitation, that is focusing efforts on coming into the market late with copycat products, and attempting to compete on the basis of price, location, or some variable other than a specific product leadership advantage.

Feig (1997) argues that the best new products are adaptations of existing products. This category offers the new product developer the best chance of success. Most successful new products are adaptations of existing products that have already established a consumer franchise (Figure 3.4). In fact, one of the ways by which a small company can compete with larger competitors is by adapting a product because the small competitor can typically react faster. One strategy is to find a category that is dominated by two or three companies and out-flank them against segments they are missing. Unless the category grows enormously, the giants will usually ignore the small competitor. When a large company enters a category with an aggressive spending plan, awareness and consumer interest rises for the category as a whole, causing the category to grow. For example, Schick has been following Gillette new product introductions for years. This allows them to save the enormous costs of advertising and R&D. The downside is that adapters (followers) are almost always number 2 or 3 or worse in the category. When you adapt a product, the change has to be both readily apparent and meaningful to the customer.

FIGURE 3.4 Modified-atmosphere-packaged fresh-cut vegetables—a convenient source for green salads—were the most successful new food product introduction of the 1990s. Among the package variations were a plastic bowl containing pouches of the green, dressing, topping, croutons, and even a fork and napkin. The consumer emptied each pouch into the bowl and ate with the supplied fork—wearing the napkin, of course.

Real innovation has become prohibitively costly according to Dwyer (1997b) and as a result, a majority of food firms are "drilling down" to a core brand emphasis. The food industry has seen a significant shift in focus to leveraging and building on flagship brands. Campbell has recently shifted to a "focus on icons" strategy; innovating by extending its current brand equity in soup and with its Prego® brands to meet consumer demand for easy dinners and retailers' demand for products that rival restaurant takeout. Characteristic of this strategy is a new 5-SKU line of family-sized Italian entrees under the Prego label and a new 5-SKU line of frozen restaurant soup. The soup launch offered potential in combining brand equity with the promise of restaurant quality. The line was positioned against restaurant soups via the tagline "Soup Du Jour to Stay Home For."

Manufacturers such as P&G are limiting new product introductions while retailers are carefully selecting products on their shelves to better target specific audiences. Because manufacturers have access to data on new product activity—sales, price, assortment, promotions, etc.—and the ability to respond quickly, retailers may see an increased focus on truly innovative products, rather than a continued proliferation of line extensions.

THE IMPORTANCE OF PLATFORM PROJECTS

Platform projects offer significant competitive advantage, because platforms can be leveraged into an entire family of products and be viable in the marketplace for several years. The technical platform that Nabisco created with its research into fat replacers not only provided the ability to create SnackWell's® but at the same time to create reduced-fat versions of many of their flagship cookie and cracker brands such as re®, Ritz®, Lorna Doone®, and others. There are several major classes of platforms: technical platforms, i.e., ingredient or process technology, packaging platforms, brand platforms, and distribution platforms.

Wheelwright and Clark (1995) discuss how Coke® and the Coke® brand name represented a single platform. Then Diet Coke was introduced, which represented a "half platform," in a sense; it was clearly a different formula but it leveraged advertising and brand identification from the parent brand. Then, responding to changes in the market and advances by "better tasting" Pepsi Cola®, Coca-Cola® introduced the much-heralded New Coke, which the company thought would be a new platform, a new-generation offering. When a major segment of the market indicated that it still wanted the old platform, the company brought back Coke Classic, the original formula, while keeping the new Coke (at least for a while). From a market leverage point of view, the company has a brand platform (Coke) that now comes in multiple formulas and variations—Coke® Classic, Diet Coke, caffeine-free Coke, and caffeine-free Diet Coke, Cherry Coke, and so on.

PROJECT PORTFOLIOS

New product strategy also involves managing a portfolio of projects. Medium and large companies typically will have multiple new product projects proceeding simultaneously. These projects are related to the extent that they compete for the same organizational resources and are linked with respect to timing. Portfolio theory suggests that firms will have projects at different stages of completion and featuring different levels of innovation, for purposes of balance as well as achieving both short-term and longer-term returns.

We have already seen that there is a range of innovation opportunities from continuous innovation (product reformulations or improvements) to restaging or replacing brands to new products in current categories (line extensions) (Figure 3.5) to new-to-the-company categories (e.g., through category extensions) to fundamental innovation (new-to-the-world categories). Using the innovation spectrum, we need to consider the portfolio of existing projects with the goal of diversification, that is, not all projects of the same type. Therefore, a company such as Gillette may be investing heavily in a fundamental new shaving technology leading to the Mach3 razor, while at the same time reformulating certain products and launching line extensions in other product categories with the intention of achieving some quick and low-risk market hits.

FIGURE 3.5 Pickles to Go! A single serving of pickles in a convenient, portable, easy-open flexible barrier package. The technology was difficult because of the high-acid content and corrosivity of the contents plus the need for a brine to maintain the crisp bite.

TABLE 3.1. **Criteria for Investment in New Product Projects.**

(1) Does the project have a sponsor or champion?
(2) Do we have the technology, processes, and financial, equipment and human resources to support the project?
(3) Is the project consistent with overall strategic objectives of the company or division?
(4) Does the project proposal achieve key financial targets and marketing objectives? For example, our product innovation charter may state that a new product must achieve a number one or number two position in the market, or we have a pre-tax ROI target of 15% within three years of launch.
(5) Do we have both the distribution capabilities and the trade relationships necessary to successfully get the product to market?
(6) Have we properly defined our target customers?
(7) Is there a roll-out plan which includes a timetable and measurement of both retailer and consumer response?

PROJECT SELECTION

Which ideas deserve review? Which projects should we initiate? Given limited resources plus the pressure to generate new revenues, how do we determine which projects deserve investment capital and assignment of a team. There are several criteria that should be applied. These are listed in Table 3.1.

STRATEGIC APPROACHES IN THE FOOD INDUSTRY

We can identify several examples of the impact of having some type of strategic guidance for new product development. Nabisco has a stated goal of doubling its business from 1994 ($3 billion) to 2000 ($6 billion). Nabisco owned 47% of the $7 billion biscuit (cookies and crackers) business in 1995. In order to achieve its very ambitious growth objectives, the company had to change its vision: while Nabisco Biscuit has always been first in biscuits, the company now seeks to be first in a bigger arena, snacking, which is a $39 billion business. One effect of this strategy on new product introductions was the launch of SnackWell's Cereal Bars, which do not fit a biscuit definition. This product sprang from NBC's "breakfast initiative," which is a conscious and coordinated effort to move Nabisco products out of the confines of the cookie/cracker aisle and into the breakfast aisle.

The major thrust of Nabisco's research and innovation strategy (Dwyer, 1997b) is designed to leverage capabilities and resources, helping to drive strategic growth and develop key technical platforms. Nabisco has reviewed basic competencies to better understand which technical areas they need to own and which they should outsource. The strategic focus is on finding the right balance between investing for the future and working shorter term to keep current brands healthy and moving forward. Lately, Nabisco has renewed emphasis on investing in and growing core brands. The basic strategy involves pursuing a few big ideas, coupled with support for core brands such as Oreo, Ritz, and Chips Ahoy®. The emphasis is on combining an understanding of consumer needs and marrying that understanding with new technologies to create the next big growth engines. SnackWell's provided such a growth engine in the mid-1990s but sales have fallen as consumer desire for taste has overtaken the desire for low fat.

As profiled in *Brandweek* (Thompson, 1997), Borden Foods Company (BFC) will attempt to build its business in grain-based meal solutions. In practical terms, that means focusing on the stable of pasta, pasta sauce, dry soup, and bullion brands, and future acquisition of aligned brands. Borden has decided to spin off "unaligned" brands, including Cracker Jack®, Cremora®, Eagle Brand®, Borden® brand processed cheese, with the emphasis on returning pasta to profitability. Borden has looked to withdraw from the private label business and find ways to consolidate branded regional businesses to make them more national in scope.

Pillsbury (Dywer, 1997b) has looked at both technology and marketing in an attempt to combine new or enabling technology with marketing or consumer insights to create new product platforms (Figure 3.6). Business teams are mandated to search for revolutionary new products and income streams, and then deliver their wish lists to the technology group, who prioritize requests based on importance and technical feasibility. Pillsbury new product strategy involves applying Pillsbury expertise to acquired brands with the objective of improving base product quality and lowering cost. The idea is to reinvent the product portfolio to deliver value-added benefits more quickly. For example, Pillsbury used a combination of consumer insights and different core technologies to create Old El Paso® low-fat cheese and salsa in a microwaveable bottle with an indicator, and One Skillet Mexican (kits) for making tacos and burritos. These included dough technology from Old El Paso (tortillas, taco shells), vegetable process technology from Progresso®, and microwave-packaging technology from Hungry Jack® Pancake Syrup. These products were built on a market opportunity for Mexican food made easy at home. Pillsbury positioned One Skillet Mexican as a meal solution, using the Nacho Man commercial to build awareness and equity.

Procter & Gamble's food and beverage division achieved a performance turnaround due to cost cutting plus revitalization of mature brands through product improvements and reformulations. For example, P&G has introduced low fat versions of Crisco® and Pringles® using olestra. They have become an ingredient supplier, selling olestra (under the brand name of Olean®) as an ingredient to Frito Lay for use in its Wow!® brand fat-free chips. P&G has also changed the market focus of certain existing brands such as Sunny Delight®

FIGURE 3.6 Recognizing the desire of many consumers for hot syrup to pour over their pancakes and waffles, Pillsbury and their suppliers developed a unique "microwave heatable" package. The bottle was reshaped into a short, squat version for more efficient microwave heating—and a very large content label was added to ensure consumers that this short bottle contained as much (or more) syrup than the more traditional tall bottles. The hollow handle was pinched off to block any hot syrup from burning consumers' fingers. A temperature-sensing label was affixed to the face of the package to signal that the microwave heating was complete.

to more profitable segments. Sunny Delight (which is 5% juice) had previously been targeted to parents of young children. However, this segment prefers 100% fruit juice. P&G shifted focus to teens, for whom taste and advertising have a bigger influence, using rap music and the catchy "Sunny D" nickname to capture 54% of the share of the U.S. market for refrigerated fruit drinks, tripling in sales from 1989 to 1996.

NEW PRODUCT OPPORTUNITY ANALYSIS

Strategic planning for new product development must involve an established and organized process for identifying new product opportunities. Opportunity identification and analysis involves ongoing environmental scanning to identify emerging consumer trends and changes in consumer habits which drive opportunities, plus a systematic procedure for evaluating whether the opportunities fit our objectives, strategic direction and resource base at that particular point in time. New product opportunity analysis flows from the business plan. Like other steps in the development process, opportunity analysis must be disciplined and directed, with specific deliverables and a procedure for evaluation.

The best categories in which to begin opportunity analyses are those in which the business unit is already involved. The reason is obvious—the risk of launching a new product in a familiar category is lessened by our ability to leverage existing brand names, distribution channels, technology and production facilities, plus our knowledge of consumers and the trade. However, launching new products in present categories typically results in reformulations or line extensions that provide low risk but minimal potential return. To the extent that the category is already crowded, contested, fragmented, and saturated, the prospects for anything more than a short-term burst of new revenues are limited. In addition, it is now generally accepted that product proliferation has a downside. It creates higher production, distribution, and marketing costs, out of stocks at the wholesale and retail level, and customer confusion. Recent evidence from P&G demonstrates that market share and sales revenue can actually increase by reducing the number of SKUs in a category.

Breakthrough products and substantial new revenues require locating emerging needs and/or emerging technologies that allow us to meet consumer needs significantly better than extant competitors. To find these emerging needs and technologies, and to find market areas which offer real growth potential, we often have to expand the scope of our category scanning beyond those categories in which we currently operate. The trade-off is potentially higher investment and greater risk, for higher returns. Finding categories that allow us to use our competencies in research and development, production, distribution, and marketing limits risk exposure and enhances prospects for success. Therefore, it is vital that we focus our environmental scanning on categories that meet the criterion of fit with organizational competencies.

One possible approach to new product search is built on the notion that competition for a brand, which is defined by what consumers perceive to be viable substitutes in a particular consumption or usage situation, ranges from very direct to much less direct. For instance, Grape Nuts® competes with Corn Flakes for breakfast, but for some people and in some instances, Grape Nuts competes with toaster pastries or even a box of raisins. Broadening the search means looking for new, novel, and emerging ways to satisfy important needs. Kellogg's launched Nutrigrain Bars because this type of convenient, healthful product fits breakfast demands for a particular group of target consumers.

INDICATORS

What factors do marketers consider when looking for new product opportunities? What causes marketers to consider a category attractive? We need to assess several factors and conditions as listed in Table 3.2.

The first key to uncovering new product opportunities is to understand the

TABLE 3.2. Factors to Consider in Evaluating Market Opportunities.

- Increasing interest in new product benefits
- Category size, as measured in sales
- Category growth rate, as measured by the rate of increase in sales
- Category margins
- Number and rate of new buyers in the category
- Degree of brand switching (a measure of customer loyalty and product appeal)
- Stage of the category life cycle, indicated by rate of growth in sales, and nature and extent of competition
- Amount of product/brand differentiation
- Recent innovation in the category
- Innovations in related fields that have potential application in the category
- Number of competitors
- Market shares of competitors
- Technology position of competitors
- Barriers to entry and barriers to exit: access to raw materials, capital requirements, proprietary technology, economies of scale, customer-switching costs
- Advertising and promotional expenses to sales ratios in the category
- Trade channels and access to consumers
- Seasonality of sales

consumer and what benefits consumers are seeking. Many of the successful new products of the late 1990s have built on the emerging consumer demand for foods that have a particular health benefit, such as those containing antioxidants or other healthful ingredients. Earlier in the decade, the opportunity area was low fat.

Categories that exhibit high growth and high margins represent opportunity areas. Categories in the early stage of the life cycle, with many new buyers and relatively few competitors, represent opportunity areas. Categories in which the amount of product and brand differentiation is minimal, in which few recent product innovations have taken place, and in which the advertising and promotional expenses represent a low percentage of sales also offer potential opportunities. Categories without entrenched and dominant competition, and without competitors with leading edge and proprietary technical advances, may represent opportunities, provided other favorable conditions exist. Categories in which it may be possible to exploit technical innovations from other categories may be attractive candidates for new product search. Generally, it is best to avoid categories with high barriers to entry and/or exit, unless you can leverage a significant technical or cost advantage. Access to trade channels is important, but a company lacking a particular type of distribution competence, such as direct store delivery (DSD), may still be able to partner with a company that possesses such a competence.

Powers (1997) reported on the launch of Tropicana® Pure Premium Bursters, a line of Tropicana Pure Premium Orange Juice targeted to children. The

FIGURE 3.7 Campbell V8 vegetable juice in injection-blow-molded polyester bottle. Generally polyester in economic weights does not have sufficient oxygen barrier to retain product quality for the requisite Campbell "shelf life." By increasing the wall thickness and reducing the distribution time requirements, this lightweight, almost unbreakable bottle, containing the 46 ounces of the traditional can could be introduced in commercial distribution channels.

opportunity for Tropicana was based on the fact that sales in the refrigerated juice category, in which Tropicana is a major player, stood at $3.5 billion at the time, which represented a 5.6% annual rate of growth. In a category this big, there is sufficient opportunity to carve out a new niche, and a niche built around a nutritious kids product makes sense on two levels. The first is the sheer size and growth of the children's market. Second, as the children of the echo boom reach school age, they exert more influence over household shopping patterns. Twenty-two percent of the U.S. population is under the age of 15, with the peak of the "echo boom" babies now entering grade school. A group called Kid Think, Inc. reports that one third of all processed foods are purchased for children, and Tropicana reports that the market for chilled juice multipacks targeting children has grown 139% since 1994. Campbell took advantage of the same opportunity with its highly successful launch of V8® Splash, in fall 1997, which provided vitamin A, an antioxidant, in a mix of good-tasting tropical fruit flavors (Figure 3.7).

SUMMARY

Economic and competitive circumstances in the food industry make it both necessary and extremely difficult to launch successful new products. Firms must understand the role of new products in business strategy and ensure that new product projects are consistent with and emerge from goals and strategy for a business. Having a consistent approach to new products, including appropriate organizational support, a charter to guide new product development efforts, management of a portfolio of development projects, and specific projects selection criteria, enhances new product success. A critical step in new product strategy is opportunity analysis, which uncovers new product opportunity areas that fit the firm's competencies and offer strong prospects for growth.

BIBLIOGRAPHY

Aaker, D. A. 1998. *Strategic Market Management.* 5th Ed. New York: John Wiley & Sons.

Anonymous. 1998. "New Product Intros Fall Flat in 1997," *New Product News,* 34(1):24–25.

Cooper, R. G. 1993. *Winning at New Products,* 2nd Ed. Boston: Addison-Wesley.

Cooper, R. G. 1996. "New Products: What Separates the Winners from the Losers," in *The PDMA Handbook of New Product Development.* Rosenau, M. D., A. Griffin, G. A. Castellion, and N. F. Anschuetz, eds. New York: John Wiley & Sons, pp. 3–18.

Crawford, C. M. 1997. *New Products Management.* Boston, MA: Irwin McGraw-Hill.

Dimancescu, D. and K. Dwenger. 1996. *World Class New Product Development.* New York: AMACOM.

Dwyer, S. 1997a. "Hey, What's the Big Idea? Pillsbury Knows," *Prepared Foods,* 166(5):8–20.

Dwyer, S. 1997b. "Taking Core of Business," *Prepared Foods,* 166(8):12–16.

Feig, B. 1997. "Success-Radical," *Food & Beverage Marketing,* 16(8):10.

Fuller, G. W. 1994. *New Food Product Development: From Concept to Marketplace.* Boca Raton FL: CRC Press.

Fusaro, D. 1995. "Kraft Foods: Mainstream Muscle," *Prepared Foods,* 164(5):8–12.

Gill, B., B. Nelson, and S. Spring. 1996. "Seven Steps to Strategic New Product Development," in *The PDMA Handbook of New Product Development.* Rosenau, M. D., A. Griffin, G. A. Castellion, and N. F. Anschuetz, eds. New York: John Wiley & Sons, pp. 19–33.

Hill, J. 1997. "The New Product Hurdle That May Cost You the Race," *New Product News,* 33(8):13.

Hoban, T. J. 1998. "Improving the Success of New Product Development," *Food Technology,* 52(1):46–49.

Hoch, J. G. 1998. "Slash-and-Burn Strategies," *Food Processing,* 59(4):32–35.

Kuczmarski, T. D. and T. Shapiro. 1996. "Measuring Your Return on Innovation," *Marketing News,* 30(20):17.

McCarthy, M. J. 1997. "Food Companies Hunt for a 'Next Big Thing' But Few Can Find One," *Wall Street Journal,* May 6, pp. A1, A8.

Mehegan, S. 1997. "Cutting Edge," *Brandweek* 38(35):1, 6.

Meyer, A. 1998. "1998 Top 100 R&D Survey," *Food Processing,* 59(8):32–40.

Neff, J. 1997. "New Products: Narrowing the Focus," *Food Processing,* 58(11):71–72.

O'Donnell, C. D. 1997a. "Nabisco's A1 R&D Strategy for Global Growth," *Prepared Foods,* 166(8):51–58.

O'Donnell, C. D. 1997b. "Campbell's R&D Cozies Up to the Consumer," *Prepared Foods,* 166(10):26–30.

Peterson, R. 1993. "Speed Is Critical in New Product Introductions," *Marketing News,* 27(5):4.

Powers, N. 1997. "Kids Get Juiced," *Food Processing's Rollout!* 2(19):1.

Schiller, Z., G. Burns, and K. L. Miller. 1996. "Make It Simple," *Business Week,* pp. 96–104.

Thompson, S. 1997. "Reborn," *Brandweek,* 38(18):28–32.

Wheelwright, S. C. and K. B. Clark. 1995. *Leading Product Development.* New York: The Free Press.

CHAPTER 4

New Product Failure and Success

JOHN B. LORD

Too many new food product introductions fall far short of the developer's expectations, visions and hopes. To minimize disappointments, the new food product development process should be a disciplined exercise in which the market environment is truly and fully comprehended. Opportunities in consumer desires and needs should be identified and clearly defined. Marketing research should be employed as a tool to qualitatively and quantitatively specify the product's role among consumers. However important is information derived from marketing research, enough data to render perfect decisions are never available. Thus, the developer should never be paralyzed by either a dearth of information or a plethora of data.

INTRODUCTION

NEW products play several roles for the organization. They help maintain growth and thereby protect the interests of investors, employees, suppliers of the organization. New products help keep the firm competitive in a changing market (Patrick, 1997). The consequences of product development have a direct impact on competitiveness. They mean the difference between falling behind a leading competitor in the marketplace and being the competitor who provides leadership, compelling others to meet similar standards (Wheelwright and Clark, 1995). Finally, new products spread the marketing risk. The investment community values new products; new products affect the top line and therefore enhance the value of the firm and shareholder value (Patrick, 1997).

The academic and the business periodical literature are replete with detailed listings and explanations of both why new products fail and what factors are

55

related to success. There is no shortage of guidance available to those interested in achieving the revenue growth, profit growth, and reputation for innovation and leadership associated with successful new product launches. Organizations invest many human, material, and monetary resources in new product development. In addition, much research, by both the academic and industry sectors, has been conducted regarding the factors involved in new product failure as well as success. Yet, the statistics that we frequently hear cited about product development and the rates at which new food products fail are frightening. How can these seemingly contradictory facts be reconciled?

The key driver of the efficient consumer response (ECR) initiative is an industry estimate that the excess cost to the grocery system in the product development and introduction process ranges as high as 4% of net sales. These costs include both:

- all development and introduction costs associated with failed products, including products canceled before introduction as well as products withdrawn after launching
- excess costs incurred in launching successful new products, principally excess manufacturing costs due to an initial massive inventory buildup needed for introductory deals and special offers, e.g., free goods

Industry data from a study commissioned by the Joint Industry Task Force on New Products and conducted by Deloitte & Touche Consulting Group in 1995 suggests that new product introductions cost the food system (manufacturers, brokers, wholesalers, and retail grocery stores) approximately $252 per SKU, per store. It is important to note that the study was conducted using 1988 data. Assuming an industry inflation rate of only 2.5% over the past 10 years, this figure now approximates $320 per SKU per store. The industry clearly spends a great deal of money on products that are introduced but do not succeed.

GETTING THE NUMBERS STRAIGHT

A key aspect of ECR, efficient product introduction, addresses the concern about the alarming number of new products launched each year, and the fact that most of these are line extensions (Kahn and McAlister, 1997). The July 1997 issue of *Progressive Grocer* included a supplement entitled "Efficient New Product Introduction." This report was intended to "describe techniques for new product introduction . . . advancing the understanding of distributors, brokers and manufacturers within the grocery industry" and cited a project undertaken by Ernst & Young who provided the data cited in the report. Prime

Consulting Group, Inc., as part of the study, computed product introduction, success, and failure rates.

Defining what constitutes a new product success or failure is a critical first step in computing and assessing success and failure rates. If a product concept demonstrates enough strength during early stage testing to warrant investment in product development, but fails to survive beyond product or market testing, is this a new product failure? If a new product is launched and gains retail distribution and generates revenues but those revenues fail to meet stated targets, is that new product a failure? If a new item is launched and generates significant first-year distribution and revenues, but loses distribution and revenues after the first year, is that a new product failure? How do we account for products that are seasonal or are merely replacements for other products in our existing line?

Clearly, therefore, the industry needed to develop a consistent definition of what constitutes new product success or failure. However, even more basic was the need to clearly define what constitutes a new product. (Figure 4.1).

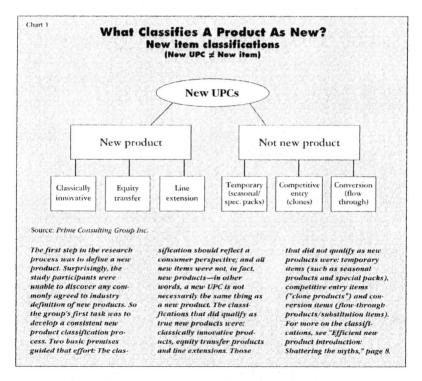

FIGURE 4.1 What classifies a product as new? New item classifications (New UPC ≠ new item). *Source:* Prime Consulting Group, Inc. Reproduced with permission from *Progressive Grocer*, July 1997, pp. 44.

Is a product new because it is new to the consumer? Most industry observers agree that a product new to the consumer is a new product. But how about a product that is new to the company? Or a product that represents improved performance?

The *Progressive Grocer* report is notable for its specification of (1) a classification scheme for new food and allied products which differentiates new products from new UPCs; (2) a specific definition of product failure which is used as a criterion to determine whether a new product is classified as a success or a failure; and (3) empirical data on failure and success rates. Previously, with the exception of data provided by Information Resources Incorporated in their annual "New Product Pacesetters" reports, little actual data has been provided to either confirm or contend the conventional wisdom that 4 out of 5 (or worse) of all new products fail.

NEW PRODUCT CLASSIFICATION

New Product News (Dornblaser, 1998a) has reported that the number of new food products introduced in the past three years ranged from 13,000–16,000. On the other hand, Marketing Intelligence Service Ltd. has declared, based on its own study, that of 11,072 new food products introduced in 1996, only 7.2% featured innovations in formulation, positioning, technology, packaging, or creating a new market (Messenger, 1997).

The Ernst & Young project group's first task was to develop a consistent new product classification process. The basic premises (Anonymous, 1997) were that "(a) the classification should reflect a consumer perspective and (b) that all new items are not . . . new products . . . that a new UPC is not the same thing as a new item." A following classification scheme was developed and is shown in Figure 4.1.

A new UPC is not necessarily a new product, and not all products new to a company are new to the consumer. In fact, for purposes of the study, only the first three categories were considered new products. According to this classification, instead of the 20,000 plus new food and allied products reported annually by *New Product News,* only 1,100 to 1,200 new products are introduced each year, of which 22% are new brands and 78% are line extensions. Of the items introduced to the trade each year, 56% represent UPC changes on existing items; 28% are test market items and new PLU (price look-up) items; 10% are seasonal items; and only 6% represent true new product introductions. Similarly, a presentation made by AC Nielsen at the 1997 *Prepared Foods New Product Conference* included an estimate that just 7% of the new products introduced in 1996 represented innovations in formulation, positioning, or technology. The data from Prime Consulting Group published in *Progressive Grocer* are shown in Figure 4.2.

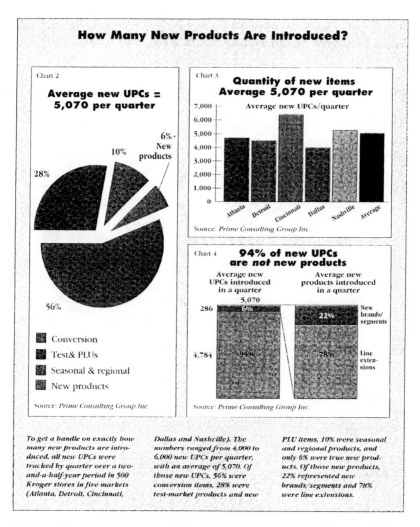

FIGURE 4.2 How many new products are introduced? *Source:* Prime Consulting Group, Inc. Reproduced with permission from *Progressive Grocer,* July 1997, pp. 45.

For purposes of its analysis, IRI has a very narrow definition of new product. The 1996 Pacesetters Report (IRI, 1996) defined a new product as one that had achieved less than 20% ACV distribution in 1995 or year 1 and had achieved at least 30% ACV distribution in year 2. This led to a list of 440 new products to be studied, with 75% of those achieving introductory sales of less than $7.5 million. Of those 440 new products, 15% featured new brand names and 85% were considered line or category extensions. There are rewards for innovation: innovative products averaged $63 million in first year sales and non-innovative products averaged $28 million.

SUCCESS AND FAILURE DEFINED AND MEASURED

The 1996 IRI Pacesetters Report also included estimates of the rates of new product success and failure. Based on Infoscan data, which provides an analysis of product movement at retail, IRI calculated that approximately 72% of new products and 55% of line extensions fail (IRI, 1996). Linton, Matysiak and Wilkes, Inc. conducted a study of new products introduced by the 20 U.S. food companies with the most new product introductions in 1995. Of the 1,935 new products introduced by these companies, 174 were "new" and 1,761 were line extensions. These new items experienced a success rate of 52% while line extensions had a 78% success rate, combining for an overall success rate of 76%. Unfortunately, companies other than the leaders introduced 14,298 products and achieved a success rate of only 11.9%, leading to the conclusion that the biggest companies succeed with new products more/better than small companies (Dornblaser, 1997a).

Why the disparity in these reported rates of success and failure? It all depends on the definition of "failure" and of "success." This problem is not easily solved. Obviously, a new item's sales cannot be tracked unless the product is launched, at least into a test market. The task of measuring the number of ideas, concepts, prototypes, and other forms of new products which never see the light of the market, in order to establish a more theoretically appealing rate of new product failure, is daunting if not impossible.

In light of this challenge, the Ernst & Young project team established the following definition of success: "A new product is considered a success if it achieved at least 80% of 26-week sales per distributing store after two years" (Mathews, 1997).

Performance data are displayed in Figure 4.3. These data were derived by applying the above-referenced definition of success to both new products and line extensions and using information provided by Efficient Marketing Services Inc. (EMS), which measured sales, distribution, and sales per distributing store. Charts 5 and 6 show sales patterns for new products, demonstrating different sales patterns after launch for successful products and those that fail. Chart 7 illustrates success rates as measured by Prime Consulting.

One of the caveats mentioned in the study is that while distribution of new products is historically been driven to achieve overall ACV penetration, this may not be appropriate for new items that are targeted to specific market segments. This problem takes on more importance in an era of more and more precise target-marketing, with product variations created to serve very specific consumer (and even trade) niches in search of very specific product benefits.

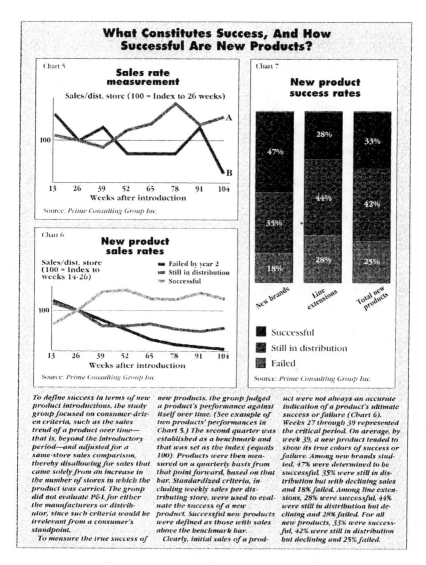

What Constitutes Success, And How Successful Are New Products?

Chart 5

Sales rate measurement

Sales/dist. store (100 = Index to 26 weeks)

A

100

B

13 26 39 52 65 78 91 104
Weeks after introduction

Source: *Prime Consulting Group Inc.*

Chart 6

New product sales rates

Sales/dist. store
(100 = Index to
weeks 14-26)

- Failed by year 2
- Still in distribution
- Successful

100

13 26 39 52 65 78 91 104
Weeks after introduction

Source: *Prime Consulting Group Inc.*

Chart 7

New product success rates

New brands	Line extensions	Total new products
47%	28%	33%
35%	44%	42%
18%	28%	25%

■ Successful
▨ Still in distribution
▨ Failed

Source: *Prime Consulting Group Inc.*

To define success in terms of new product introductions, the study group focused on consumer-driven criteria, such as the sales trend of a product over time—that is, beyond the introductory period—and adjusted for a same-store sales comparison, thereby disallowing for sales that came solely from an increase in the number of stores in which the product was carried. The group did not evaluate P&L for either the manufacturers or distributor, since such criteria would be irrelevant from a consumer's standpoint.

To measure the true success of new products, the group judged a product's performance against itself over time. (See example of two products' performances in Chart 5.) The second quarter was established as a benchmark and that was set as the index (equals 100). Products were then measured on a quarterly basis from that point forward, based on that bar. Standardized criteria, including weekly sales per distributing store, were used to evaluate the success of a new product. Successful new products were defined as those with sales above the benchmark bar.

Clearly, initial sales of a prod-uct were not always an accurate indication of a product's ultimate success or failure (Chart 6). Weeks 27 through 39 represented the critical period. On average, by week 39, a new product tended to show its true colors of success or failure. Among new brands studied, 47% were determined to be successful, 35% were still in distribution and 18% failed. Among line extensions, 28% were successful, 44% were still in distribution but declining and 28% failed. For all new products, 33% were successful, 42% were still in distribution but declining and 25% failed.

FIGURE 4.3 What constitutes success, and how successful are new products? *Source:* Prime Consulting Group, Inc. Reproduced with permission from *Progressive Grocer,* July 1997, p. 46.

NEW PRODUCT FAILURE

Hoban (1998) conducted a survey of food and allied products manufacturers and distributors to ascertain both positive and negative influences on new product success. Respondents were asked how much impact each of 11 different barriers or negative influences had on the success of their firm's new product development efforts. Results indicated that lack of strategic focus, limited understanding of the market, priorities not set or communicated, lack of financial resources, and focus on short-term profitability were rated as having the highest negative impact. Poor product quality, limited creativity or vision, and lack of support for risk-taking were also rated as high impact by more than half of the respondents.

Hoban's survey respondents were also asked to state in their own words why some new products are not successful. Answers were coded into eight main categories. The top four reasons listed by manufacturers were insufficient product marketing, duplication or lack of innovation, lack of support during rollout, and lack of a compelling consumer benefit. The top four reasons listed by distributors included: lack of a consumer benefit, duplication or lack of innovation, insufficient product marketing, and inadequate market research (Hoban, 1998).

Lots of food products get to market, obtain early distribution, and fail to sustain it. Before too long, distribution is waning, and the product is being "swapped out" for another new item. Although there are examples of new food product failures almost too numerous to mention, a couple of names are Campbell's® Souper Combos and Crystal Pepsi®. Why did these and numerous other products fail? Crawford (1997) has identified a fairly complete set of reasons including poor planning, poor management, poor concept, and poor execution. Virtually every reason for failure falls into one of these categories.

Poor planning incorporates issues such as developing a product that doesn't fit a company's strategy, competencies and/or distribution strength. Bic makes great disposable pens and razors, but not perfume. Anheuser-Busch felt it could make Eagle® Snacks a success because snacks are complementary to beer and can be distributed using the same channels, but snack manufacturing and distribution relies on different competencies than beer brewing and distribution. Poor planning also includes failure to properly analyze the market to understand whether and what type of opportunity may exist in a category and what specific unsolved problems consumers have. Also included in this category of reasons for failure is not understanding the stakes of the game, that is, the cost to enter and sustain a position in a given product category. Finally, products can fail because firms did not perform due diligence in discovering patent and copyright issues and/or understanding the potential impact of regulations such as those governing the use of certain types of advertising claims or labeling information. These failures are all attributable to poor planning.

Poor management is all about the organizational culture, support and resources for new product process, management's expectations and focus, the process used to develop products, and the process of deciding upon which projects are funded and which are not. New product development inherently involves investment, risk, and rewards that may not be immediately evident. Firms that don't encourage entrepreneurial behavior will not observe any. Firms that discourage or penalize taking risks will find that no one in the firm is willing to take risks. Firms that don't provide adequate investment capital can never achieve big winners. Firms that don't put the best people in effective cross-functional teams, on new products, will succeed only rarely. Firms that demand immediate returns will have only limited successes with line extensions but no innovation. Projects which lack the individual—the new product champion—willing to see the project through, and willing to fight the battles to obtain necessary resources, will wither for lack of support. Organizations lacking clear goals and direction for new product efforts will find a real difficulty in achieving a coherent and coordinated effort which is both effective and efficient. The bottom line is that firms that do not invest in innovation won't experience any innovation.

A poor product concept lacks a compelling consumer benefit, is a simple me-too item with no real and relevant difference from items already available, does not have a defined market target with adequate sales potential, or is introduced at the wrong time. Products lacking a significant consumer benefit as well as competitive clone items simply do not provide consumers a reason to buy. The trade may take on such an item if the up-front financial incentives are adequate to pay for the shelf and display space, but such products cannot sustain their market position and will quickly lose distribution.

Poor concepts also are those that are positioned incorrectly. A product's position is the place it holds in the consumer's mind, relative to the positions of other brands in the category, and based on the characteristics and benefits which consumers associate with the brand. What the marketer says about the product, how it is priced, packaged, merchandised, and how it performs create a product's position. A poor position is one which associates benefits with a product that are irrelevant to the consumer, or alternatively one by which the consumer is unable to perceive a meaningful difference between brands. In either case, the consumer sees no reason to buy the product.

One of the big issues in food product distribution is that of variety versus duplication. A new product possessing unique characteristics and benefits adds to the variety offered by a retailer and available to the consumer. This uniqueness adds to the appeal of the product to both the trade and the household consumer. On the other hand, me-too products, which merely duplicate products already on the shelf, add to shelf clutter and to consumer confusion and serve merely to further fragment the product line. Fragmentation of the product line adds operating costs in the areas of production, distribution, and inventory control, and can threaten the existing product base by increasing potential stock-outs and reducing on-shelf visibility.

Poor execution covers a multitude of sins. Included in this category are products which fail to deliver on the promise, that is, they simply do not perform to customer expectations as well as products that simply do not work as intended or expected. Food products that lack taste or texture, packages that are difficult to open or do not dispense properly, and products which deteriorate before going out of code are all examples of products which do not execute correctly.

Poor execution extends to all areas of the marketing program. Failure to properly execute the sales plan and achieve targeted levels of distribution, as well as failure to achieve appropriate display and retailer support, all contribute to new product failure. Advertising that is of insufficient weight to build awareness and trial according to plan, is poorly scheduled relative to seasonal sales factors or when the product is available, fails to reach the target consumer, or fails to communicate clearly the new product's brand name, benefits, and availability are all examples of poor execution. Products that are mispriced or have a poor price-value relationship, as perceived by the target customer, will likely fail.

Products that are introduced too late, either because they have been pre-empted by competitors or because the market has moved on to something else, are poorly executed. The early 1990's saw the introduction of many low-fat products; however, the late 1990s have witnessed a return to indulgent products. Products that are introduced before their time are poorly executed. Campbell's Intelligent Cuisine was a product line with a very strong concept and rationale, but the idea of a medically prescribed, fully prepared diet may have been a bit ahead of its time. Of course, another execution issue with Intelligent Cuisine is that Campbell was not set up to detail the product to physicians, which is something for which pharmaceutical companies but not food companies have a competency.

ACHIEVING SUCCESSFUL NEW PRODUCT LAUNCHES

Mathews (1998) reported on a food industry task force that was commissioned in 1997 to study new products and how they function in the industry. The work of the task force has resulted in some key findings. Only one in 20 U.S. brands have increasing sales in their second and third year in the market. There is a direct relationship of consumer trial to sales volume. Sales volume for new brands seems to be almost completely dependent on sustained advertising, and a reduction of advertising after the introductory period almost invariably leads to dramatic sales declines. Study results indicated that 69% of new

products have sales declines after the first year. As a result of their resea.
the task force concluded that there are four basic consumer truths associated
with new products:

- The product needs to deliver on the concept promise. Products with
  high consumer acceptance, as measured by after-use purchase intent,
  are likely to succeed. Those with low acceptance are likely to fail.
- Advertising quantity and quality matter. Advertising drives awareness,
  which drives trial, which drives sales.
- Distribution drives sales.
- New brands need long-term support if they are going to succeed.

Understanding what truly drives new product success requires evaluating
new product activity from the point of view of the entire food distribution
channel. From this vantage point, other "truths" about new products are (1)
a new item or line needs to add incremental dollars and profit for both the
category and the brand, (2) new products should enhance the manufacturer's
and the retailer's competitive position in the market, (3) launches should be
carried out with minimum disruption to the distribution system as a whole;
(4) a launch should be accompanied by both a product service plan and a
consumer target plan defined by both the manufacturer and distributor, and
(5) there should be an identification of activity-level costs (Mathews, 1998).

While useful to the industry, these findings are incomplete because the
focus is on the back-end of the development process, specifically, new product
launch. Cooper (1996) reported the results of various studies which concluded
that the most important discriminators between new product winners and
losers were, in rank order:

(1) Understanding of user's needs
(2) Attention to marketing and launch publicity
(3) Efficiency of development
(4) Effective use of outside technology and external scientific communication
(5) Seniority and authority of responsible managers

These factors have much more to do with the front-end of the process.
Wheelwright and Clark (1995) maintain that the entire range of functions and
activities, every dimension, of the organization drives the success or failure
of product development. The development of a new product is the development
of every aspect of the business that the product needs to be successful. And
consistently successful new products need every aspect of the business working
in harmony.

Thus, it may be instructive to construct a listing of new product success
requirements, which integrates the findings of numerous industry experts and

observers, and will provide guidance to firms looking to enhance their success in new product development. These success requirements can be developed into a 13 point listing which is shown in Table 4.1.

AN APPROPRIATE ORGANIZATIONAL ENVIRONMENT AND TOP MANAGEMENT SUPPORT

An organizational environment that is right for successful new product development is one in which there is a nurturing of innovative thinking, such as the case with a leading innovator like 3M, and a culture that rewards creativity. Firms must also support risk taking and take a long-term perspective, which involves supporting new products through the challenging "early years." Firms must recognize that while there are corporate hurdles for any investment, really new products often show losses during the first year to three years because of high introductory marketing costs, and this effort must be sustained. Organizations with a strong track record in new products invest in continued innovation. Organizations must clearly communicate that it is acceptable to take risks and even make mistakes or else they will clearly wind up with a "nothing ventured, nothing gained" type of situation. Finally, firms should commit to putting their best people on new products.

Cooper (1996) discusses the importance of top management support in driving projects to market, that is, seeing the project through to completion.

TABLE 4.1. **New Product Success Requirements.**

(1) An appropriate organizational environment and top management support
(2) Use of a disciplined new product development process
(3) Dedicated development teams plus the willingness and ability to partner and outsource
(4) Product development activities that start with and flow from business unit strategy
(5) Understanding the external environment and identifying market opportunities that fit core competencies
(6) Identification and specification of what is driving the consumer and what the consumer wants
(7) Processes and techniques for keeping the pipeline filled with a wide variety of product ideas
(8) Clear and focused product definition early in the process before development work begins
(9) A superior and differentiated product and package
(10) Use of research to measure reaction to the product and all elements of the program throughout the development process
(11) Use of category management philosophy to align manufacturer and retailer focus on the consumer
(12) A well-executed launch
(13) Ability to adapt, grow, and improve as market and competitive conditions evolve

Top management's main role in supporting innovation is to set the stage: commit to a game plan and make available the right resources. Top management must define project boundaries, missions, and charters. Top management must pick the right kind of teams and create job opportunities and career paths that result in qualified team leaders and qualified core team members, and that support individuals when, where, and in quantities needed. Top management must also ensure that teams are not only effective in achieving their individual project goals, but also fit well with the overall strategy of the business (Wheelwright and Clark, 1995). Good managers use objectivity to deal with failures, correct problems, and celebrate successes.

One of the most important characteristics of successful product development is achieving speed to market, but with quality of execution. The right management approach to new products can help achieve this. For example, Pillsbury launched Totino's Stuffed Nachos in late summer 1996. The idea for the product emerged in early 1995 and original projections were that it would take a full year to bring the product to market. However, brand management realized that in order to be in full distribution prior to January, the top snacking month, the product had to be ready for shipping by late summer. Top management accomplished this by telling the development team for Stuffed Nachos that the management group would assume all risks associated with the accelerated timetable, allowing the development team to focus on the task at hand. The team delivered (Dwyer, 1997a).

USE OF A DISCIPLINED NEW PRODUCT DEVELOPMENT PROCESS

Successful product launches involve a number of different activities and an increasing amount of investment as the project proceeds closer to launch. A disciplined process with defined stages and gates such as the Arthur D. Little process (Rudolph, 1995) (see Chapter 5) provides the basis for objectivity in decision making because deliverables and metrics are defined for each stage. Such a process helps to ensure that investment capital will be expended wisely and on the best projects. Cooper (1996), the originator of the stage-gate process, notes the importance of the quality of execution of technological activities such as preliminary technical assessment, product development, in-house product or prototype testing, trial or pilot production, and production start up. There is a need for completeness, consistency, and proficiency which is much more likely with a well-planned and disciplined process for which all constituent activities are defined and sequenced at the front end.

A disciplined process should contain realistic but aggressive timetables. Speed to market can be improved by up-front planning of our projects to identify which activities must be performed in sequence and which can be performed in parallel. Being first in the market can be very important, particularly when the technical differences between products are slight or nonexistent. For example, in the razor category, we had Gillette's Sensor versus Schick's Tracer. Both products represented innovative enhancements; both were introduced with huge spending and incentives. The major difference is that Sensor beat Tracer to market by one year. For Sensor, trial build was substantial and consistent. Sensor achieved a trial rate of 40.3% during the first six months of introduction and 66.6% during the first 12 months. On the other hand, Tracer achieved trial rates during comparable periods of 14.1% and 49.7% respectively. Part of the performance difference is attributable to Gillette's market leadership; part was attributable to the fact that Tracer's introduction lagged Sensor's by one year.

DEDICATED DEVELOPMENT TEAMS PLUS THE WILLINGNESS AND ABILITY TO PARTNER AND OUTSOURCE

The use of cross-functional teams encourages communication across functions and enhances speed to market. Interaction among team members who are co-located and share a common set of objectives facilitates the implementation of projects. Dwyer (1998b) described how Dean Foods revitalized the milk category with the launch of Milk Chugs in 1996 (Figure 4.4). The firm's ability to mobilize into teams allowed it to scale high hurdles during the 2-

FIGURE 4.4 Fluid milk packaging has finally departed from that of commodity and into more market-oriented packaging. These plastic bottles are decorated with reverse-printed shrink film labels that conform to the bottle shape. The labels provide differentiating color that can be seen at a distance from the shelf. Note also that multipacks of unit-portion sizes are in printed shrink film.

year development cycle. A flat organization such as Dean Foods uses cross-functional teams so that individuals can contribute from their own perspective and then cross over and contribute from the perspective of other functions. The team approach allowed engineers to come up with the best and most economical choice for machinery, with finance determining how to pay for equipment and marketing to create programs so that volume projections could be met (see Figure 4.1).

Cross-functional teams make it easier to marshall necessary resources from the many disparate functions required to successfully launch a new product. Pillsbury used the team approach to develop and launch low-fat Häagen Dazs (Dwyer, 1998b). The Haagen Dazs team stressed a process in which there was a mutual ownership of the project, delegation, and goal-setting, serving to empower the team. The team included as many as twelve individuals that worked in two integrated groups—a business team and a project team.

At Nabisco, a cross-functional new products team generally consists of 8 to 12 people from commercialization, marketing, product development, engineering, quality assurance/quality control, packaging, sensory services, logistics, and finance, and may include a representative from a supplier in the very beginning of the project (see Chapter 9). The team is almost certain to interface with the company's legal, regulatory, and public relations personnel as well as with top management. In addition, every new product project needs a champion and a sponsor to compete for scarce organizational resources and see the project through to completion. At Nabisco, each team has one person assigned in the following roles: sponsor, team leader, phase leader to motivate and manage the project.

As a strategic response to intensifying global competition, food processors are reassessing and redefining core competencies, then downsizing and restructuring to focus on these competencies that add value to the business. Core competencies increasingly are being focused on consumer marketing and product innovation, not plant and process design. As a result, although most medium- to large-sized companies have at least the minimal in-house capabilities to generate new products, few possess all of the necessary skills to efficiently develop the new products which today's market demands. This situation creates the need to outsource certain new product development activities.

Increasingly, food processors take advantage of supplier expertise in the realms of ingredients, equipment and research/laboratory facilities. Services provided by suppliers may involve one or more of chemistry, microbiology, food technology, processing (including scale-up), or sensory evaluation/market research (Morris, 1996b). Some manufacturers are partnering or joint venturing with each other and with suppliers to retain core competencies, pool resources so they can access new markets or new technologies, cut costs, leverage manufacturing resources, and optimize assets and gain a competitive advantage (Morris, 1996a).

According to Ferrante (1997), there are four basic reasons to turn to co-packing:

- when there is a short term need for emergency or additional capacity
- to leverage technology and expertise that the company does not own internally
- to gain speed to market; to expedite a product line to market with low risk by using a third party manufacturer
- to reduce up-front costs and investment (capital investment, fixed and variable costs, transportation of ingredients, supplies, finished product, etc.)

Similarly, Berne (1996) maintains that outsourcing is useful in certain circumstances. These include when firms

- do not have the internal resources or time to accomplish all necessary development work
- do not have the needed technology in-house
- have a blue sky idea so far out they do not know where to start
- have had core competencies weakened due to downsizing
- need some expert advice

According to Demetrakakes (1996), the biggest reason for the increased reliance on outside help is money. Results of *Food Processing's* "The 1998 Top 100 R&D Survey" (Meyer, 1998) indicated that 76% of R&D personnel respondents reported their department's budgets were stagnant or declining. Downsizing is the inevitable result of these budget pressures. But along with downsizing comes pressure to bring new products to market ever faster. When 10 employees are doing the benchtop work that used to be done by 40, a boost from the outside is often welcome.

There are four basic reasons why Kraft turns to co-packing: (1) a short-term need for additional/emergency capacity, which may result from a seasonal spike in sales or an internal manufacturing issue; (2) to leverage technology and experience which Kraft does not own internally; (3) risk management and speed to market (to avoid the risk of investing in new equipment when someone else already has the needed equipment and the market outlook is uncertain and/or to reduce cycle time; (4) cost, because a co-packer can occasionally produce for less (Ferrante, 1997).

PRODUCT DEVELOPMENT ACTIVITIES MUST START WITH AND FLOW FROM BUSINESS UNIT STRATEGY

In its most basic form, strategy involves choices of products and markets, and how product-market combinations will be exploited for the purpose of achieving business objectives. Product development cannot be isolated from overall business unit strategy; on the contrary, it is essential that new product efforts emanate from a clearly articulated set of objectives and strategy for business growth. Consistent with the notion of building on and leveraging core competencies, this "rule" stipulates that our new product efforts must be clearly focused and must start with those business processes and assets which are the bases for the firm's competitive advantage.

Hawkins (1996) states that one of the "eight ways to win in the marketplace" is building new products on the company's strongest assets, which might be a technology platform, a particular production process, or brand equity. According to Cooper (1996) firms should attack markets from a position of technological strength, and elements of technological synergy, including fit with research and development, and engineering and production skills, should be critical criteria in the prioritization and evaluation of new product projects. Some basic rules for new product success include developing a sound strategy that leverages core capabilities and clearly differentiating new products from those of competition through carefully conceived, developed, and designed brand identities.

Cooper (1996) also notes the need for marketing synergies which means a strong fit between the needs of the project and a firm's marketing competencies and assets: brands, research and market intelligence capabilities, distribution channels, sales force, advertising resources, and so on. When firms launch products that sell into markets that are familiar to the firm, and which utilize a firm's existing distribution channels and sales force, prospects for success are heightened considerably.

We must also consider financial and human resources since both are quite obviously essential aspects of new product development. Projects must be both adequately funded and properly resourced with key people from a variety of disciplines. The number, type, and timing of projects must be set with reference to available financial and human resources.

UNDERSTANDING THE EXTERNAL ENVIRONMENT AND IDENTIFYING MARKET OPPORTUNITIES

One of the most important guidelines for new product development voiced by both experts and practitioners is the need to consider the attractiveness of the market. Opportunities for successful new products exist when firms can leverage their technical, marketing, financial and human resources against marketplace gaps. These gaps represent combinations of benefits sought by groups of consumers who can be identified, reached selectively, of adequate size to represent market potential, and who have not found an existing product that provides these benefits. What creates a gap? A gap is created by changing consumer tastes that are driven by changes in the marketing environment.

Of primary significance for marketers are demographic, socioeconomic and lifestyle changes which drive changes in the way consumers eat and the types of products that fit their consumption system. For example, the aging population has created increasing interest in foods that deliver specific health benefits to provide a better overall health profile for older consumers. The increasingly hectic pace of society combined with a growth in the number of working women and single parent households has created an increasing time pressure that has led to changes in the way we prepare food and an emphasis on convenience and time-savings. Thus, the hottest topics in food product development today are nutraceuticals (wellness) and home meal replacement (convenience).

In addition, advances in ingredient, process and packaging technology, and changes in the regulatory environment affect both the manner and the economics of delivering relevant consumer benefits, and provide food processors with capabilities of delivering taste, convenience, and health benefits in ways previously not practical or even possible. New product success rates improve significantly when companies monitor the environment and become aware early on of changes that create marketplace gaps and define attractive markets.

Market attractiveness is a strategic variable of increasing importance. Attractive markets have at least some of the characteristics listed in Table 4.2.

IDENTIFICATION AND SPECIFICATION OF CONSUMER DRIVERS AND CONSUMER WANTS

Unsatisfied consumer needs represent potential new product opportunities. Thus, effective and successful new product development has to start with the customer. Having identified the market opportunity, the next step involves studying and understanding the dynamics of the consumers in the category, using extensive market research and working closely with customers/users to identify customer needs, wants, and preferences, and to define what the cus-

TABLE 4.2. **Characteristics of Attractive Markets.**

- Sufficient size
- High growth rates
- Few barriers to entry, such as patents, very high advertising to sales ratios, restricted access to raw materials, and so on
- Few barriers to exit, such as restrictive service contracts with customers or fixed assets with no alternative uses
- Few recent product or technical developments in the category
- Categories that are not dominated by a strong competitor or set of competitors
- New emerging needs driven by demographic and lifestyle changes.

tomer sees as a "better product." According to Margo Lowry, vice president for new business development at Campbell Soup, "effective new product development means extensive research to understand the consumer plus intuition. You must understand why people do what they do and say what they say. This is where data and consumer research ends and intuition begins" (O'Donnell, 1997b).

One of the most important principles of marketing is that successful marketing involves truly understanding your customers. Understanding your customer requires market research, which provides data about the market in the aggregate. However, astute marketers recognize that truly understanding your customers requires that you hold regular individual and small group discussions with your customers. Qualitative research, typically utilizing focus groups and sometimes depth interviews, can be utilized to uncover and define customers' underlying needs and motivations, preferences, likes and dislikes for current products, and changing tastes and habits. Quantitative research is also useful in quantifying market opportunities by providing marketers with such information as market size, market structure, and awareness and usage patterns in the category. Quantitative research also can be used to develop demographic and lifestyle profiles plus media usage patterns that can be correlated with product usage to yield valuable market insights. Finally, there are a large number of "trendmeisters" who provide insight into major consumer trends, knowledge that is necessary to lead instead of merely to follow the market.

A great example of a new product which was clearly driven by an understanding of the customer is pre-cut, packaged salads (Figure 4.5). Consumers want to eat healthy, at least some of the time, but lack both the time and the energy to wash, peel, cut, core, dice and slice several different vegetables and fruits to make up that healthy lunch or dinner. Understanding the consumer helped to create a product idea that revolutionized the produce category (see Figure 4.2).

PROCESSES AND TECHNIQUES FOR KEEPING THE PIPELINE FILLED WITH A WIDE VARIETY OF PRODUCT IDEAS

Although there is typically no shortage of new product ideas, they do not appear magically. Firms must have a systematic process and set of techniques for uncovering new product ideas. Ideation cannot occur only periodically or on a pure project-by-project basis. Through competitive intelligence, market research, and creative techniques, firms need to consistently identify and screen a large multitude of new product ideas. One of the keys to success for Kraft Foods is the process of exploration, whereby Kraft identifies literally hundreds of new product ideas. For every 50 new product ideas identified, approximately six survive initial screening. For every 25 to 30 concepts screened, five concepts are refined, three developed, two go to in-home use and one to test

FIGURE 4.5 Fresh-cut salad vegetables, one of the successful new food product introductions of the 1990s, are generally packaged in flexible pouches. The flexible materials are engineered to provide a very high oxygen permeability which permits oxygen to enter and minimize the probability of respiratory anaerobiosis which can adverse affect the respiring product's flavor.

market. These numbers emphasize the need to be very productive in the ideation process, which is a function of developing a superior understanding of the consumer. Ideation also requires both a good invention system that encourages and rewards creativity as well as excellent market intelligence to uncover what is happening in the marketplace. Great ideas require breaking away from the traditional and letting wild ideas fly without succumbing to the temptation to evaluate or judge.

Beyond the traditional and not-so-traditional ideation techniques of problem detection, brainstorming and others, firms must also keep in mind that successful new products can often come from other parts of the food industry, including foodservice and niches such as gourmet and health foods. Successful new products also come from creating new categories by shooting the gaps between products. Combination products that solve consumer problems include combination glass and household cleaner; analgesics and cold medicines in the same pill or liquid, and so on. New categories can also be created through value-added bundling (e.g., Healthy Choice Salad Bar Select). Sometimes ideas previously rejected might merit consideration if market or competitive conditions have changed. According to Dornblaser (1997b), "Although there seem to be almost no truly new product ideas, sometimes products hit the market at the wrong time, marketed the wrong way, or without the needed support. That's why a failed product concept from the past may be a success in the present." For example, Dean has recently launched Birds Eye Side Orders microwaveable side dishes, which are vegetable and sauce combinations that serve two. Previously, Birds Eye (then Pillsbury) introduced Birds Eye for One side dishes, which is essentially the same concept but in a single serving portion. The new version provides more convenience and fits perfectly with the trend to convenience or "home meal replacement" foods.

CLEAR AND FOCUSED PRODUCT DEFINITION EARLY IN THE PROCESS PRIOR TO DEVELOPMENT WORK

Cooper (1996) notes that a pivotal step in the new product process is to define the target market, product concept and positioning, and the consumer benefits to be delivered prior to the development phase. A fully articulated product concept represents a customer "wish list," which is in turn a protocol for product development, and gives the development team a set of requirements that guide technical development, package development, process development, equipment specifications, and preparation of the marketing program. Without this focused and agreed-to definition, the prospects for an efficient and effective process that minimizes time to market are adversely affected.

Pillsbury's motto (Dwyer, 1997a) is to "test early, fail early and eliminate options that won't work." Initial screening, preliminary market and technical assessments, a detailed market study, and an early stage business analysis are included among predevelopment activities that are pivotal to new product success and must be built into the process. Similarly, Kraft Foods includes among its principles for new product success early development work on positioning, which includes specification of the target audience, key benefit(s), point of difference and source of business (Fusaro, 1995).

SUPERIOR AND DIFFERENTIATED PRODUCT AND PACKAGE

Ultimately, we have to give the consumer a product that promises and delivers meaningful, unique, and superior benefits. Kraft achieves new product success by developing a product bundle that solves an important customer problem in a demonstrably superior way. The product itself—its design, features advantages, and benefits to customer—is the leading edge of new product strategy, according to many experts. According to Cooper (1996), six items comprise a superior product: innovative, possessing unique features, meeting customers' needs better, delivering higher relative product quality and solving the customer problem better (than competitive products), and reducing total customer costs. The message is that products succeed in spite of the competitive situation—because they are superior, well defined, and executed well, and have certain synergies with the firm. Cooper (1996) argues that perhaps there is too much preoccupation in today's firms with competitive analysis, and not enough focus on delighting the customer.

Food Processing magazine (Bennett, 1997) highlighted "the 15 most interesting new products of 1997." All of these products share at least one common feature—a significant difference that drives a consumer benefit. Several of these 15 most interesting new products are shown in Table 4.3.

The success of Frito-Lay's® Baked Lay's salted snacks (Mathews, 1997) highlights the ability of a new brand with a significant and meaningful point of difference to drive top-line growth. Baked Lay's built upon a consumer trend toward eating "better-for-you" foods, and delivered a key benefit, lower fat, without sacrificing taste. An aging and increasingly health-conscious population, but one which wanted taste sensation and an ability to engage in "low guilt indulgence," was willing to pay more for the product because it delivered on the promised benefit. The objective was to develop a line that would grow the category and offer higher margins for retailers through providing consumer value in the way of variety and taste satisfaction. Early versions of the product did not meet taste expectations, requiring breakthroughs in both formulation and processing. Sustained consumer testing and in-market testing was necessary to insure that the significant investment in process and equipment was wise. Ultimately, Baked Lay's delivered the highest trial rates in the Frito-Lay history and a brand that was the largest-selling new item in the 1990's, through 1996, surpassed as the largest selling new item by Frito-Lay's Wow brand of fat-free salted snacks in 1998.

TABLE 4.3. Food Processing's Most Interesting New Products of 1997.

- *Milk Chugs,* from Dean Foods Company, is milk packaged to compete with bottled waters and iced teas by making the product more portable and more hip; the package is the difference, and builds on the heritage of the old round milk bottle, but is resealable and fits in automobile cup holders.
- *Blast O Butter Ultimate Movie Theatre Butter Microwave Pop Corn,* from American Pop Corn Company, is high indulgence (double butter) with only a little more guilt (only more gram of total fat per serving than regular butter flavor).
- *Häagen Dazs Low Fat Ice Cream,* from Pillsbury, provides full indulgent high-fat ice cream taste but with only 3 grams of fat or less per serving, based on a patented production process using lactose-reduced skim milk with a high milk protein content.
- *The UnHoley Bagel,* from SJR Foods, is a bagel without a hole but already filled with cream cheese, so you can enjoy a bagel with cream cheese but you don't have to come to work with cream cheese smeared all over your clothing and the steering wheel of your car.
- *Uncle Ben's Calcium Plus Rice,* from Uncle Ben's, Inc., is rice fortified with calcium to help address the epidemic levels of osteoporosis affecting the increasingly elderly population in this country and builds on the trend of nutritionally-enhanced foods.
- *Cooking Made Easy* refrigerated dinners, from Mallard's, are cooked-chilled dinners, with a shelf-life of 45 days, that provide consumers meals like Homestyle Pot Roast or Chicken Scaloppini at a price of $6 for two people, or less than half of what people would pay for the same items in one of Mallard's restaurants, these meals give consumers an alternative to either cooking or going out to eat.

Source: Bennett, 1997.

General Mills (Kroskey and Krumrei, 1995) developed Betty Crocker® Whipped Deluxe Frosting as a response to consumer complaints about RTS frosting—that it was too sweet and difficult to spread. The development team recognized the need to create a frosting with a completely different taste and texture, with breakthrough new ingredients required that reduce the density of the frosting without compromising shelf life. The point of difference, delivered by R&D, is a texturally different frosting that used a proprietary, patented production process that served to slow copycat efforts. Unique advertising, which communicated the idea of spreadability, used a strawberry as the frosting-spreading device that clearly demonstrated one advantage of the product. The successful launch of Whipped Deluxe Frosting grew the category by 6% and Betty Crocker business by 9%.

USE RESEARCH TO MEASURE REACTION TO THE PRODUCT AND ALL ELEMENTS OF THE PROGRAM THROUGHOUT THE DEVELOPMENT PROCESS

Listening to the customer does not end with the birth of the product idea. Successful product development must involve talking to the consumer early and often. The product development process is essentially a series of screens and evaluations, starting with preliminary market assessment to determine market size, market potential, and the competitive situation. Detailed market studies are used to determine customer needs, wants and preferences that serve as input to product definition and product design. Pushing the numbers early and often is key; quantitative research must be employed to gauge market size, growth and trends. Early stage volumetric estimates can be generated from concept tests that help to measure likely market acceptance of the product and provide an intent-to-purchase score that can be normed to provide estimates of trial. It is vitally important to establish the correctness of the proposition with the consumer early in the process: engage the consumer to help define and refine concepts.

Throughout the development process, the concept must be continually refined and the physical product evaluated to insure that sensory attributes and nutritional characteristics match the product concept. Rigorous testing of the product, including shelf life, quality, and safety elements must be undertaken under normal and extreme customer conditions. Successful firms make sure the customer is exposed to the product as it takes shape, and throughout the entire new product project. All elements of the program must be evaluated as they are created: packaging (structure and graphics), labels, and advertising (copy and media plan), promotions, display, and shelf sets. Even after the product and package survive testing during the early and intermediate stages of development, and after all elements of the marketing program have been identified, refined, and tested, the reaction of both the trade and the target household customer must be assessed. This assessment typically involves some type of market testing or field trial.

Companies that use disciplined processes that feature regular and rigorous evaluation have better overall new product performance than companies lacking such evaluative processes. Campbell Soup attributes improved recent new product performance to the establishment of disciplined development processes. In the early 1990's Campbell was launching an average of two new, single SKU products with more than $10 million sales per year. By the mid-nineties, Campbell was launching more than 20 new products per year that surpassed the $10 million threshold. Tremendous upfront discipline is the key, with most new products spending up to 12 to 18 months in rigorous testing (Dwyer, 1997b).

USE OF CATEGORY MANAGEMENT PHILOSOPHY TO ALIGN MANUFACTURER AND RETAILER FOCUS ON THE CONSUMER

Ernst & Young has identified eight critical success factors in new product launches which can be applied to both the manufacturer and the retailer. The key is partnering with a focus on the consumer and on achieving objectives for the category. These success factors are shown in Table 4.4.

A WELL-EXECUTED LAUNCH

The launch is where "the rubber hits the road" and is all about execution and proper timing. The launch consists of setting specific goals and targets for market performance, running initial production, distribution and selling it to the trade, implementing the advertising and promotional programs we have planned for the trade and for the household consumer, and tracking all of these activities relative to goals and targets. Success is measured by performance in four key areas:

- production and distribution—having the product available in the correct quantities and at the right times and places
- sell-in—executing the sales activities to the trade and achieving the targeted levels of distribution penetration, including the targeted shelf sets, displays and numbers of facings
- consumer reaction, as measured by awareness, trial, repeat purchase, and customer satisfaction
- overall performance, as measured by sales revenue and sales growth, market share and share trends, and source of volume—cannibalization versus category growth versus taking sales from competition

One of the key challenges of the launch is coordinating sourcing and production, and availability of the product, with both the sales presentations to the trade and the breaking of the initial consumer advertising and promotional activity. Product must be in the store before advertising breaks and coupons are dropped. When planning the launch, seasonal sales patterns must be taken into account. It should seem obvious that seasonal items cannot afford to miss the major selling season. The firm needs to set the launch date and then back up from there. It must be determined, primarily from experience, when the product needs to be in wholesale warehouses, when it has to be available to ship, when production must be run, and ultimately when ingredients and raw materials need to be acquired. For consumers to buy the new product it must be on the shelf and we must build awareness through advertising and motivate purchases through promotion.

Finally, we need to use a combination of syndicated data and market research to track the launch. We need to obtain data on retail distribution and overall market penetration [e.g., using Efficient Marketing Services (EMS) data]. We need to gather sales and share data at the retail level (typically through IRI or Nielsen scanner data). We need to monitor purchase dynamics including patterns of substitution and the incentives associated with purchase (household

TABLE 4.4. Evaluating New Product Introductions (NPI).

Manufacturers Does the Manufacturer . . .	NPI Critical Success Factors	Retailers Does the Retailer . . .
• Practice or plan to implement ECR, category management and efficient assortment concepts • Involve the retailer in test marketing and collecting and analyzing data at the SKU level when determining the optimal number of SKUs to introduce • Focus on bringing real value and innovation to the category • Strive to match technology innovations with unmet customer needs • Evaluate the number of UPC changes that occur within the category	Category Management Match Technology with Unmet Consumer Needs	• Have a category strategy that details the Category's role Relative size of category Optimal number of category SKUs Full P&L product evaluations for each product line in a category Category teams with well-defined goals who are empowered to make decisions on all aspects of marketing the category
• Perform extensive consumer research, with early retailer involvement, which drives a formal business case evaluation, including volume projections • Have a documented NPI process with formalized review points and identified critical-path model of the process	Fact-Based Analysis and Planning (doing your homework) A Formal NPI Process with Pre-Established Criteria	• Have a fact-based evaluation process as a baseline for all introduction evaluations regardless of category • Have a documented monitoring process to ensure successful introduction, where problems are investigated to determine underlying cause and appropriate corrective actions
• Plan and develop effective programs with respect to consumer marketing/promotions, strong advertising and retailer programs • Ensure early retailer and consumer involvement to greatly improve the odds of execution-level success • Work with the retailers tying the critical-path model and planned introduction time to largest cluster(s) of customer re-planogram time(s)	A Strong Implementation Program A Strong Advertising Plan Interaction, Coordination, and Communication	• Participate with the manufacturer in the development and execution of promotions and retailer programs • Get involved early in the NPI process to greatly improve odds of execution-level success • Work with manufacturer to optimize introduction dates to retailer schedules

TABLE 4.4. (continued).

Manufacturers Does the Manufacturer . . .	NPI Critical Success Factors	Retailers Does the Retailer . . .
	Store Support	• Work with the manufacturer in evaluating category performance so the manufacturer can suggest new assortments • Partner with key manufacturers to improve process, share data and category insights and help test-market products
• Plan for and support execution activities in conjunction with the retailers early in the process		• Support introductions with adequate inventory, facings, and effective joint promotions • Employ computer-based planograms and stocking level/facing optimization programs to improve execution
• Evaluate category-level rewards and compensation based on profitability of the category, along with individual recognition	Aligned Performance Metrics and Rewards	• Have fact-based performance evaluations of buyers on volume projections, contribution, traffic, growth, and profit, depending on category role • Rotate key employees through the various roles in stores, merchandising, and buying to ensure a broad perspective
• Have detailed cost and target pricing policies in the NPI process	Profitability	• Have NPI evaluations based on category growth, sales volume and long-term profit potential

81

panel data), and we need to track awareness, trial, repeat purchase, and customer satisfaction through consumer surveys.

ABILITY TO ADAPT, GROW, AND IMPROVE AS MARKET AND COMPETITIVE CONDITIONS EVOLVE

Proper execution of the development process and the launch does not guarantee the long-run success of a new product. We do not exist in a competitive vacuum; there is a need to adapt and grow as we adjust to realities of the marketplace and competition. In most food product categories, new items, including those that feature a technical advancement, can be easily duplicated. A competitive difference lasts but a short time if competitors note that the new product has a significant impact on consumer purchase patterns. Moreover, the trade behaves impatiently with respect to new products, and if sales objectives are not achieved, there is always another new item to fill the shelf. Thus it is essential that firms plan to (1) ensure market success via contingency planning and (2) to continuously improve the product bundle.

Contingency planning involves establishing alternative scenarios of market and competitive reaction to the new product launch, setting up a monitoring system to track market and competitive reaction, and having alternative strategies to implement for different market and competitive conditions. For example, what do we do if the reaction to our new product launch is for a competitor to lower price? We need to be proactive instead of purely reactive to minimize the potentially negative impact of a competitive countermove.

Kraft Foods is one practitioner of the model of continuous improvement. A key part of Kraft's new product strategy is to stay on the offensive by value engineering to continuously improve, to add value to products, to delight the customer and to drive down costs. Another aspect of staying on the offensive is to flank the new brand with line extensions to meet the needs of a market with needs and wants that become more refined and more differentiated over time. The original Kraft Oscar Mayer® Lunchables® brand has been modified several times during its highly successful life. First, the container was changed to lower unit costs. Then, items such as branded drinks and desserts were added to create additional customer value. Ultimately, Kraft rolled out versions such as nachos and pizza targeted specifically at kids, and more recently has added pancakes and waffles to extend the "day-parts" for consumption.

FINAL THOUGHTS

The Leo Burnett Company New Product Planning Group presented a paper at the 1995 *Prepared Foods Annual New Products Conference* which presented "15 Principles" to guide success in new product development (Leo Burnett U.S.A., 1995). These principles, for the most part, are similar to what much

TABLE 4.5. Leo Burnett's 15 Principles to Guide New Product Success.

(1) Distinguish your product from competition in a consumer-relevant way
(2) Capitalize on key corporate competencies and brand strengths
(3) Develop and market products to people's needs and habits
(4) Market to long-term trends, not fads
(5) Don't ignore research, but don't be paralyzed by it
(6) Make sure your timing is right
(7) Be a marketing leader, not a distant follower
(8) Offer a real value to consumers
(9) Determine a product's short- and long-term sales potential
(10) Gain legitimacy and momentum for the brand
(11) Give the trade as good a deal as the consumer
(12) Clearly define, understand, and talk to your target
(13) Develop and communicate a distinctive and appealing brand character . . . and stick to it
(14) Spend competitively and efficiently, behind a relevant proposition
(15) Make sure the consumer is satisfied . . . and stays that way

of the literature contains. The agency recommends that new product marketers follow the principles listed in Table 4.5.

Although much of what Burnett recommends is consistent with what has been published both in the academic and the trade literature, it is interesting to note the focus on advertising and franchise-building, which is quite natural considering the source of Burnett's revenue. Research results, previously cited by Hoban (1998), bear out the Burnett contention that there must be clearly understood market targets, a distinctive positioning and brand character, and competitive spending, primarily advertising, behind the proposition. And the information derived from distributors and retailers clearly points toward the need for aggressive trade programs to access some of the food retailers' most valuable resource—shelf and display space.

BIBLIOGRAPHY

Anschuetz, N. F. 1996. "Evaluating Ideas and Concepts for New Consumer Products," in *The PDMA Handbook of New Product Development*, Rosenau, M. D., A. Griffin, G. A. Castellion, and N. F. Anschuetz, eds. New York: John Wiley & Sons, pp. 195–206.

Anonymous. 1996. "Betty Crocker Whips Up a Winner by Combining Technological Savvy and Old-Fashion Attention to the Consumer," *Food & Beverage Marketing*, 17(4):24.

Anonymous. 1997. "Getting the Numbers Straight," in *Efficient New Product, Introduction*. Supplement to *Progressive Grocer*, pp. 44–46.

Bennett, B. 1997. "The 15 Most Interesting Products of 1997," *Food Processing*, 58(11):21–33.

Berne, S. 1996. "Options in Outsourcing," *Prepared Foods*, 165(5):24–26.

Bierne, M. 1998. "Feeding Frenzy," *Brandweek*, 39(37):3.

Blaesing, D. 1997. "Lunchables Makes a Run for the Border," *New Product News*, 33(8):35.

Buss, D. 1998. "The Dynamic Dozen," *Brand Marketing*, 5(5):14–20.

Clancy, K. and R. Shulman. 1995. "Test for Success," *Sales and Marketing Management*, pp. 111–114.

Cooper, R. G. 1993. *Winning at New Products,* 2nd Ed. Boston, MA: Addison-Wesley.

Cooper, R. G. 1996. "New Products: What Separates the Winners from the Losers," in *The PDMA Handbook of New Product Development,* Rosenau, M. D., A. Griffin, G. A. Castellion, and N. F. Anschuetz, eds. New York: John Wiley & Sons, pp. 3–18.

Cooper, R. G. and E. J. Kleinschmidt. 1990. *New Products: The Key Factors in Success.* Chicago: American Marketing Association.

Crawford, C. M. 1997. *New Products Management.* Boston, MA: Irwin McGraw-Hill.

Demetrekakes, P. 1996. "Take Out Technology," *Food Processing,* 57(6):53–62.

Dimancescu, D. and K. Dwenger. 1996. *World Class New Product Development.* New York: AMACOM.

Dornblaser, L. 1997a. "Success at the Top," *New Product News,* 33(4):8.

Dornblaser, L. 1997b. "This Glass Is Half Full," *New Product News,* 33(11):3.

Dornblaser, L. 1998a. "Out with the Good Stuff," *New Product News,* 34(1):14.

Dornblaser, L. 1998b. "Only the Value-Added Survive," *New Product News,* 34(4):6–7.

Dwyer, S. 1996. "Rookies of the Year," *Prepared Foods,* 165(11):13–14.

Dwyer, S. 1997a. "Hey, What's the Big Idea? Pillsbury Knows," *Prepared Foods,* 166(5):8–20.

Dwyer, S. 1997b. "Red Alert: The Soup's Back On," *Prepared Foods,* 166(10):14–23.

Dwyer, S. 1998a. "Home, Sweet HMR," *Prepared Foods,* 167(1):14–26.

Dwyer, S. 1998b. "Inextricably Linked," *Prepared Foods,* 167(2):21–22.

Dwyer, S. 1998c. "Hain's 'Natural' High," *Prepared Foods,* 167(4):12–22.

Dwyer, S. 1998d. "Delicious Diversity," *Prepared Foods,* 167(6):4–27.

Dwyer, S. 1998e. "Bye-Bye Lowfat, Hello Big Taste," *Prepared Foods,* 167(6):29–32.

Dwyer, S., and B. Swientek. 1998. "A-1 Food Teams," *Prepared Foods,* 165(6):12–20.

Erickson, P. 1994. "It Works, but Does It Taste Good?," *Food Product Design,* pp. 71–74.

Feig, B. 1997. "Success-Radical," *Food & Beverage Marketing,* 16(8):10.

Fellman, M. W. 1998. "New Age Dawns for Product Niche," *Marketing News,* 32(9):1, 14.

Ferrante, M. A. 1997. "Outsourcing: Gaining the Manufacturing Edge," *Food Engineering,* 69(4):87–93.

Findley, J. 1998. "New Product Pacesetters 98." Information Resources, Inc., presented at the *Annual Convention of the Food Marketing Institute.*

Fusaro, D. 1995. "Kraft Foods: Mainstream Muscle," *Prepared Foods,* 164(5):8–12.

Fusaro, D. 1996a. "Food Products of the New Millennium," *Prepared Foods,* 165(1):28–30.

Fusaro, D. 1996b. "Food by the Numbers," *Prepared Foods,* 165(2):28–30.

Hawkins, C. 1996. "Rookies of the Year," *Prepared Foods,* 165(11):13–17.

Hoban, T. J. 1998. "Improving the Success of New Product Development," *Food Technology,* 52(1):46–49.

Hustad, T. P. 1996. "Reviewing Current Practices in Innovation Management and a Summary of Selected Best Practices," in *The PDMA Handbook of New Product Development,* Rosenau, M. D., A. Griffin, G. A. Castellion, and N. F. Anschuetz, eds. New York: John Wiley & Sons, pp. 489–510.

IRI. 1996. "New Product Pacesetters 96." Chicago: Information Resources, Inc., presented at *Saint Joseph's University Executive Food Marketing Program.*

Kahn, B. and L. McAlister. 1997. *Grocery Revolution: The New Focus on the Consumer.* Reading, MA: Addison-Wesley.

Kane, C. 1996. "In Search of the Magic Formula," *Brandweek,* 37(17):17.

Khermouch, G. 1998a. "Triarc's Smooth Move," *Brandweek*, 39(25):26–32.

Khermouch, G. 1998b. "New Dairy Drinks Set to Go," *Brandweek*, 39(33):4.

Kroskey, C. M. and D. Krumrei. 1995. "Project Everywhere," *Bakery Production and Marketing*, pp. 26–43.

Leo Burnett U.S.A. 1994. "15 Principles," Presented at the *Prepared Foods Annual New Products Conference*.

Maremont, M. 1998a. "Gillette Finally Reveals Its Vision of the Future, and It Has 3 Blades," *Wall Street Journal*, April 14, pp. A1, A10

Maremont, M. 1998b. "How Gillette Brought Its MACH3 to Market," *Wall Street Journal*, April 15, pp. B1, B10.

Mathews, R. 1997. "Efficient New Product Introduction," July 1997 Supplement to *Progressive Grocer*, pp. 8–12.

Matthews, R. 1998 "What's New in New Products?" *Grocery Headquarters*, 64(8):18–19.

Melcher, R. A. 1997. "Why Zima Faded So Fast," *Business Week*, March 10, pp. 10–11.

Messenger, B. 1996. "Companies to Watch," *Food Processing*, 57(11):69–74.

Messenger, B. 1997. "Bob Messenger's Food Trends Newsletter," 10(240).

Messenger, B. and K. Kevin. 1997. "The Future According to Pillsbury," *Food Processing*, 58(3):22–28.

Meyer, A. 1998. "The 1998 Top 100 R&D Survey." *Food Processing*, 59(8):32–40.

Morris, C. 1996a. "Beyond the Box with New Manufacturing Alliances," *Food Engineering*, 68(4):63–70.

Morris, C. 1996b. "Cut to the Core: Food Companies Reassess Engineering Competencies," *Food Engineering*, 68(11):57–67.

O'Donnell, C. D. 1997a. "Nabisco's A1 R&D Strategy for Global Growth," *Prepared Foods*, 166(8):51–58.

O'Donnell, C. D. 1997b. "Campbell's R&D Cozies Up to the Consumer," *Prepared Foods*, 166(10):26–30.

Parker-Pope, T. 1998. "The Tricky Business of Rolling Out a New Toilet Paper," *Wall Street Journal*, January 12, pp. B1, B8.

Patrick, J. 1997. *How to Develop Successful New Products*. Lincolnwood, IL: NTC Business Books.

Perlman, A. 1997a. "1997 New Product Resolutions," *Food Processing's Rollout!* 2(1):4.

Perlman, A. 1997b. "Evolutionary, Not Revolutionary," *Food Processing's Rollout!* 2(1):4.

Reynolds, M. 1997. "Let's Make a Meal," *Food & Beverage Marketing*, 16(6):7–8.

Rudolph, M. T. 1995. "The Food Product Development Process," *British Food Journal*. 97(3):3–11.

Spethmann, B. 1995. "Big Talk, Little Dollars," *Brandweek*, 36(4):20–29.

Stein, J. 1998. "The Men Who Broke Mach3," *Time*, 151(16):4.

Stinson, W. S., Jr. 1996. "Consumer Packaged Goods (Branded Food Goods)," in *The PDMA Handbook of New Product Development*, Rosenau, M. D., A. Griffin, G. A. Castellion, and N. F. Anschuetz, eds. New York: John Wiley & Sons, pp. 297–312.

Swientek, B. 1998a. "A New Formula for Growth," *Prepared Foods*, 167(1):11.

Swientek, B. 1998b. "Toasts of the Town," *Prepared Foods*, 167(5):21–26.

Symonds, W. C. and C. Matlack. 1998. "Gillette's Edge," *Business Week*, January 19, pp. 70–77.

Symonds, W. C. 1998. "Would You Spend $1.50 for a Razor Blade?" *Business Week*, April 27, p. 46.

Thompson, S. 1998. ''Minute Maid Aims Tangerine Blend with Calcium at OJ Resistant Kids,'' *Brandweek,* 38(30):4.

Von Bergen, J. M. 1998. ''Newest Tasty Treats Have Tropical Flavor.'' *Philadelphia Inquirer,* May 12.

Wheelwright, S. C. and K. B. Clark, 1995. *Leading Product Development.* New York: The Free Press.

CHAPTER 5

The Food Product Development Process

MARVIN J. RUDOLPH

Although no guarantees can ever be offered for a new food product's success, the implementation of a carefully orchestrated plan significantly increases that probability. Planning is a series of well-considered steps to be taken from gap analysis through concept generation and evaluation to prototyping and assessment, positioning through mapping, optimization, market testing, and scale-up for launch. The emerging discipline of quality function deployment may have an integral role to play in establishing the phases and benchmarks before products are actually formulated.

INTRODUCTION

THE current process for developing new food products is seriously flawed. Of the 8,077 new food products (stock keeping units or SKUs) introduced to U.S. retail markets in 1993, only about one quarter of them were novel not simply line extensions (Morris, 1993). Although there are no published data on successes and failures of new food products, it is estimated that 80 to 90% of them fail within 1 year of introduction. These are just the products that made the retail cut; consider all the products whose efforts fell short and retail introduction never took place. This means that about 200 novel food products introduced in 1993 made it after their first year on the U.S. retail shelves in 1994. The failure cost to the U.S. food industry is estimated at $20 billion (Morris, 1993). This cost results from missed sales targets, lost revenues, and postponed profits in addition to wasted development resources.

Reprinted with permission from *British Food Journal,* Vol. 97, No. 3, 1995, pp. 3–11, MCB University Press Limited, 0007-070X.

Being complex and iterative, the food product development process has proved difficult to define and model. It begins with a concept and ends with either the entry of the product in the market or the maintenance of the product in the marketplace, depending on whose model is studied. Barclay's review of research work into the process of product development (not only food) and the way in which it has progressed over 40 years shows that much of the work studying the new product development process is unknown to product development managers (Barclay, 1992a). After investigating current practices in new product development at 149 U.K.-based companies, only 78 were found to have some form of a new product guide to help manage the process, and 65 of the 78 stated their guide had originated through experience. Only one company based its new product development process on literature describing process models.

Traditional approaches to managing the product development process often fail because they result in unbalanced milestone structures. Typically there is no preplanned structure of milestones that identifies the deliverable for each functional group at each step of the program. Project managers apply their often-limited experience in developing a program plan that emphasizes their area of expertise. Often, milestones for following product performance are absent. For example, production data are not fed back to the team members who developed the product. Furthermore, management traditionally concentrates on the process only when authorization for large amounts of money is requested. Management attention, therefore, is focused typically on the purchase of equipment, not on concepts or brainwork.

We believe that Arthur D. Little, Inc. (ADL) has developed a comprehensive philosophy to guide food product development activities. It is based on establishing clear, consistent milestones for the entire development process and identifying the required deliverables by each of the functional elements contributing to product development with the firm. Milestones are viewed as an opportunity to monitor progress against a planned set of goals, to review the next tasks and anticipate problems, and to initiate program changes. This approach to milestones is analogous to the use of bivouac site by a mountain climbing team to regroup and make adjustments before proceeding on the journey. Figure 5.1 depicts the milestone structure graphically, highlighting the natural and effective shifting in the level of activity for each functional area.

While the peak activities shift, it is critically important that each group provide input to all of the milestones throughout the process. As Barclay states, "The [food product] development process needs to be linked with the corporate objectives and to the external environment to allow new ideas into the organization" (Barclay, 1992b).

The ADL milestone-driven product development process recognizes that a good process is flexible and continuously evolving (Figure 5.1). The many advantages of this process for project leaders, team members, and management are based on the following features:

All organization functions are involved throughout the project, but the level of activity varies for each function

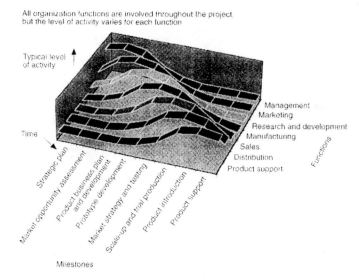

FIGURE 5.1 Organizational involvement in the product development process.

- use of common, project-specified vocabulary facilities communication among the case team, management, and project reviewers
- development of a standard framework of milestone deliverables reduces project start-up time
- a consistent definition of milestone structure that allows for internal benchmarking
- a proven methodology that allows for more accurate project planning, including allocating resources, establishing budgets, and scheduling tasks

The total milestone driven food product development process under investigation is illustrated in Figure 5.2. It can be envisioned as a skier racing in a three-phase giant slalom, the phases defined as product definition, product implementation, and product introduction.

PHASE I: PRODUCT DEFINITION

STRATEGIC PLAN

The skier (product developer) begins the race by proceeding form the starting gate under a strategic plan implemented by what we at ADL refer to as "third generation R&D" (Roussel et al., 1991). Third generation R&D is simply a holistic linking of business and technology goals.

Product definition phase

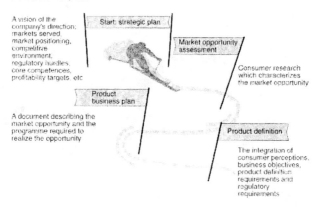

A vision of the company's direction; markets served, market positioning, competitive environment, regulatory hurdles, core competences, profitability targets, etc

Start: strategic plan

Market opportunity assessment

Consumer research which characterizes the market opportunity

Product business plan

A document describing the market opportunity and the programme required to realize the opportunity

Product definition

The integration of consumer perceptions, business objectives, product definition requirements and regulatory requirements

Product implementation phase

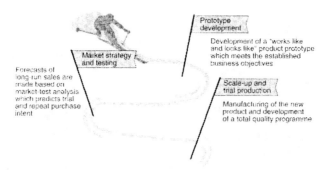

Prototype development

Development of a "works like and looks like" product prototype which meets the established business objectives

Market strategy and testing

Forecasts of long-run sales are made based on market-test analysis which predicts trial and repeat purchase intent

Scale-up and trial production

Manufacturing of the new product and development of a total quality programme

Product introduction phase

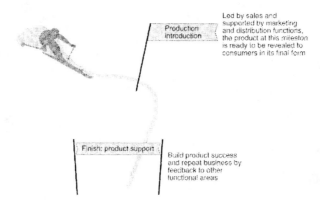

Production introduction

Led by sales and supported by marketing and distribution functions, the product at this mileston is ready to be revealed to consumers in its final form

Finish: product support

Build product success and repeat business by feedback to other functional areas

FIGURE 5.2 Milestone structure.

90

Prior to the early 1960s, there was first generation R&D, which can be defined as the strategy of hope. A research center was established, usually separated in distance from the main activities of the corporation, whose mission was to come up with interesting phenomenon. Work was self-directed, with no explicit link to business strategy. R&D activities were a line item in the corporate budget. If something came out of the effort, so much the better. However, the attitude was fatalistic—expectations usually were nonspecific.

From the early 1960s to today, we have second generation R&D. It is characterized by a mutual commitment to goals by upper management and R&D, consideration of the corporate strategy when setting these goals, and the implementation of a control system to track progress toward the goals. There is, however, no corporate-wide integration of R&D activities.

The future belongs to third generation R&D, emergent in leading companies. It articulates the issues that the firm must consider as it decides how to define overall technology strategy, set project goals and priorities, allocate resources among R&D efforts, balance the R&D portfolio, measure results, and evaluate progress. The crucial principle is that corporate management of business and R&D must act as one to integrate corporate, business, and R&D plans into a single action plan that optimally serves the short-, mid-, and long-term strategies of the company. A major output of third generation R&D is a vision of the company's direction. It characterizes the markets served and that competitive environment; details regulatory hurdles; and identifies the company's market positioning, core competencies, and profitability targets.

MARKET OPPORTUNITY ASSESSMENT

Once the strategic plan is in place, the second gate to traverse is the characterization of the market opportunity. This means that market requirements must be defined. In the food industry, this usually means consumer research. Frequently, focus groups (real-time knowledge elicitation) are conducted to identify potential opportunities for new products. The appeal of focus groups is in the "free format"—usually resulting in qualitative (anecdotal) comments that may be misinterpreted when a strong observer bias exists. Lack of a rigorous method for sifting through consumers' verbal comments may result in a misunderstanding of the real trade-offs that consumers are making or a failure to uncover an opportunity because of insufficient probing.

On the other hand, quantitative methods, like conjoint analysis (Rosenau, 1990), are criticized because of the limited and unrealistic nature of the options that can be presented and evaluated by the consumer.

The Arthur D. Little approach to conducting real-time knowledge elicitation combines the advantage of the unrestricted choices offered by free-choice profiling with the powerful collection of quantitative data for statistical analysis. We use a software tool for real-time data capture and analysis during the focus group. The online analysis provides the moderator with feedback on how well the consumer descriptions differentiate the samples. The moderator can then pursue further probing to uncover subtle yet important differences among samples. Subsequent statistical analysis is used to reduce the tension set to those sensory characteristics that account for the majority of the variance.

The method is effective in explaining the full range of sensory dimensions where individuals are able to discriminate effectively between items. As alternative features are defined, these consumer opportunities can then be translated into product concepts for further testing during consumer real-time knowledge elicitation activities. This can improve the success rate of the concepts as they move further through the milestone gates.

THE BUSINESS PLAN

The output of consumer real-time knowledge elicitation is the identification of new consumer needs and product concepts that can be incorporated in a product business plan (Hopkins, 1985); a document that describes a market opportunity and the program required to realize the opportunity. The business plan, usually written for a 12-month period, does the following:

- defines the business situation—past, present, and future
- defines the opportunities and problems facing the business
- establishes specific business objectives
- defines marketing strategy and programs needed to accomplish the objectives
- designates responsibility for program execution
- establishes timetables and tracking mechanisms for program execution
- translates objectives and programs into forecasts and budgets for planning by other functional areas within the company

PRODUCT DEFINITION

The last step in this phase is product definition. The key to product definition is the integration of multiple, and often conflicting, objectives. The integration of consumer requirements, business objectives, product delivery, requirements, and regulatory requirements is illustrated in Figure 5.3.

Arthur D. Little has pioneered a structured approach to integrating research of consumer needs and descriptions of the competitive environment with technical realities into a unique product specification. We use a product definition process based on quality functional deployment (QFD) to help us combine

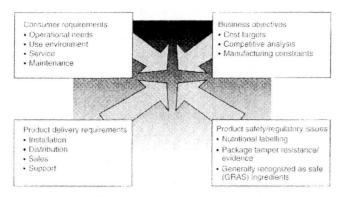

The key product definition is the integration of multiple
and often conflicting, objectives

Consumer requirements
• Operational needs
• Use environment
• Service
• Maintenance

Business objectives
• Cost targets
• Competitive analysis
• Manufacturing constraints

Product delivery requirements
• Installation
• Distribution
• Sales
• Support

Product safety/regulatory issues
• Nutritional labelling
• Package tamper resistance/
  evidence
• Generally recognized as safe
  (GRAS) ingredients

FIGURE 5.3 Integration of objectives.

and translate consumer requirements into product specifications. (See also Chapter 14 for the application of QFD for package development.) QFD (Sword, 1994) was developed in Japan as a tool to provide designers with an opportunity to consider the qualities of a product early in the design process. QFD is a method that allows us to consider the qualities of a product, process, or service. It helps us to focus our activities on meeting the needs of the customer:

- Who are the customers?
- What is it they want?
- How will our product address those wants?

Using QFD leads to a better understanding of customer needs that the product must meet to exceed competition.

QFD methodology evolves around the "house of quality" (see Figure 5.4, a graphical representation of the interrelationships between customer wants and associated product characteristics. It maps product requirements, helps to identify and understand requirement trade-offs, and predicts the impact of specific product features. Additionally, it provides a team-building tool for interdisciplinary product planning and communication. It is a method that is an important part of the process to develop successful products that fit the strategic and tactical needs of the business.

PHASE II: PRODUCT IMPLEMENTATION

PROTOTYPE DEVELOPMENT

Once the food product is defined, a "works like, looks like, tastes like"

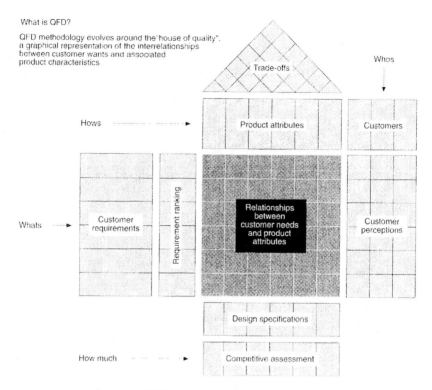

What is QFD?

QFD methodology evolves around the "house of quality", a graphical representation of the interrelationships between customer wants and associated product characteristics

FIGURE 5.4 The house of quality.

product prototype is constructed or formulated (see Chapter 10). To demonstrate that the product prototype in its conceptualized final form will meet the technical and business objectives established, Arthur D. Little staff use their profile attribute analysis (PAA) method (Hanson et al., 1983). PAA is an objective method of sensory analysis that uses an experienced and extensively trained panel to describe numerically the attributes of the complete sensory experience of a product. These attributes are a limited set of characteristics which provide a complete description of the sensory characteristics of a sample. When properly selected and defined, little descriptive information is lost. PAA is a cost-effective tool for product development that takes advantage of the use of powerful statistical techniques, such as Analysis of Variance.

PAA is used in product prototype development in two ways, competitive product evaluations (benchmarking) and product optimization (see Chapter 12).

BENCHMARKING

Competitive product evaluations provide formulators with objective information regarding the flavor quality of competitive products and the areas of flavor opportunity. The following example demonstrates a competitive product evaluation of several nationally branded oatmeal cookies. The products sampled (four full-fat oatmeal cookies, two reduced-fat cookies, and a no-fat oatmeal cookie) are identified in Table 5.1.

The evaluation was conducted to determine whether reduced-fat and/or no-fat products provide sensory characteristics similar to those of their full-fat counterparts.

- A panel of trained sensory analysts evaluated two replications of each of three lots of each type of cookie. The samples were presented in a random order with no visible identification of the brand.
- The panel used PAA to evaluate the products against the 13 sensory attributes shown in Figure 5.5.
- Summary indices for texture and flavor were developed using principal components analysis to summarize the attribute data, and were interpreted as shown in Table 5.2.
- The resultant flavor map of the flavor and texture indices (see Figure 5.6) shows that the full-fat Pepperidge Farms Santa Fe oatmeal cookie is not significantly different from the Nabisco® SnackWell's reduced-fat oatmeal cookie.

When the balance and flavor indices are shown on the flavor map (see Figure 5.7), the reduced- and no-fat products, as well as two out of the four full-fat cookies (Pepperidge Farms, Pepperidge Farms Santa Fe), exhibited a thinner, less blended flavor. The Archway and Nabisco full-fat products had a fuller, more blended flavor but differed significantly in texture. Archway is a soft, fragile type of cookie and Nabisco is a moister, chewier cookie.

The results of this study indicate formulation improvements to the flavor of the reduced- and no-fat cookies should be focused on improving the balance and fullness of flavor.

TABLE 5.1. **Profit Attribute Analysis (PAA): Competitive Product Evaluation.**

Full-fat cookies	Archway oatmeal cookies
	Nabisco⁺ oatmeal cookies
	Pepperidge Farms oatmeal cookies
	Pepperidge Farms Santa Fe cookies
Reduced-fat cookies	Pepperidge Farms wholesome choice oatmeal cookies
	Nabisco Snack Well's⁺ reduced-fat oatmeal cookies
No-fat cookies	Entenmann's no-fat oatmeal cookies

The sensory analysis evaluated two replications of each of three lots of each oatmeal cookie product. The samples were presented in a random order with no visible identification of the brand.

The panel used profile attribute analysis (PAA) to evaluate the products against
13 attributes that best describe important sensory characteristics

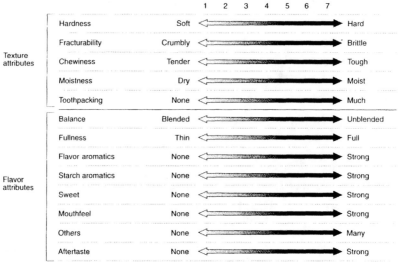

FIGURE 5.5 Sensory attributes of oatmeal cookies.

PRODUCT OPTIMIZATION

Response surface methodology (RSM) can be used to achieve product
optimization. In experimental food product formulations with multicomponent
mixtures, the measured response surface, usually a flavor attribute, can reveal
the "best" formulation(s) that will maximize (or minimize) the attribute
(Cornell, 1990). An RSM experimental design for optimizing a ketchup formulation is illustrated in Figure 5.8.

TABLE 5.2. Summary Indices for Oatmeal Cookies.

	Higher Scores	Lower Scores
Texture index	Less crumbly	More crumbly
	Harder	Softer
	More moist	More dry
	More chewy	More tender
Flavor index	More starch aromatics	Less starch aromatics
	More others	Fewer others
	Less blended	More blended
	Thinner flavor	Fuller flavor
	More mouthfeel	Less mouthfeel

Summary indices for texture and flavor were developed using principal components analysis to summarize the attribute data.

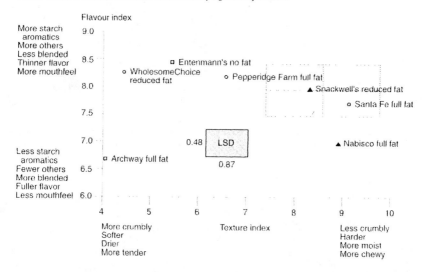

On flavor and texture, the Nabisco Snackwell's reduced fat and the Pepperidge Farms Santa Fe full fat products are not statistically significantly different

FIGURE 5.6 Texture index versus flavor index, oatmeal cookies.

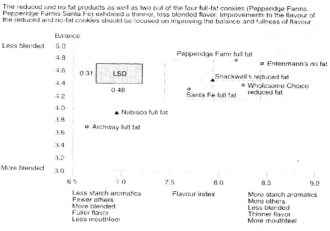

The reduced and no-fat products as well as two out of the four full-fat cookies (Pepperidge Farms, Pepperidge Farms Santa Fe) exhibited a thinner, less blended flavor. Improvements to the flavour of the reduced and no-fat cookies should be focused on improving the balance and fullness of flavour

LSD = Least significant difference

FIGURE 5.7 Flavor index versus balance, oatmeal cookies.

After identifying three or four of the most important factors using a screening design, we use response surface methods to optimize the formulation, for example, we helped to optimize a ketchup product

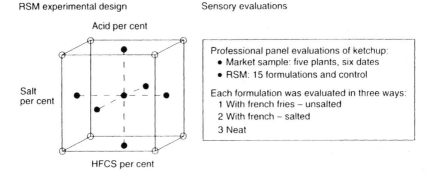

RSM experimental design   Sensory evaluations

Professional panel evaluations of ketchup:
- Market sample: five plants, six dates
- RSM: 15 formulations and control

Each formulation was evaluated in three ways:
1 With french fries – unsalted
2 With french – salted
3 Neat

FIGURE 5.8 Response surface methodology experimental design.

Professional panelists characterized the initial market samples from five plants with six code dates and the subsequent reformulated experimental products. The three most important variables were identified as percentages of salt, acid, and high fructose corn syrup. The experimental design called for the manufacture and sensory analysis of 15 ketchup formulations and a control. Each formulation was evaluated three ways: with unsalted french fries, with salted french fries, and neat. The RSM results revealed that our client's original evaluation design (without including different "carriers") was misleading.

- Ketchup evaluated neat has different flavor characteristics from ketchup evaluated with french fries.
- The original "optimized" formulation was acceptable in limited use situations.
- The resultant optimized formulations were acceptable in all ketchup uses when a standard manufacturing process was, followed.

When RSM is utilized, product optimization time is greatly reduced from traditional "cook and look" optimization techniques that depend on subjective formulation and evaluation procedures and often stop short of fully realized product improvements.

MARKET STRATEGY AND TESTING

At this point in the product development process, the organization has invested time and money in developing a new product from the initial concept to product optimization. If marketing forecasts look good, the temptation is to prepare for a full-scale launch. However, the product definition phase is based on models that are reasonably accurate representations of market response, but not of reality (Chapter 11). Things can still go wrong, as witnessed by the failed national introductions of "New Coke," Milky Way® II candy bars (25% fewer calories and 50% fewer calories from fat), and ConAgra's Life Choice Entrees (low-fat diet regimen).

Long-run sales are based on two types of consumer behavior, product trial and repeat purchase. Forecasts of long-run sales can be made if market test analysis can predict the percentage of consumers who become repeat users as well as those who will try the product. Numerous models have been developed that present the new product to consumers in a reasonably realistic setting and take direct consumer measures leading to the forecasting of cumulative trial and repeat purchases. Although they will not be discussed in the detail here, some of these models are based on previous purchasing experience, the so-called stochastic, or random models, and a combination of trial/repeat and attitude models (Urban and Hauser, 1980). New services are constantly being developed commercially, and it is clear that technically strong models and measurement systems will be widely available to forecast sales of new packaged food products.

SCALE-UP AND TRIAL PRODUCTION

Ultimately, the new food product has to be manufactured to meet the needs of the consumer. Early involvement of the manufacturing function in the product development process helps avoid problems that invariably surface when consumer expectations conflict with engineering constraints. The product's success is often linked to the level of compromise that is reached between the R&D and manufacturing functions (Chapter 11).

Implicit in the scale-up and trial production of the new food product is a total quality program that continuously identifies, analyses and controls risk. The risk controlling process, as shown schematically in Figure 5.9, begins with the identification of all potential hazards and proceeds through the screening, analysis ranking, quantification, and evaluation stages, ultimately to the controlling of the risks.

A hazard analysis critical control point (HACCP) matrix (Chapter 7) is a useful tool for identifying and prioritizing hazards which may affect food product quality. Such a matrix has the following elements:

- identification of critical control point
- evaluation of hazard potential
- assignment of degree of concern (low, medium, high)
- development of criteria for hazard control
- preparation of monitoring/verification procedures
- designation of corrective action alternatives that may be required

Total quality implies the continuous identification,
analysis and control of risks

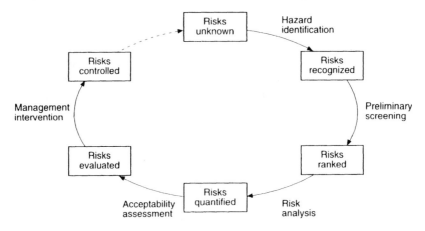

FIGURE 5.9 Food industry risk controlling process.

Because food safety is always of paramount concern, new products often linger or die at this point in the process if the issues cannot be resolved satisfactorily

PHASE III: PRODUCT INTRODUCTION

The product introduction milestone (Chapter 17) is led by sales but supported through all other functional areas, especially marketing and distribution. Field trials have been completed and the product is designed to meet the needs of the consumer. The product has been packaged and priced appropriately to convey the correct messages of quality and value. Packaging for transport has been tested and the product has been distributed in a timely and correct fashion so that it flows through the distribution system without impediments.

This phase is perhaps the most exciting and anxious, where customers see the product for what it is. Their initial response generally reveals the potential for success or failure of the product.

PRODUCT SUPPORT

Product support is a complementary milestone that builds product success and repeat business because it feeds back valuable information to other functional areas that can lead the process for line extensions, product upgrades, and the creation of all new opportunities. Product support is the ''infantry'' for the battle at the retail shelf; the first line of communication from the point-of-sale back to the organization.

CONCLUSIONS

It is worth assessing how effectively your company controls the basic product development process. Symptoms of a "broken" process are:

- longer development time than competitors
- missed targeted introduction dates
- significant number of "crash projects"
- a succession of stop/go decisions

If your company exhibits these symptoms, you may need to overcome a common misconception about control: milestones are not an administrative burden imposed by senior managers eternally worried about the team's capacity to deliver. Rather, they are a self-help tool without which a team can feel hopelessly lost. Ultimately, the mountain-climbing team and the base camp are jointly responsible for the success of a climb (Vantrappen and Collins, 1993).

BIBLIOGRAPHY

Barclay, I. 1992a. "The New Product Development Process: Past Evidence and Future Practical Applications, Part I," *R&D Management,* 22(3):255–263.

Barclay, I. 1992b. "The New Product Development Process: Part 2. Improving the Process of New Product Development," *R&D Management,* Vol. 22, No. 4, pp. 22(4):307–317.

Cornell, J. A. 1990. *Experiments with Mixtures.* New York: John Wiley & Sons, p. 8.

Hanson, J. E., D. A. Kendall, and N. F. Smith. 1983. "The Missing Link," *Beverage World,* 112:1–5.

Hopkins, D. S., 1985. *The Marketing Plan.* New York: The Conference Board, Inc.

Morris, C. E. 1993. "Why New Products Fail," *Food Engineering,* 65(6):132–136.

Rosenau, M. D. 1990. *Faster New Product Development.* New York: American Management Association, p. 239.

Roussel, P., K. N. Saad, and T. J. Erickson. 1991. *Third Generation R&D.* Boston, MA: Harvard Business School Press.

Sword, R. 1994. "Stop Wasting Time on the Wrong Product!" *Innovation,* Winter: 42–45.

Urban, G. L. and J. R. Hauser. 1980. *Design and Marketing of New Products.* Englewood Cliffs, NJ: Prentice-Hall.

Vantrappen, H. and J. Collins. 1993. "Controlling the Product Creation Process," *Prism,* Second Quarter: 59–73.

CHAPTER 6

Product Concepts and Concept Testing

JOHN B. LORD

Concepts are essentially means to fill real and perceived consumer needs and wants. They are meaningful descriptions of the augmented product as the target consumer— and not the developer—perceives it. Without good concepts, nothing effective can be prepared, although, of course, a good concept is only a first step in food product development. Concept generation emerges from consumer inputs, marketing research inputs, technological feasibility studies, and thoughtful analyses. Concepts should be objectively evaluated against pre-established criteria in order to filter through those which appear to hold the highest probability of success.

INTRODUCTION

THE new product development process starts with strategic business planning and opportunity analysis, followed by idea generation. Having established both a strategic direction and identified a category in which there seems to be a gap that offers opportunities for a new product or a new product line, the objective of the ideation phase is to create, discover, and uncover as many new product ideas as possible. During ideation, we suspend evaluation so as not to hamstring the creative process. Whatever the number of new product ideas put forth, many will not be feasible. In addition, the firm, regardless of size, has limited resources for development. Thus, the next step after ideation involves screening ideas with respect to technical, manufacturing, distribution, marketing, and financial feasibility and passing along to the next stage only those few ideas with the best prospects for success based on all of these dimensions. The next stage involves turning ideas into concepts and then

103

using marketing research to elicit consumer reaction to and the level of interest in concepts before proceeding to prototype development.

Some of the most necessary and most effective research for product development can be conducted at the concept development and testing stage. Early in the development process, we need to prioritize concepts according to which are most worthy of development. Successful new products come from winning ideas and the firm must test propositions with consumers to minimize the chance of proceeding with marginal ideas and wrongly positioning a winning new product idea. Because the firm must allocate scarce resources to competing projects, we need to perform some financial analysis early on, and the measurements of purchase interest and intent we obtain from consumer research on concepts helps us to generate early estimates of revenues. Separating the potential losers from the potential winners early on also lessens the risk associated with the later stages of product development.

In this chapter, discussion will center on product concepts and concept testing. We must keep in mind the fact that, although firms must elicit early consumer reaction to new product ideas, the depth and complexity of the development process is related to the degree of newness, and of risk, of the product to the company and to the market. The greatest source of information is past experience. Thus, firms developing line extensions rely on experience and these line extensions tend to be tested in a more finished stage. This is done as a way of shortening time to market, which is crucial, and is possible because producing a prototype of a variation of an existing product tends to be relatively easy and inexpensive. Firms testing new products with which they have less experience rely on concept testing to a much greater degree.

PRODUCT CONCEPTS

A product concept is a printed, pictorial, or mocked-up representation and description of a new product. The concept provides a device for communicating to both consumers and the development team the nature of the new product, how it will work, the product's features and characteristics, benefits, reason for being, and what problems it will solve for the user. A concept also represents a protocol that provides the development team a specific set of expectations as well as direction for both the development process and the launch strategy. The concept provides the basis for positioning strategy and product formulation, plus packaging, advertising, pricing, and distribution.

According to Patrick (1997), concepts evolve during development. At the earliest stages, a concept may be a simple standard description of the product that will include its major benefits, features, package size, and price; a simple drawing may accompany the description. At later stages, the concept is enhanced and may resemble a rough print ad with an interesting headline focusing on the key consumer benefit, attractive graphics, and copy with style and personality.

CONCEPTS CAN BE PRESENTED IN DIFFERENT FORMATS

Depending on both the relative newness of the concept and the stage of development, there are multiple formats in which concepts can be presented. These include

- simple verbal descriptions
- more elaborate verbal descriptions incorporating advertising language
- storyboards including product descriptions and typically an illustration of the package and/or product
- a mock-up of the package
- a product prototype

Since concept testing, like the development process itself, differs in complexity and depth across different products, the format for the concept often is made more elaborate as more learning takes place, allowing for more detailed information from concept tests.

The development team must take into account the fact that as we add more detail, particularly creative execution, to the concept, we risk testing reaction to the advertising instead of reaction to the concept. This will confound the feedback we are gaining from consumers and may result in both unreliable and invalid data on market reaction to the concept. As Patrick (1997) advises, because we use concept boards to garner reaction from consumers during interview situations, both care and creativity must go into the development of a concept board so we avoid displaying pure advertisement, which copywriters and art directors are trained to create.

The objective of the board is to present one major benefit, simply stated. Both the headline and copy should be simple, direct, and easy to comprehend. The concept statement must communicate the main point, the benefit, right up front, and simply stated so the respondent knows immediately *what's in it for me.* Any graphic or illustration must support the main product idea. The results of research in the social sciences indicate that any type of concept is communicated more effectively if presented in an environment that is familiar, and the graphics should help create that familiar environment. Make certain respondents can easily recognize the main benefit. Remember to make sure that concept statements are clear and understandable as well as realistic.

CORE IDEA CONCEPTS AND POSITIONING CONCEPTS

Schwartz (1985) differentiates core idea concepts and positioning concepts.

A *core idea concept statement* is usually quite short, just a few sentences or brief paragraphs. The core idea concept focuses directly on the product's main benefits, puts little emphasis on secondary features, and avoids persuasive communication. Core idea concepts can be supported by rough artwork to help communicate how the product might look or function. Core idea concepts are used as building blocks. In other words, the development team creates alternative concepts that can be screened, elaborated, and fleshed out on the way to becoming a positioning concept. From the Gibson Consulting Group, we have these examples of core idea concepts (Gibson, undated):

Here's a new line of snack dips for vegetables from French's that comes in convenient dry mix form.

Here's a new product from French's that does for your hot vegetable side dishes what salad dressings do for salad vegetables.

Here's a new product from French's that adds new taste variety to your everyday chicken dinners.

A *positioning concept* statement is longer and more detailed than a core idea concept (Figure 6.1). Positioning concepts can run several paragraphs, and attempt to communicate all of the product's main as well as secondary benefits. The positioning concept incorporates language that compares the

FIGURE 6.1 Heater meals: fully prepared, ambient-temperature, shelf-stable entrees in a barrier plastic tray. Incorporated into the secondary folding carton is a device to which the consumer adds water to generate heat to raise the temperature of the tray of food which is placed in the water. This product/package is particularly suited to truckers, campers, boaters—and to those others who are hungry and have no access to an oven.

advantages of the new product to other products, and is often supported with a high-quality photograph or illustration. Positioning concepts typically use both visual *and* verbal stimuli to describe the product, its benefits, and its end-use applications. Following on our earlier example from Gibson Consulting, the full positioning concept for the new French's chicken flavoring product is

French's Introduces Chicken Dippin' Sauces.

Chicken is chicken, right? Broiled, fried or baked, it all tastes the same, right?

Not any more!

Now French's brings you new Chicken Dippin' Sauces for delicious tasting, moist chicken, right at the table.

Chicken Dippin' Sauces come in two tasty varieties that complement the taste of chicken: there's Mild—a light fruity flavor; and, Zesty—a blend of rich robust flavors.

Regardless of how each member of your family eats chicken—with their fingers or a fork—every bite will be flavorful and moist with new Chicken Dippin' Sauces.

New Chicken Dippin' Sauces from French's . . . add flavor and moistness to chicken right at the table.

Concept statements provide the basis for gaining feedback from consumers to help measure the strength, viability, and sales potential of a new product. A poorly worded concept can lead to inaccurate data, which in turn can lead to development mistakes and wasted investment capital, not to mention the waste of the development team's time and effort. Good concept statements clearly communicate the nuances of the product idea but are realistic. Realism is often a major deficiency in writing concepts. Concept statements must reflect reality, both in terms of the capabilities of the product and in terms of the realities of the competitive environment.

Some other issues with writing concepts include clarity of language, focusing on what the consumer needs to know, proper length, and appropriate degree of emotion. Frequent errors in writing concepts include placing too much emphasis on the headline and illustration at the expense of poor body copy, making the concept too long, and overstatement of what should be obvious points. Writing a concept is an art, not a science.

CONCEPT TESTING

Concept testing is defined as a marketing research technique that is used to evaluate a concept's market potential and provide information useful in strengthening the concept and developing introductory marketing strategy. Testing a concept involves exposing a product idea to consumers and getting their reaction to it, using a predetermined series of questions designed to measure various reactions, feelings and opinions.

Concept testing has several important purposes. First, a concept test should enhance the efficiency of the development process because only concepts judged to have sufficient competitive strength and market potential pass this particular gate and go on to the next stage, thus conserving product development resources. Second, the fully elaborated concept that emerges from the test provides direction in terms of a specific protocol for the development team to follow through subsequent development stages. Third, the concept test helps the development team gain understanding of consumer reaction to the concept and its components. Fourth, concept testing allows an estimate of purchase intent, and thus provides data for early stage sales volume estimation. Fifth, concept test results provide diagnostic data as well as guidance for positioning strategy.

Concept testing can use either qualitative or quantitative techniques. Qualitative techniques, such as in-depth interviews and focus groups, can provide a great deal of useful knowledge regarding consumer reaction to the concept, the concept's strengths and weaknesses, and recommendations for improvement. Qualitative research, however, cannot provide data necessary to make early sales volume estimates.

The most important purpose of a positioning concept is to present the product idea realistically to learn whether consumers will eventually buy the product. The emphasis should be on clear and accurate product description communications rather than persuasion so as to portray the product as it will eventually be presented to the market. Sometimes copywriters can get carried away with a concept description; we must remember to test the concept and not the advertising.

A concept test can yield numerous important classes of information, which are listed later. Not all of this information will be yielded by any one test, and what information the test provides will be dependent on the research methodology, specifically the types of questions asked, as the proper execution of the test. The types of data provided by a concept test are listed in Table 6.1.

Virtually every quantitative concept test will contain an intention to buy or purchase interest question. Most will attempt to measure uniqueness and price-value, and most will attempt to elicit consumer reaction not only to the concept as a whole but to the elements or characteristics that comprise the new product. (Figure 6.2).

According to Schwartz (1985), a properly planned, executed, and analyzed concept test can provide several important pieces of information:

- An early read on sales potential. If potential is adequate, test results should aid in identifying the segments of the market that are potentially most responsive to the new product. If results are below

TABLE 6.1. **Concept Test Data.**

- Consumer attitudes and usage patterns in the product category
- The competitive setting or market structure in the category
- Size of the potential market
- Segments and characteristics of the consumers making up each segment
- The main idea being communicated by the concept
- The importance of the main idea to consumers
- The concept's relevance to consumer needs
- Consumer reaction to product attributes and features, and the relationship between consumer reaction to product attributes and overall concept rating
- Purchase interest and intent
- Reasons consumers give for purchase interest
- Major strengths and weaknesses of the concept
- Strength of the concept versus other concepts tested
- Perceived advantages and disadvantages relative to competitive products
- Uniqueness—a measure of competitive insulation and a strong indicator of concept strength if combined with high purchase interest
- Believability, the extent to which consumers feel that the concept can deliver the benefits promised
- Perceived price-value—consumers expect good value so they are not necessarily turned on by concepts which provide a good value, but a poor price-value relationship can dramatically depress overall appeal
- Usage situations and frequency of usage
- The expected frequency of purchase
- Potential source of volume (incremental category sales, from competitors, cannibalization of the developing firm's product line)

expectations, the test results should permit determination of any potential targets of opportunity if the market population were to be segmented into different or narrower demographic, behavioral, or attitudinal groups. Alternately, we may find simply that the concept lacks sufficient appeal to be a viable entry into the category.
- The degree to which the message that consumers receive from the concept (in both objective and subjective dimensions) is consistent with the message intended. This is diagnostic element that lets the development team know if the concept's level of success or failure is attributable to the attractiveness and strength of the new product idea, or is compromised in some way by a biased or inaccurate execution. This is especially important in the case of a poor-performing concept when a decision must be made to either kill the idea or make another attempt at positioning and describing the concept.
- Identification of the individual strengths and weaknesses of the concept, providing insight into the relationship between consumers' overall evaluation of the concept and the role of each of the concept

FIGURE 6.2 Dole cut fruit in an oxygen barrier plastic bowl that is hot-filled. The use of the oxygen barrier, coupled with clarifying agents in the polypropylene for the cups, provides a contact clarity to permit visibility of the fruit.

elements in that evaluation. The test should relate which elements of the concept contributed to positive evaluation and which contributed nothing or to negative evaluation. This insight aids in strengthening the concept via heavier emphasis on positive characteristics and elimination or downplaying of neutral and negative elements.

METHODOLOGY

The first decision a firm makes about concept testing is whether to outsource. ACNielsen BASES (formerly Bases, Inc.) has several alternative testing designs and is a frequent research supplier to the food industry. We will discuss BASES tests later in the chapter. Of course, there are numerous other research suppliers that offer alternative research designs. One of the reasons so many food processors use BASES is the large database of test results they have assembled which allows for fine-tuning the model and more accurate sales forecasts.

The methodological issues for concept testing are essentially the same as for any marketing research project, and include type of sample, sample size, method of administration, questionnaire design, and analytical techniques.

TYPE OF SAMPLE

The type of sample can be either probability (random) or non-probability (judgment, quota, or convenience). Random samples use alternative sampling methodologies, including simple or one-stage random samples and multi-stage sampling designs, including stratified random samples, area samples, and cluster sampling. The major advantage of random samples is the ability to use techniques of statistical inference, and to determine the precision of our sample estimate or the amount of sampling error. Random samples require sampling frames (lists of potential sample members) and the use of specific and correct techniques of sample selection to guarantee true randomness. The disadvantage of random sampling is high cost relative to other sampling designs.

Non-probability (or purposive) samples are of three major types. The first type is the quota sample, whereby we use quotas to draw a sample that matches, on certain predetermined dimensions, those same characteristics of the population. The second type of nonprobability sample is called a convenience sample. A very typical convenience sample design is administered using mall intercepts (central location testing), where the focus is on conveniently (and efficiently) reaching a large number of potential respondents. The third type is the judgment sample, which involves selecting sample members who, because of their background and/or experience, should be a useful source of information. The purpose of a "purposive" sample is to draw a sample which is representative of the population of interest, which in concept testing is the potential market for the new product.

Companies conducting their own tests typically use either central location tests using a combination of convenience and quota sampling, or use mail surveys sent to households or individuals who are members of a preselected and demographically balanced panel. However, since concept testing, like the entire process, is iterative, one project may feature multiple concept tests using different types of samples with concepts that become more elaborate and more focused with repeated testing.

SAMPLE SIZE

There are no hard and fast rules governing sample size. The old rule in basic statistics is that if we are using a probability sample, we can use a z-test instead of a t-test if the sample size is 120 or above. The tradeoff in sample size is, of course, cost versus representativeness and accuracy. A more representative sample requires a larger sample size as the population becomes more heterogeneous. Most concept tests use samples consisting of 300 to 500 potential respondents. This number is necessary to provide adequate size of each cell when doing cross-tabulations. Smaller sample sizes can be used for relatively homogeneous populations. For instance, the Tasty Baking Company launched Tropical Delights, which is targeted to Hispanics and specifically Hispanics of Puerto Rican extraction. The relative homogeneity of that population permits a smaller sample size to yield useful and projectable data.

ADMINISTERING THE TEST

Concept tests can be administered in person, typically via the use of central location tests or focus groups; by the use of mail questionnaires; and by the use of telephone surveys. The relative advantages and disadvantages of these methods of gathering data are well-documented in marketing research texts. In-person interviews allow a great deal of range and flexibility in the types of questions researchers can ask and the visual stimuli that can be employed, and typically allow for greater amounts and depth of data to be gathered. In-person interviews are expensive and require trained interviewers as well as a location in which the test is administered. Central location tests require careful screening of potential respondents to ensure sample representativeness. In a depth interview situation, researchers strongly prefer personal interviews. However, concept tests in many cases use closed-end questions and scaling questions, which can be effectively and efficiently handled via telephone or by the use of mail surveys.

Telephone interviews allow a great deal of control, access to a greater range of respondents in a shorter period of time than central location tests, and, through the use of random digit dialing, can be used in a random sampling situation. Limitations include the ease with which respondents can refuse to be questioned or terminate the interview before completion as well as the lack of any visual stimuli. Telephone interviews can be accomplished very quickly but the range and depth of data gathered is somewhat limited. Telephone interviews are most useful when the objective is to gather some basic consumer response data in a short period of time, which is typical of new product development projects.

Mail surveys tend to be the least expensive, provide an opportunity for the respondent to ponder questions and even gather information before answering questions, allow for visual stimuli to be presented, and allow for careful targeting of households or at least neighborhoods. Response rates tend to be lowest for this type of questionnaire administration; however, this disadvantage is negated if we have access to a panel of households who have been recruited and screened.

DESIGNING THE QUESTIONNAIRE

Just as there are no hard and fast rules governing sample size, there is no single approach to the number, type, and sequence of questions in a concept test. However, a typical concept test questionnaire would begin with an assessment of the respondent's current experience and practice in the product category, demonstrate the concept, elicit reaction to the concept, measure intent to purchase, and then ask for relevant demographic and other data useful in classifying respondents. Table 6.2 lists several dimensions on which reaction to the concept can be assessed.

The single most important question in concept test is the purchase intent question; this is the tool most frequently used to measure concept success, both as a decision-making variable, and as a key element in various volume-prediction models. By asking this question, we not only measure the overall strength of the concept, but we also can pinpoint the type(s) of individuals most likely to buy the product. A typical statement of the purchase intent question is:

If (this product) were currently available in your local store, how likely is it that you would buy/try this product?

Since most available normative data are based on a five-point scale, the purchase intent question usually employs such a scale with the following scale points:

- would definitely buy
- would probably buy
- might or might not buy
- would probably not buy
- would definitely not buy

The percentage of respondents who respond "definitely would buy" is called the *top box* score. The percentage of respondents who respond "definitely would buy" plus the percentage that respond "probably would buy" is a measure of positive purchase intent and is called the *top two box* score. According to Schwartz (1985). On average, the typical concept stimulates a "definitely will buy" score of 19% among target group consumers and a *total*

TABLE 6.2. Dimensions of Concept Tests.

- Uniqueness
- Believability
- Level of consumer interest
- Whether the new product will solve the consumer's problem
- Practicality and/or usefulness of the new product
- An affective or relative liking dimension
- Reaction to price or a price-value rating
- Strengths and weaknesses
- Problems with the concept

positive interest score ("definitely will buy" plus "probably will buy") of 65%.

While gauging purchase intent is critically important to a concept, the development team must consider several caveats. First, every respondent who indicates "definitely will buy" will not buy. Over time, marketers develop norms for a category whereby we can hypothesize a certain numerical relationship between positive purchase intent and projected trial rate. When you analyze the results of a concept test, it is helpful to have scores available from previous concept tests to act as a point of reference or benchmark, especially when these scores have been related to actual in-market performance. In effect, these data serve as a historical perspective against which the current concept's result can be compared. Accurate forecasts of trial and first-year sales volume are completely dependent on these norms (Figure 6.3).

Second, many marketers simply place too much value on purchase intent. New product ideas often live or die on the basis of one score. Blind reliance on any single question is dangerous, and purchase intent is no exception. Third, if the primary objective is to predict actual volume once the product

FIGURE 6.3 One of the most successful new product developments of the 1980s and 1990s is moist pasta designed for rapid and easy preparation by the consumer. The development was one of the first employing "hurdle" or "combined" technologies to deliver a microbiologically safe and high-quality convenience product. The formulation often contains natural antimicrobials, as well as being low water activity to retard anaerobic pathogenic microorganisms. Processing and packaging are performed under clean and cold conditions. The air is removed from the oxygen barrier package in which an oxygen scavenger is carefully adhered, and the package is sealed for refrigerated distribution.

is introduced to the marketplace, make sure to tell respondents the price *before* asking about their interest in purchasing the product. This is critically important to the accuracy of the forecast of sales volume.

Schwartz (1985) makes several other suggestions regarding the purchase interest question:

- For diagnostic purposes, that is to learn whether the basic idea of the product has merit or needs improvement, the purchase intent question should be asked twice, once at the beginning of the survey with no price and again toward the end with price. This is based on the premise that the respondent can provide better diagnostic information without reference to price.
- Because the reasons consumers give for having high or low levels of purchase interest often provide valuable insights about a concept's strengths and weaknesses, we should assess reasons for purchase interest. This can aid in spotting winners and losers, and in understanding the factors responsible for their performance.
- Analyze the reason for purchase interest separately among individuals with positive or negative purchase interest. Also, respondents with neutral purchase interest often like the product on some dimensions and dislike it on others, and the elements they dislike may be different from the things disliked by those with negative purchase interest. Therefore neutral respondents should be analyzed separately provided that the number of people who fall into this neutral group is large enough to be significant analytically.

Purchase Intent alone is not sufficient to properly evaluate the concept. There are, according to Schwartz (1985), other key indicators that can make a big difference in the value of your test data.

- the concept's uniqueness
- the concept's appropriateness or relevance to the consumer's needs
- the main idea being communicated by the concept
- the importance of the main idea
- the expected frequency of purchase

Uniqueness represents an important dimension for an obvious reason: successful new products must be meaningfully different than competitive products. Two elements of uniqueness—of the overall product and of individual product characteristics—should be studied. Uniqueness is an excellent diagnostic tool to help pinpoint how to improve a deficient concept. Cross-tabulating uniqueness by purchase interest can help pinpoint factors that are responsible for weak concept ratings.

Determining whether and to what extent consumers perceive a concept as delivering important benefits or satisfying recognized needs is particularly important in understanding whether consumers might adopt a new product. Often, we are willing to buy a new product once or twice, because novelty is an important motivation, but also because new products often come with significant incentives that lower the effective price. But one or two purchases do not sustain a new product. Unless we are specifically planning to launch a ''fad'' item, such as a ready-to-eat breakfast cereal using a licensed character, or a seasonal item, such as a Christmas snack cake, we need consumers to make the product part of their regular lifestyle and consumption system. This can only happen if consumers perceive the item as having the ability to meet ongoing and important needs.

Virtually every concept test questionnaire includes a question that asks respondents what they consider to be the main idea of the product. Purchase intent and attribute ratings can only be valid if consumers understand the concept. There is also value in discovering how consumers evaluate the importance of the key benefit. This can be particularly valuable when testing a concept for a breakthrough product—something completely different from products currently available. Early stage consumer evaluation of a ''new to the world'' product presents certain challenges. It is crucial to determine the importance of the concept to the consumer to properly measure the strength of breakthrough concepts. This type of concept, even if it has high appeal, may score low on purchase intent due mainly to the fact that consumers may not believe that the product can really deliver the promised benefit.

Purchase intent provides a basis for estimating trial. In order to estimate overall sales volume, we also need to assess potential purchase frequency, and make some projections about quantities purchased per purchase occasion. Of course, products that have high purchase frequency may represent a high-volume opportunity. Schwartz (1985) points out that questions that measure intended purchase frequency are often omitted from concept tests because the data may prove to be unreliable—consumers often significantly overstate their intended purchase or usage frequency. However, eliminating the question is not the solution. It is better to ask the question and use some care in interpreting the result, assessing data in a relative sense by using norms from previous introductions in the category or a similar category. We can also employ consumer panel data to deflate responses to purchase frequency questions to bring them into line with actual purchase patterns. It is insightful, with a sufficiently large sample size, to cross-tabulate intended purchase frequency with different population characteristics such as household size, household income, and so on. This will be particularly helpful in identifying the heavy user segment if one exists.

As we have previously discussed, concept testing plays several roles, including providing not only consumer reactions to the concept but also a diagnosis of those reactions. In order to understand why concepts are perceived as acceptable or unacceptable, strong or weak, and so on, we can ask respondents to rate the concept on specific product attributes. Ratings on individual product attributes represent a "disaggregation" of the overall concept rating. Schwartz (1985) provides a series of questions that can be used to measure consumer reaction to product attributes.

I would like you to rate this product on several different characteristics. Although you have evaluated the product on an overall basis, you may feel differently about it on some of these characteristics. Since you may not have used this product before, please base your answers on your impressions from what you've just read. After you read each characteristic, rate the product either as excellent, very good, good, fair, or poor. Pick the choice that best describes how well you think the product would perform on that characteristic.

Now I'd like to find out how important each of these characteristics is to you. After you read each characteristic, rate it as either very important, somewhat important, neither important nor unimportant, somewhat unimportant, or very unimportant.

Plotting the concept ratings on each attribute and the importance ratings of each attribute on a on a two-dimensional scale provides a clear visual presentation that highlights the strengths and weaknesses of the concept. The importance ratings are plotted on the vertical axis, highest at the top and lowest at the bottom. The concept ratings are plotted on the horizontal axis, highest on the left and lowest on the right. The upper left quadrant contains attributes that are important to the respondent and that the test concept scores well on. The upper right quadrant contains attributes that are important but low-rated. This provides direct and dramatic insight for the development team regarding how the concept can be improved.

Other useful findings from concept tests include identifying the user and the usage occasion(s), and how the concept might be improved. The user is the type of person that respondents associate with your product. This information is important primarily in planning the advertising because user imagery is a key element of creative strategy, particularly in categories such as beverages, cosmetics and fragrances, and tobacco products. Usage occasions not only help to identify the competitive set, which may and often does differ according to usage occasion, but also provide clues for our positioning and marketing program. Consumers often can provide very good insight into how to improve a concept. Simply include the question, "In what ways, if any, could this product be changed or improved?"

ABSOLUTE VERSUS RELATIVE CONCEPT TESTS

Concepts can be tested in two ways: either individually or with control concepts. Most experts prefer absolute or monadic tests as opposed to relative or sequential tests. This is because absolute measures are less biased and more amenable to sales volume estimation.

VALIDATION

We have previously discussed the notion of norming concept test results, particularly purchase intentions. This is one type of validation procedure. The major premise of any concept testing procedure is that consumer's reactions to a concept are valid and reliable indicators of their likely future behavior in the market. To establish the validity of purchase intent scores as predictors of trial requires the analysis of a significant number of cases wherein we can compare purchase intent scores with subsequent performance of the product in actual market conditions, either a sell-in test market or after launch. Without this information, we can have little confidence in the validity of our purchase intent data and the ability to predict trial. One technique used by some analysts is to develop a trial estimate based on a weighted average of our purchase intent scores. For example, the "top box" score is weighted by 80, the second box score is weighted by 60 and so on so that the percentage of respondents who checked "definitely would not buy" is weighted by zero. It is important to remember in this context that high scoring concepts may be successful and low scoring concepts will most likely not be successful.

VOLUME ESTIMATION

One of the most critical outcomes of the concept testing phase is an estimation of first-year sales volume. Sales volume estimates require the data shown in Table 6.3.

Market size estimates are based on category data that measure the number of households purchasing items in the category plus a projection of the geographic scope of the launch. For instance, canned soup has close to 100% household

TABLE 6.3. **Data Requirements for Estimating Sales Volume for a New Product.**

- Estimated market size, usually stated in terms of number of households in the target market
- Projected level of awareness of the new item which will be created primarily through consumer advertising plus consumer and trade promotions
- Projected level of distribution penetration, typically stated in terms of a percentage of ACV distribution to be attained within a given period of time after launch, and based on both the strength of the concept and the strength of the trade program
- Estimated percentage of households that will try the product plus the average trial volume per household
- Estimated percentage of households that will buy the product at least once after trial (repeat purchase) plus the projected number of purchase cycles and the average volume per purchase during the introductory period
- The price of the product to the trade

FIGURE 6.4 Ever since man began to eat, soup has been among the more popular food forms. Since the last century, soup has been offered commercially in dry form for reconstitution with hot water, in canned form for ambient temperature shelf stability, and in canned condensed form for reconstitution with hot water. Frozen single-strength and concentrated soups have been in and out of commercial markets over the years. An interesting format is chilled concentrate—in stand-up flexible pouches. The chilled distribution channel suggests a higher-quality product than canned or dry, and much more convenient than frozen.

penetration and there are slightly more than one hundred million households in the U.S. Thus, Campbell Soup Company might identify the potential market size to be 50 million households in a launch that will reach about 50% of the population during the introductory period (Figure 6.4).

Awareness levels are projections based upon relationships between the strength of the introductory marketing campaign and the percentage of households in the target market that will become aware of the new item. Introductory marketing campaigns, for purposes of these projections, typically consist of several different elements. Media advertising campaigns help build awareness. The strength of an advertising campaign or program is usually measured by the number of gross impressions delivered (i.e., the number of exposures of the message to consumers, including multiple exposures to the same consumer) or by gross rating points. Gross rating points (GRPs) represent a measure of the weight of an advertising campaign. GRPs are calculated by multiplying the net unduplicated reach of the campaign, stated in terms of a percentage, by the frequency, or average number of times each household is exposed to at least one message. Thus, a campaign that reaches 80% of the target audience an average of 3 times generates 240 gross rating points (Figure 6.5).

FIGURE 6.5 Wraps may represent the food product star of the turn of the millennium. Derived from cultures around the world, wraps are being adapted to as many market segments as food processor/marketers are able to identify. A competition has developed between wraps prepared on demand or at hotel/restaurant/institutional sites and those prepared by food processor/marketers for grocery store distribution. Technologically, it is difficult to main textural integrity between the wrap and its contents.

Introductory marketing campaigns also include consumer promotions, such as coupons, and trade promotions, which are monetary incentives provided to the trade to encourage the stocking and displaying of the new item. In many cases, the retail trade will pass at least some of the savings along to consumers in the form of lower introductory prices. There is no way to determine the relationship between the strength of the introductory marketing campaign and the awareness levels created without reference to historical launch data. One of the major reasons many food processors outsource this type of research is because of the large number of cases research companies, such as BASES, Inc., have developed that provide reasonably accurate and dependable mathematical relationships.

Projections of distribution penetration rely on the same type of historical database identified above. Wholesalers and retailers will "cut-in" new items if there is money to be made. Thus, the ability of the concept to sell to consumers, trade margins, and trade incentives will impact distribution penetration. We need reference to historical data to make accurate projections.

Trial can be measured empirically through the use of a simulated test market (which will be discussed in more detail in Chapter 17 on market testing and launch), which provides a sample of consumers with an opportunity to purchase the product under simulated market conditions. Alternatively, trial can be projected from purchase intent scores. This is where norms and validation of these scores comes heavily into play. Again, historical data that allow us to determine some sort of probability relationship between "top box" or "top two box" scores and trial rates, are a necessary component of the process. The average quantity purchased at trial may simply be projected as one unit. However, if we are launching a line of items simultaneously, and we incent the consumer to buy more than one item at a time, we have to adjust the quantity estimate accordingly.

Repeat purchase can be measured empirically, again using simulated test market methodology, or projected from past comparative introductions in the company, or better yet, the category. It is vitally important to note that repeat purchase is a function of satisfaction with the product, so that we can use ratings of concept strength, in the absence of any empirical data, as a surrogate for satisfaction. The resulting deflation factor is applied to the test concept. The repeat purchase estimate is a weak link in the predictive process, regardless of how it is derived. You never know for sure if your concept will stimulate more or less than average repeat buying until it reaches the market (Figure 6.6).

FIGURE 6.6 In the realm of new food products, few are more difficult to deliver than egg-based. This kit was composed of aseptically packaged liquid egg (in the sealed cup in the center); a small cup of cheese, ham or bacon resting on top of the container; plus a paperboard tub that acted as a carrier, a cooking vessel, and a unit from which to eat. The consumer poured the liquid egg and the inclusion into the tub and heated the mass in a microwave oven until the egg coagulated (and the cheese melted).

How does one estimate repeat purchase? Repeat volume is a function of how many times a household purchases a new item after trial and the average volume at each purchase occasion. Historical data from launches of products in the same category, as well as the specifics of both the category and the launch program, must be considered in estimation of purchase frequency and the average quantity purchased. According to Schwartz (1985), most models take data on purchase frequency and amount purchased directly from questions included in the concept test, adjusted in some manner. The reason for the adjustment is that consumers usually cannot accurately provide volume and frequency information; they tend to overstate their intended consumption. Adjustment factors are calculated by examining historical relationships. One technique is to correlate responses to similar questions in previous concept surveys versus actual diary panel or store scanner data for those same products (Figure 6.7).

The price of the product to the trade comes from the launch team. We must consider both normal and introductory prices and trade margins. We also need to consider both seasonal sales patterns and the timing and pattern during the introductory period of volume build, which more sophisticated models allow. A procedure for estimating first-year sales volume is shown in Table 6.4.

It should be obvious to even the casual observer that these computations will be extremely sensitive to the values we assign to the parameters and

FIGURE 6.7 Piggybacking on the fresh-cut vegetable growth has been unit-portion packaging of "finger food" vegetables such as carrots—a "healthy" snack replacement for salty cereal-based snacks.

TABLE 6.4. A Procedure for Estimating First-Year Sales Volume.

- Start with the number of households in the target audience
- Multiply by the level of awareness, the product of these two numbers is the percentage of target households who become aware of the existence of the new item
- Multiply the aware households by distribution penetration because one cannot purchase the product if it is not available in stores they shop
- Multiply the number of aware households with access to the product by the projected trial rate; you have now computed the number of trial households
- Multiply the number of trial households by the average volume purchased at trial; you now have trial volume in units
- Multiply the number of trial households by the percentage of repeat purchasers; you now have the number of repeat purchase households
- Multiply the number of repeat purchase households by the average number of purchase events, based upon the repeat purchase cycle which is the average amount of time between purchases, and the average number of units purchased per purchase event; you now have the repeat purchase volume in units
- Add trial volume plus repeat volume, multiply by the price, and you have projected first year sales volume in dollars

variables, so realistic numbers must be assigned to awareness, distribution penetration, average quantities purchased, and number of purchase events. Plus, the estimates of trial and repeat, if overstated, can mislead companies into spending money to develop losers. Finally, one should note that we could perform sensitivity analyses with models of this type. If projections do not meet critical levels, we can adjust the strength of our consumer or trade marketing campaigns to create greater awareness or distribution penetration, to make the numbers move. Clearly, we can also adjust components of trial and repeat purchase estimates.

Table 6.5 provides an example of a simulation leading to a sales forecast. Two treatments are simulated: a low spending introductory plan and a high spending introductory plan, yielding dramatically different sales forecasts:

TABLE 6.5. Sample Simulation.

Assumptions	Low Spend	High Spend
Distribution	80%	90%
FSIs	2	4
GRPs	1,500	2,500
		Yes
Sampling	No	
Output/Forecast		
Yr 1 Trial	4.2%	9.5%
Yr 1 Repeat	41%	43%
Yr 1 Volume (000 cases)	1,700	4,000

CAVEATS

Regarding concept testing, Fitzpatrick (1996) notes several important caveats to keep in mind:

- The concept statement is a description written to explore potential consumer interest; include some ''sell,'' i.e., how the product is different, makes life easier, etc.
- Include some sort of price context to avoid major misunderstanding.
- Don't make materials too elaborate; retain flexibility in presentation.
- Be prepared to modify the benefits as you get consumer reaction.
- Consider ''layering'' different options on the basic concept.
- Elicit individual as well as group opinion.
- Make sure you understand the difference between lack of consumer understanding (i.e., they understand the concept differently than us) and consumer confusion (they don't know what we are talking about) (Figure 6.8).

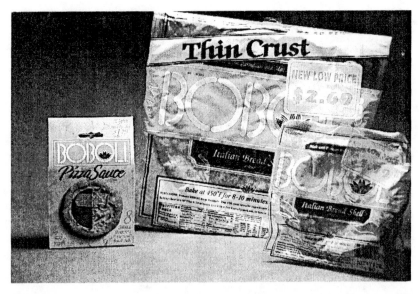

FIGURE 6.8 Pizza, one of the largest single American food categories, is offered in a wide variety of forms, from home delivery through refrigerated. Among the more interesting is ambient-temperature shelf-stable kits. Each component is preserved by ''hurdle'' or combined technologies. The par-baked shell is modified atmosphere packaging in gas barrier flexible materials. The high-acid sauce topping is hot-filled into heat-resistant, gas barrier flexible pouches.

BASES

[Author's note: Most of the material in this section is taken from two sources, a BASES, Inc. product brochure, and HBS Case Services case number 595-035 Nestle Refrigerated Foods: Contadina Pasta & Pizza (A) Rev. 1/97.] ACNielsen BASES (formerly the BASES Group) specializes in estimating and analyzing sales potential for new products, line extensions, and restaged brands, using market simulation models and analytical tools. BASES tests, like most concept testing methodologies, attempt to help decision makers reduce risks by providing accurate estimates of consumer sales volume at the earliest possible stages of product development, well before major resources are invested in production, packaging, marketing, or management time. BASES, Inc. offers alternative testing products, including

- Pre-BASES is a concept test with only rough volume estimates.
- BASES I is a concept test with volume estimates generally thought to be within a 25% accuracy range. The methodology employed attempts to assess the current level of awareness and usage in the category, and gain insight into consumer perception of alternate possible positioning statements with respect to competitive brands. A BASES I test provides an estimate first year trial volume, simulates total year 1 sales volume, and provides some understanding of the likely effect of alternate positionings.
- BASES II is a concept test in combination with a product taste test, with first year sales volume estimates reliable within 20% of actual. In addition, BASES II tests measure claimed source of volume.
- BASES II Line Extension is a test for new products that are part of a pre-existing product line.

BASES is the industry leading simulated test market product. Introduced in 1978, BASES has been around for approximately twenty years. Most companies use BASES tests as screeners for products to be test marketed or rolled out. A BASES test starts with consumer attitudinal data and reactions to concept boards, including purchase intent, that are adjusted in order to provide accurate predictions of consumer behavior, specifically trial of a new product. Repeat purchase rate is a function of product satisfaction. Different repeat purchase rates can be projected for products that prove to be, respectively, mediocre, average, and excellent.

These adjusted consumer data are combined with client-provided assumptions of distribution, media plans, and trade and consumer promotional events in order to estimate trial and repeat purchase potential for the new brand. Given their extensive product category experience, with results of tests of close to 6,000 new concepts and 3,000 new products during the past ten years,

BASES has in-market databases which help ensure the reasonableness of inputs and outputs to the model. BASES' clients routinely share their in-market experience with BASES for products that were tested in the BASES system and subsequently introduced in-market, and BASES buys IRI panel data. These data allow BASES models to be adjusted and fine-tuned so that about 90% of its forecasts fall within plus or minus 20% of actual market results.

Major factors contribute to the reliability of a BASES test. These include the sample size, the similarity of the estimated marketing plan to the actual plan used to launch the product, and the BASES experience with similar product categories. A BASES II test also employs in-home product usage tests to derive more accurate assessments of product satisfaction and repeat purchase.

SUMMARY

Product concepts are verbal or pictorial representations of products. Concept statements provide a description of a new product's attributes and benefits, as well as a stimulus to which consumers can respond to assess the overall strength of the concept and the level of consumer interest. Concept statements also provide a protocol to guide the work of the project team as the concept is developed into a physical product. Concept tests employ a variety of research methodologies to measure consumer reaction to the concept, determine the concept's strengths and weaknesses, and measure purchase intent, and serve as a gate through which only high-scoring concepts pass to the next stage of development. Purchase intent, along with other data, is used to create early stage sales volume estimates. Purchase intent scores must be validated, usually through comparison with previous test scores combined with actual market performance. This requires that a substantial database of concept test results be maintained. BASES, Inc. offers concept and simulated test market products frequently used by food processors.

BIBLIOGRAPHY

Clancy, K. and R. Shulman. 1995. "Test for Success." *Sales and Marketing Management*, pp. 111–114.

Clancy, K. J., R. S. Shulman, and M. M. Wolf. 1994. *Simulated Test Marketing*. New York: Lexington Books.

Crawford, C. M. 1997. *New Products Management*. Boston, MA: Irwin McGraw-Hill.

Feig, B. 1993. *The New Products Workshop*. New York: McGraw-Hill.

Fitzpatrick, L. 1996. "Qualitative Concept Testing Tells Us What We Don't Know," *Marketing News*, 30(20):11.

Fuller, G. W. 1994. *New Food Product Development: From Concept to Marketplace.* Boca Raton, FL: CRC Press.

Gibson Consulting Co. brochure.

Gill, B., B. Nelson, and S. Spring. 1996. "Seven Steps to Strategic New Product Development," in *The PDMA Handbook of New Product Development,* Rosenau, M. D., A. Griffin, G. A. Castellion, and N. F. Anschuetz, eds. New York: John Wiley & Sons, pp. 19–33.

HBS Case Services case number 595-035, Nestle Refrigerated Foods: Contadina Pasta & Pizza (A) Rev. 1/97.

Patrick, J. 1997. *How to Develop Successful New Products.* Lincolnwood, IL: NTC Business Books.

Schwartz, D. 1985. *Concept Testing.* New York: AMACOM.

Stinson, W. S., Jr. 1996. "Consumer Packaged Goods (Branded Food Goods)," in *The PDMA Handbook of New Product Development,* Rosenau, M. D., A. Griffin, G. A. Castellion, and N. F. Anschuetz, eds. New York: John Wiley & Sons, pp. 297–312.

CHAPTER 7

Food Science, Technology, and Engineering Overview for Food Product Development

ROMEO T. TOLEDO
AARON L. BRODY

Development of new food products requires an understanding and appreciation for the scientific and technological principles of foods and food processing. Food serves the basic functions of providing nutrition and psychological satisfaction. Almost all foods are complex biochemical and biophysical structures that inevitably are vulnerable to deterioration. Vectors of deterioration include microbiological, enzymatic, biochemical, and physical. Food science and technology is the discipline that strives to retard food deterioration by applying measured heat, temperature reduction, water removal, blending, and/or chemical additives. Microbiological food safety demands that all products be developed and handled so that the probability of an adverse public health incident is minimized. Food engineering is a subdiscipline in which the processes and equipment to effect the thermal, physical, and/or chemical changes are optimized. Food product development must take into account the ability to translate concepts and prototypes into commercially viable entities.

FOOD is the most primal of human needs. The need for food is manifested by physiological changes in the body, primarily low levels of blood sugar and a near empty stomach which signals the brain to make a person feel hungry. Visual and olfactory signals to the brain stimulate or depress appetite. Food, in the most basic sense, provides the nutrients needed for the body to have the energy to perform muscular movement, sustain growth, or maintain health. The study of nutrient needs to maintain a healthy life is the science of nutrition.

In affluent societies, food also has a social role in addition to nutrition. Eating is a social event, a pleasant activity in an atmosphere where business decisions may be easily made or where friends and family may strengthen

129

the closeness of relationships. To a food scientist, food is more than just a collection of nutrients. Food should provide the consumer with the enjoyment and satisfaction associated a pleasing visual, oral, and olfactory stimuli. Food technology in product development involves manipulation of the chemical constituents of food and ingredients to maximize the positive sensory perceptions by consumers of the product.

Unlike freshly prepared foods in restaurants or domestic kitchens, commercial food products are consumed after they have spent a period of time in the distribution and retail system. Although commercially processed foods must attract the consumer by the appearance of the package, repeat purchases will depend on how pleasing a sensory experience is perceived when the product is eaten. Thus, food scientists must not only be concerned with nutrient content and sensory attributes of food immediately after preparation but also with how changes which occur during storage and the method of preparation prior to serving affect product sensory properties.

PRIMARY FOOD COMPONENTS

The primary food nutrients arranged in the order of their contribution to the total dietary energy intake are carbohydrates, fat, and protein. These nutrients provide 4, 9, and 4 kcal per gram of food, respectively. Metabolism of these major nutrients produces the chemical energy that powers muscular contraction, allows brain cells to function, and permits the synthesis of compounds the body needs from simpler compounds in the diet. Regulations that govern the labeling of food products as reduced calorie, "lite," or low calorie, specify the caloric density of food and the amount of fat present per serving. The digestible carbohydrates are the preferred energy-contributing component of food formulations. Their metabolism produces carbon dioxide and water, which are eliminated through the respiratory system. In contrast, metabolism of fats and proteins involve the liver to safely convert metabolic intermediates into compounds that are further metabolized or eliminated through the urinary system

CARBOHYDRATES

Fiber

Carbohydrates constitute the structural and storage organelles of plants. The structural carbohydrates in plants are complex compounds of repeating five carbon sugars called pentosans or repeating six carbon sugars called glucans. The former, called hemicellulose, and the latter, called cellulose, are nondigestible and, along with another complex noncarbohydrate compound called lignin, constitute most of what is considered dietary fiber. Cellulose, hemicellulose, and lignin are fibrous. They do not soften appreciably on hydration and heating, and therefore they often impart a rough mouthfeel to food products. Another plant component, pectin, is soluble in hot water but is nondigestible. Pectin, a soluble dietary fiber and a component of cell walls of plants, is responsible for the softening of vegetables on cooking. Changes in the pectin molecular structure causes the softening of ripened fruits. Commercial pectin preparations are used as gelling agents for jams, jellies, and preserves.

Some plant seeds, exudates from the bark, or the stem and leaves of aquatic plants contain soluble complex carbohydrates called gums. Gums dissolve in water to produce viscous solutions and at specific concentrations, or in the presence of specific ions or sugar, they would form gels. Gums provide food technologists with the tools needed to modify body, firmness, and mouthfeel of food products.

Sugars

The simple sugars are the immediate products of photosynthesis. They are also the end product of the complete digestion of carbohydrates. Simple sugars cannot be further broken down and still remain a sugar. Sugars are sweet soluble carbohydrates. The end products of digestion of complex carbohydrates in the alimentary tract, sugars are rapidly absorbed from the intestines into the blood circulatory system. This characteristic of easy absorption of simple sugars make them a rapid energy source for some individuals, but it could also be detrimental to diabetics who are susceptible to hyper/hypoglycemic swings with intakes of simple sugars. Refined sugars are the common carbohydrate sweeteners such as corn syrup, high fructose corn syrup, and cane sugar. Some consumers may tend to consume these sugars with minimal intake of the other food nutrients, and so when refined sugars are consumed in excess, there might be inadequate intake of other dietary nutrients. One health concern with refined sugars in the diet is their ability to support the growth of bacteria responsible for tooth decay. These health concerns must guide a food technologist in the choice of carbohydrate used in food formulations.

Starch

Starch is the storage carbohydrate in grains and root crops. It is a complex molecule which consists of a long chain of repeating units of the simple sugar, glucose. Since plants also utilize starch as they respire during their resting period in the absence of sunlight, starch is easily digestible by humans. Starch is the complex carbohydrate recommended to provide the majority of dietary calories. In addition to being a source of calories, starch plays a significant role in determining the texture of foods.

Native starch occurs as granules in a cellular matrix within grains, and storage roots or tubers of plants. Macerating the root or milling the grain releases starch as granules which can be easily separated from the other components to produce a purified starch. Wheat flour is a mixture of starch and other components of the grain. Corn starch is pure starch.

Within the starch granule are two types of polymers, a straight chain called amylose and a branched chain called amylopectin. These large complex molecules are tightly coiled within the granule. Heating starch in the presence of water permits the granules to absorb water, swell, and eventually release the amylose and amylopectin into solution. When intact starch granules are no longer visible, the hydrated starch molecules in solution are called gelatinized starch. The gelatinization process is manifested by an increase in viscosity. Different starches will have different gelatinization temperature and different viscosities at specific concentrations. When starch at an appropriate concentration is gelatinized and allowed to cool, a firm gel may form. The gelling properties of the starch determine the ability of that starch to bind water and impart firmness to food products.

Amylose and amylopectin are present in different starches in different proportions. Gelatinized amylose and amylopectin have different characteristics. Gelatinized amylopectin solutions are opaque and do not form firm gels on cooling. Gelatinized amylose solutions on the other hand are clear and form firm gels with a tendency to release free water from the gel matrix on cooling. Amylose has a higher gelatinization temperature than amylopectin.

Although most natural starches consist of a mixture of these two polymers, some starches contain more of one than the other. For example, waxy maize starch contains practically all amylopectin. Regular corn starch on the other has more amylose than amylopectin.

When gelatinized starch is stored, starch molecules in the solution slowly lose water of hydration and those trapped within the spaces between adjacent molecules are slowly squeezed out resulting in agglomeration of starch molecules. The solution turns turbid and becomes less viscous so that, eventually, starch agglomerates precipitate. This process of conversion of gelatinized starch from a soluble to an insoluble form is called retrogradation. Starch retrogradation is a reversible process that may be reversed by heating. In low moisture cooked starchy products such as bread, retrogradation is manifested by a stiffening of the structure or staling, making the product appear hard and dry. Starch retrogradation rate increases with decreasing moisture content and reduced temperature. In frozen foods, however, water is immobilized by freezing, and so there is little movement from entrapment within the starch molecular network. Thus retrogradation in frozen foods is minimized if both freezing and thawing processes are carried out rapidly. Foods containing starch stored under refrigeration are most susceptible to retrogradation. In frozen foods, the time of holding at low temperature just above the freezing point is the critical time for development of retrograded starch. When starch is used in a food formulation as an emulsifier or a suspension medium for critical ingredients, starch retrogradation can irreversibly alter the desirable product quality attributes.

Because of its straight molecular structure, gelatinized amylose is more susceptible to retrogradation than amylopectin. Starch manufacturers have developed a number of products called modified starches, which are designed to alter the gelatinization temperature, viscosity enhancing properties, and retrogradation tendency of natural starches. A number of these modified starches are available commercially and starch manufacturers are a good resource for food scientists needing starch with specific functional attributes desirable in a formulation.

FATS AND OILS

Fats or oils, also called triglycerides, are compounds that consist of three long chain fatty acids attached to a glycerol molecule. Fats and oils have a similar chemical structure, but different types of fatty acids in the molecule change the physical state of the compound at room temperature. Fat triglycerides are solid at room temperature while oils are liquid at room temperature. When the fatty acids are highly saturated, each carbon atom in the molecule has its full complement of hydrogen atoms, i.e., single bonds join the carbon atoms, and the molecule is triglyceride containing these fatty acids and will be solid at room temperature. Fats made up of saturated fatty acids are stable against reacting with oxygen, and so saturated fats are less subject to oxidative rancidity development.

On the other hand, oils are made up of unsaturated fatty acids, i.e., they do not have their full complement of hydrogen atoms in the molecule; some of the carbon atoms in the chain are joined by double bonds. These triglycerides will easily react with oxygen to produce compounds which impart a rancid odor or flavor. Physiologically, in animals, saturated fats tend to be present in leaf fat or adipose tissue while unsaturated fats form a component of cellular membranes. The degree of saturation of fats depends upon the animal source. Fish has the least saturated of fats in animal foods, followed by poultry. Red meats contain the highest level of saturated fatty acids. Plant-derived oils, such as peanut, canola, corn, soybean, and cottonseed are highly unsaturated. In contrast, tropical plant oils such as cocoa, coconut, and palm oil contain highly saturated fats.

In the human body, saturated fats cannot be dehydrogenated to transform them into unsaturated compounds. Thus the need for highly unsaturated fats to form cellular membranes must be met by dietary intake of these compounds. Highly unsaturated fats, also known as polyunsaturated fats, are considered essential dietary nutrients. Saturated fats perform no major physiological function and serve only as a source of dietary calories. Fats have nearly double the caloric equivalent of the same weight of carbohydrates and proteins. Thus, high fat products are also high in calories. Therefore, the best way of reducing caloric content of foods is in formulating them to have reduced fat.

The melting point of fat used in a food formulation plays a role in how the components blend, how the product holds water and fat as it undergoes the mechanical and thermal rigors of the process, and eventually affects the texture, appearance and even flavor of the product. Fat is an excellent carrier for flavors and color. Thus, success of a food formulation may depend on the choice of fat used. Achievement of adequate storage stability against oxidative rancidity development will dictate formulation and packaging options which are necessarily highly dependent upon the type of fat in the formulation.

PROTEINS

Proteins are the structural component of animal tissue. The muscles which contract and relax to allow the body to move are composed of the proteins actin and myosin, while the connective tissue which separates muscle bundles is collagen. The protein collagen also forms the matrix into which calcium and phosphorus are deposited in bones.

Proteins are molecules which consists of a long chain of amino acids or nitrogen-containing organic compounds. There are 22 known amino acids in proteins, and the type and number of amino acids in a protein determines the protein type. Plant and animal proteins differ in the distribution of different types of amino acids in the molecule. Animal proteins most closely resemble the amino acid profile of human tissue protein, and so these proteins are best for human growth and maintenance. Plant proteins, however, may contain a low level of some of the essential amino acids, and so more of the protein must be consumed in the diet to meet the required protein intake for growth and maintenance of healthy tissue. Some proteins such as gelatin lack one or more of the essential amino acids and so have no nutritive value as a protein unless another protein is present. These will be metabolized, however, for energy and thus will contribute to dietary calories.

On digestion, proteins produce free amino acids. Commercial proteolytic enzymes from microbial sources may also digest proteins *in vitro* to alter their solubility, alter gelled product textural attributes, or to impart a characteristic flavor. Proteins are generally bland in flavor. However, as the length of the amino acid chain becomes shorter, the flavor becomes more distinct. Hydrolyzed proteins with a high fraction of free amino acids have a very strong flavor and odor. Some may even have a bitter flavor. Hydrolysis of proteins to give the desirable meaty flavor is a carefully orchestrated process of manipulating enzyme type, temperature, time, and pH to obtain the desired product attributes.

Foods that contain animal muscle as the primary ingredient, and plant protein-based foods must be carefully formulated and processed to take advantage of the ability of proteins to bind water and fat and set into a stable solid matrix. Proteins and carbohydrates have the same caloric equivalent; therefore, they may be exchanged in the food formula to produce least cost formulations. A number of commercial protein products and protein hydrolysates are available to food scientists for product formulations.

OTHER FOOD COMPONENTS

A number of food components are present at low concentrations, and yet they define the characteristic color and flavor of foods.

Acids

After the primary nutrients, acids are present at the next higher level. Acids are intermediates in plant metabolism; certain plants have the capacity to accumulate specific acids at high enough levels to give a distinctive flavor and mouthfeel. Acids impart a sour flavor. The intensity of the sourness is dependent on both the pH and the fraction of undissociated acid. Generally, strong acids give low pH solutions at relatively low acid concentrations and the mouthfeel is harsh on swallowing.

The types of undissociated acid molecules also differ in the interaction with other food components such as carbohydrates to modulate or enhance the sourness perception. In general, sugars modulate sourness, whereas mineral salts intensify the sensation. Gums such as pectin tend to coat the linings of the throat to minimize the harshness of the acid, but they also prolong the sourness sensation after swallowing. Buffering the acid by addition of one of its salts also tends to reduce the harshness of the sourness sensation. Lactic acid has a slightly bitter aftertaste, whereas tartaric acid leaves a slight scratchy feeling in the throat after the product is swallowed. The most common acidulant, citric acid, is best used in a formulation with a small amount of sugar to leave a smooth nonpersistent sourness sensation after swallowing.

Acids may be naturally present in the product or they may be an added ingredient. Citric acid is the most common plant acid and is the primary acid in citrus fruits. Malic acid is the primary acid in apples and tartaric acid is the primary acid in grapes. Lactic acid is produced by bacterial fermentation in salt-packed vegetables such as cucumbers and cabbage and is also the acid in fermented dairy products. Most of these acids are produced commercially by fermentation and are available commercially in pure form for use in food formulations.

Pigments

Color pigments comprise the next lower level food component after the acids. Plants synthesize pigments to produce the various colors and hues of plant products. The most widely distributed of the pigments is chlorophyll, the green pigment in plants. Chlorophyll is water soluble and is unstable with heat, particularly at low pH. Degradation of chlorophyll bleaches out the green color and transforms the color to brown. Care must be taken when processing products with green color or else severe color changes will develop.

The next most widely distributed water-soluble pigments are the anthocyanins. These colors range from pink to deep blue. Anthocyanin color is pH dependent, being red at the low end of the pH scale with the blue intensifying as the pH increases. Anthocyanins also change color to brown when they degrade during storage or during heat treatment of the product. The degraded pigments tend to agglomerate with surrounding degradation products of the molecule to form an unsightly precipitate at the bottom of the container. The presence of oxygen as well as elevated storage temperature tend to accelerate the degradation of anthocyanins, while the presence of sugar tends to slow down the rate of degradation. Anthocyanins do not have an attractive color at near neutral pH, and so they are most used as a colorant in frozen sliced fruit and shelf stable juices or beverages.

The natural fat-soluble colors available are yellow, orange and red. These compounds include carotene, which also is a precursor of vitamin A, and a group of similar compounds, the carotenoids, which could not be transformed into vitamin A. These fat soluble colors are heat stable, although slow loss of color can occur during storage if they are stored under conditions which promote oxidation. Natural plant derived water and fat soluble color pigments are commercially available for use as food colorants.

Meat pigments are normally not intensified by addition of artificial colorants although on some processed meat products color enhancement by the addition of colored spices such as paprika, turmeric, and annatto is practiced. Meat pigments such as myoglobin vary in concentration in the meat with the age and species of the animal. In raw meats, color changes from dull red with a purplish hue when meat is stored in the absence of oxygen and to a deep red on exposure to oxygen. Oxidized pigments turn to brown metmyoglobin. When cooked the pigment becomes a gray color. Addition of nitrite produces a stable pink color, the characteristic color of cured meats.

Micronutrients

The micronutrients needed for meeting dietary needs are normally not a consideration in food product development, except when formulating analogs using ingredients that do not contain the micronutrients in the product being simulated.

Flavor

Other compounds present in trace quantities contribute to the color and flavor of foods. When volatile, these compounds are responsible for the aroma of food and when nonvolatile they contribute to the taste sensation. Typical nonvolatile compounds are the tannins and high molecular weight alcohols. The volatile compounds have low molecular weight and may be lost during processing. An important component which participates in flavor development during heating is the group of compounds called carbonyls. They usually participate in reactions which result in roasted flavors. The flavor of coffee and cocoa, roasted nuts, and roast beef may be attributed to the reactions involving these carbonyls. Although these compounds may be naturally present, formulations may be developed where these compounds are added to intensify the desired flavor or color effect. Aroma, taste, and mouthfeel together constitute the flavor.

FUNCTIONAL PROPERTIES OF THE PRIMARY FOOD COMPONENTS

Starch, sugars, proteins, and fats interact significantly in a product to set the texture and flavor. Manipulation of these interactions to produce desirable product attributes tests the creativity of food scientists in food product development.

Immobilization of water through hydration of macromolecules or entrapment is a major goal in the design of food formulations. By virtue of their small molecular size, sugars are mobile in solution, and so the molecules may be easily positioned in intermolecular spaces in the product matrix to facilitate immobilization of water and maintain separation of macromolecular species preventing their aggregation. Thus, sugars could be used to minimize starch retrogradation. The gelling of the pectin-sugar system in fruit preserves, jams, and jellies is a good example of the water immobilizing properties of sugars in food systems.

The water immobilizing properties of starches associated with gelatinization are a well-known phenomenon, as are the water immobilizing properties of heat setting proteins. Some food ingredients such as fibers and nonheat setting proteins simply imbibe water reducing free water in the macromolecular interstices in the product.

Seldom understood is the role of fat in the immobilization of water. This role is primarily physical. Fat is hydrophobic, and so it can attach to hydrophobic moieties in macromolecules forming a water-impermeable membrane that would prevent water migration within the product matrix. Examples of protein–fat interactions that are effective in water immobilization are those that occur in processed meat products.

Because of the hydrophilic nature of starch, starch-fat interactions are not very effective in immobilizing water, but the presence of proteins in starch systems intensifies starch-protein-fat interactions to amplify the water immobilizing effect. Some modified starches have hydrophobic moieties created within the starch molecule making them effective as water immobilizing agents.

Minerals of the ionic or cationic species may enhance or reduce the water immobilizing properties or macromolecules. Mineral ions may crosslink sites in adjacent macromolecules forming interstitial spaces to trap water, or they could promote intermolecular attraction and agglomeration to squeeze water out of the system. Thus, care must be exercised in the selection of ingredients, including tap water used in the formulation to fully understand whether or not the presence of mineral ions is beneficial or detrimental in a system. The hardness of water used in a product is a commonly ignored factor in food product development. Similarly, the presence of minerals in ingredients which occurs by virtue of the original source or their addition during manufacturing, could be a factor that affects formulated product properties unbeknownst to the formulator.

Flavor and color development may also be manipulated by formulation or processing. In particular, roasted flavors develop as a result of reactions between carbonyl compounds and free amino acids in a reaction called the Maillard reaction. This reaction results in the formation of large complex molecules which have a brown color and characteristic roasted flavor. Low molecular weight volatile compounds with a pleasing odor are also formed. A pleasing roasted flavor is different from a burnt flavor. Flavor development through the Maillard reaction depends on the concentration of carbonyls and amino acids in the food, the moisture content, temperature, and time. Aldose or ketose reducing sugars such as glucose and fructose may provide the carbonyls required by the reaction. Amino acids can come from breakdown of proteins by indigenous enzymes or by added hydrolyzed proteins. Because reaction time is easily controlled, the most difficult part of ensuring proper extent of the reaction is controlling temperature and moisture content. The reaction appear to be favored by high temperature and low moisture content.

Flavor of Maillard reaction products does not appear to be as good in the presence of excess moisture compared to low moisture conditions. In high moisture products heated at atmospheric pressure, temperature is limited to the boiling point of water, 212°F (100°C). Slow moisture evaporation from the surface and adequate replacement by diffusing water from the product interior results in constant high moisture at the surface, and so Maillard reaction flavor development does not occur until a dry surface crust is formed. On the other hand, roasting in air at very high temperatures rapidly forms the dry surface crust because rapid heating vaporizes surface water faster than it is replaced by diffusion from the interior of the material. Thus, for the same time of roasting, the latter results in more intense Maillard reaction flavor development compared to low temperature roasting.

Although fat itself has no flavor, its presence in roasted foods intensifies the surface temperature from radiant heating in the oven, and the hydrophobic nature of fat isolates water from the reacting amino acids and carbonyls resulting in more Maillard reaction flavor development compared with what is formed in the same product without fat.

FOOD SPOILAGE

Food spoilage is primarily microbiological in nature, although biochemical and physical changes may also render a product less than acceptable by consumers. Physical changes such as starch retrogradation and loss of turgidity in frozen/thawed fruits and vegetables render products less desirable than their fresh counterparts, but the product is still edible. Biochemical changes in contrast occur over longer times, and so, for a period considered the product shelf life, the product may be still acceptable to consumers. Examples of biochemical changes that occur during processing are enzymatic oxidative browning, unwanted Maillard reactions, and caramelization of sugars. Examples of chemical changes that occur over a long distribution time are off-flavor development from uninactivated oxidative enzymes in plant products, ascorbic acid degradation in juices, and lipid rancidity development. Although physical changes may be minimized by careful selection of conditions during processing, prevention of chemical changes may require chemical additives.

Microbiological spoilage is the most significant type of spoilage in terms of economic loss to the industry and consumer dissatisfaction with the processed food. Microorganisms which cause spoilage are grouped into the yeasts, molds, and bacteria. Reproduction of these microorganisms to generate numbers that cause spoilage requires conditions favorable for their growth.

All microorganisms require water to grow, and an index used to determine the availability of water to support microbial growth is the water activity. Water activity is decreased by reduction of moisture content and lowered by the presence of solutes such as salt and sugar. Water activity is generally independent of temperature, although below the freezing point, crystallization of water removes water from solution as ice, and so the concentration of solutes increases, decreasing the water activity. Molds require the least water activity to support growth, generally a water activity of 0.7. Yeast are generally inhibited from growing at a water activity of 0.85. Bacteria on the other hand, are generally inhibited from growing at a water activity as high as 0.90. Bacterial spores generally do not germinate at water activity below 0.94.

The pH of food is also a factor inhibiting microbiological growth. Yeasts and molds are not inhibited at the pH levels of most foods. However, the large number of groupings of bacteria and the differences in requirements for growth by the different groups, means that only a few species are inhibited at a certain pH while some species may grow at this pH. However, spores of bacteria which are resistant to inactivation by heat or chemical disinfectants, generally do not germinate at a pH of 4.6 and below.

Some bacteria called aerobes require oxygen to grow while the anaerobes require the absence of oxygen. Between these two extremes are the facultative anaerobes that can grow under either condition and the microaerophiles that will grow even with very little oxygen present. Oxygen content of foods can be manipulated by modified atmosphere or vacuum packaging. The pH of foods can be adjusted by acidification using food grade acidulants. When combined with water activity reduction using solutes in the formulation or removal of water, an effective tool is provided the food scientist for altering the milieu to prevent microbiological spoilage.

All microorganisms respond to temperature in their rate of growth. The lower the storage temperature, the slower the rate of growth. The lowest temperature which inhibits growth classifies the microorganisms into the psychrophiles (cold loving), mesophiles (grow at room temperature), and thermophiles (grow at high temperatures). Mesophiles are spoilage organisms in dry shelf stored products. Psychrophiles are spoilage microorganisms in refrigerated products. Because low temperature is effective in slowing microbiological growth, short-term storage is possible without microbiological spoilage when foods are properly refrigerated. Freezing prevents growth by a combination of low temperature and decrease in water activity as solutes concentrate with ice crystal formation.

If pH, water activity, and oxygen concentration in the packaged food cannot effectively prevent or retard microbial spoilage, the food must be heat treated to inactivate potential spoilage microorganisms. When a food is low acid, i.e., the water activity is greater than 0.85 and the pH is greater than 4.6, heat treatments must be performed with steam or water under pressure to ensure the inactivation of bacterial spores. When the food is acid, i.e., pH below 4.6, heat treatment to stabilize food against spoilage may be adequate in boiling water at atmospheric pressure.

Although short-term storage of perishable foods may be prolonged by temperature control, reduction of the microbiological population, and modified atmosphere packaging, long-term shelf storage is possible only if all microorganisms capable of growing in the food under the conditions which exist in the package, are inactivated by heat treatment or other means.

MICROBIOLOGICAL FOOD SAFETY

As indicated in Chapter 1, microbiological food safety appears to be the most important problem in food safety in the United States, with up to 9,000 deaths and 33 million cases recorded annually. Foodborne illnesses are overwhelmingly microbiological and almost all from home-kitchen- or restaurant-prepared foods. Very few are from factory-processed foods.

As indicated above in the section on food spoilage, microorganisms use our food as their food and consume the food to produce carbon dioxide and water eventually, but can produce other undesirable end products such as acids and sulfide odors. Usually the end products are spoilage and not necessarily harmful. Spoiled foods, however, can be harmful to some persons, and so food spoilage has the potential to be unsafe, and some food spoilages are always unsafe.

Microorganisms are ubiquitous and increase their growth rate with rising temperature. Usually, but not always, the greater the numbers of microorganisms, the greater the problem with respect to spoilage and illness. Food infections are usually related to the quantities of microorganisms, e.g., *Salmonella* or *Listeria* infections, which are very hazardous to young, old, and ill persons, and pregnant women. In food intoxications, toxins are produced as microorganisms' biological end products. The toxins are very hazardous to everyone. Examples of toxin-producing microorganisms are *Clostridia* and *Staphylococci*. In addition, some "emerging" microorganisms such as *E. coli* O157:H7 can produce adverse effects with only a few microorganisms present.

Microbiological food safety issues may be retarded or obviated by one or more of several mechanisms:

- heat
  —pasteurization
  —sterilization
- reduced temperature
  —chilling
  —freezing
- reduced moisture
- chemical additives
- irradiation

Thermal energy may destroy microorganisms. If mild heat is applied, some heat-resistant microorganisms may survive, and so the product must be refrigerated in distribution. Considerable heat is required to destroy spores of low-acid anaerobic toxin formers such as *Clostridia*. If sufficient heat is applied to destroy vegetative and spore forms, the product is usually ambient-temperature shelf stable (provided the product is not recontaminated).

Refrigeration or reducing the temperature slows the rate of microbiological reproduction, but does not necessarily stop the growth of some pathogenic microorganisms. On the other hand, freezing can arrest the growth of pathogenic microorganisms. If the product is thawed, however, these microorganisms may grow.

Removal of water is an effective means to retard pathogenic microbiological growth. Included in water removal are drying and water activity adjustment.

Combination or "hurdle" technologies can reduce or, in some instances, completely stop, pathogenic microbiological growth and/or toxin production.

The issue here is that food products are at risk to a variety of pathogenic microbiological hazards which must be controlled if the product is to be distributed.

HACCP

The Hazard Analysis Critical Control Point (HACCP) system is a newly accepted preventive system to attempt to ensure safe production and distribution of foods. It is based on the application of technical and scientific principles to food processing from field to table. The principles of HACCP are applicable to all phases of food production, including agriculture, food preparation and handling, processing, packaging, distribution, and consumer handling and use, although it is mainly used today in food processing. Efforts are underway to extend HACCP to all food processing and into distribution.

The most basic concept underlying HACCP is prevention rather than inspection. Processors should have sufficient information concerning this segment of the food system, so they are able to identify where and how a food safety problem may occur. HACCP deals with control of safety factors affecting the ingredients, product, processing, and packaging. The objective is to make the product safely and to be able to prove that the product has been made safely.

The "where" and "how" are the HA (Hazard Analysis) part of HACCP. The proof of control of processes and conditions is the CCP (Critical Control Point) element. Flowing from this concept, HACCP is a methodical and systematic application of the appropriate science and technology to plan, control, and document the safe production of foods.

The HACCP concept covers all elements of potential food safety hazards that will more probably lead to a health risk—biological, physical, and chemical— whether naturally occurring in the food, contributed by the environment, or generated by a deviation in processing. Although chemical hazards appear to be the most feared by consumers, as indicated above, microbiological hazards are the most serious. HACCP systems address all three types of hazards, but the emphasis is on microbiological issues.

SOME HACCP TERMS

Control: to manage the conditions of an operation to maintain compliance with established criteria. The state in which correct procedures are being followed and criteria met.

Control Point: any point, step, or procedure at which biological, physical, or chemical factors can be controlled.

Corrective Action: procedures to be followed when a deviation occurs.

Critical Control Point (CCP): a point, step, or procedure at which control can be applied to a food safety hazard to be prevented, eliminated, or reduced to acceptable levels.

Critical Defect: a deviation at a CCP which may result in a hazard.

Critical Limit: a criterion that must be met for each preventive measure associated with a critical control point.

Deviation: failure to meet a critical limit.

HACCP Plan: the written document based upon the principles of HACCP which delineates the procedures to be followed to ensure the control of a specific process or procedure.

HACCP Plan Validation: the initial review by the HACCP team to ensure that all elements of the HACCP plan are accurate.

HACCP System: the result of the implementation of the HACCP plan.

HACCP Team: those persons responsible for developing and implementing a HACCP plan.

Hazard: a biological, chemical, or physical attribute that may cause a food to be unsafe for consumption.

Monitor: to conduct a planned sequence of observations or measurements to assess whether a CCP is under control and to produce an accurate record for future use in verification.

Preventive Measure: physical, chemical or other factors that can be used to control an identified health hazard.

Risk: an estimate of the likely occurrence of a hazard.

Verification: methods, procedures, or tests in addition to monitoring to determine if the HACCP system is in compliance with the HACCP plan and/or whether the plan needs modification and revalidation.

BIOLOGICAL HAZARD ANALYSES

A biological hazard is one which, if uncontrolled, will result in foodborne illness. As indicated above, the primary organisms of concern are pathogenic bacteria, such as *Clostridium botulinum, Listeria monocytogenes, Campylobacter,* and *Salmonella* species, and *Staphylococcus aureus* and *E. coli* O157:H7.

Physical hazards are represented by foreign objects which are capable of injuring the consumer. The HACCP team must identify both physical and chemical hazards associated with the finished product.

PRINCIPLES OF HACCP

HACCP is a systematic approach to food safety consisting of seven principles:

(1) Conduct a hazard analysis. Prepare a list of operations in the process at which significant hazards could occur and describe preventive measures that might obviate the hazards.

(2) Identify the Critical Control Points (CCPs) in the process.

(3) Establish critical limits for preventive measures associated with each identified CCP.

(4) Establish CCP monitoring requirements. Establish procedures for using the results of monitoring to adjust the process and maintain control.

(5) Establish corrective actions to be taken when monitoring indicates that a deviation from an established critical limit has occurred.

(6) Establish effective record-keeping procedures that document the HACCP system.

(7) Establish procedures for verification that the HACCP system is functioning properly.

Principle No. 1: Conduct a Hazard Analysis; Prepare a List of Operations in the Process at Which Significant Hazards Occur and Describe the Preventive Measures

The steps that precede the development of a hazard analysis are

- Assemble the HACCP team.
- Describe the food and its distribution.
- Identify intended users of the food.
- Develop and verify a flow diagram.
- Conduct the hazard analysis.

The HACCP team is responsible for conducting hazard analyses and identifying the operations in the process in which hazards of potential significance can occur. For inclusion in the hazard analysis list, the hazards must be of a nature that their prevention, elimination, or reduction to acceptable levels is essential. Hazards that are of low risk and not likely to occur would not require much further consideration when developing the HACCP plan. Low risk hazards, however, should not be dismissed as insignificant and may need to be addressed by other means.

Principle No. 2: Identify the Critical Control Points (CCPs) in the Process

A CCP is a point, step, or procedure at which control can be applied and a food safety hazard can be prevented, eliminated, or reduced to acceptable levels. An ideal CCP has

- critical limits that are supported by research and/or information in the technical literature
- critical limits that are specific, quantifiable, and provide the basis for a go or no-go decision on acceptability of product
- technology for controlling the process at a CCP that is readily available and at reasonable cost
- adequate monitoring (preferably continuously) and automatic adjustment of the operation to maintain control
- historical point of control
- a point at which significant hazards can be prevented or eliminated

All significant hazards identified during the hazard analysis must be addressed.

Principle No. 3: Establish Critical Limits for Preventive Measures Associated With Each Identified CCP

A critical limit is a criterion that must be met for each deterrent measure associated with a CCP. Each CCP has one or more measures that must be properly controlled to assure prevention, elimination, or reduction of hazards to acceptable levels. Each preventive measure has associated critical limits that serve as boundaries of safety for each CCP. Critical limits may be set for preventive measures such as temperature, time, physical dimensions, humidity, moisture, water activity (a_w), pH, acidity, salt concentration, viscosity, preservatives, etc.

Principle No. 4: Establish CCP Monitoring Requirements: Establish Procedures for Using the Results of Monitoring to Adjust the Process and Maintain Control

Monitoring is a planned sequence of observations or measurements to assess whether a CCP is under control and to produce an accurate record for future use in verification. Monitoring serves three main purposes:

- It is essential to food safety management that it track the system's operation. If monitoring indicates that there is a trend toward loss of control, i.e., exceeding a target level, then action can be taken to bring the process back into control before a deviation occurs.
- Monitoring is used to determine when there is loss of control and a deviation occurs at a CCP, i.e., exceeding the critical limit. Corrective action must then be taken.
- It provides written documentation for use in verification of the HACCP plan.

Principle No. 5: Establish Corrective Action to Be Taken When Monitoring Indicates That a Deviation from an Established Critical Limit Has Occurred

The HACCP system is intended to identify potential health hazards and to establish strategies to prevent their occurrence. However, ideal circumstances do not always prevail and deviations from established processes may occur. If a deviation from established critical limits occurs, corrective action plans must be in place to.

- Determine the disposition of noncompliant product.
- Fix or correct the cause of noncompliance to assure that the CCP is brought under control.
- Maintain records of the corrective actions that have been taken.

Principle No. 6: Establish Effective Recordkeeping Procedures That Document the HACCP System

The approved HACCP plan and associated records must be on file at the food establishment. Generally, the records utilized in the total HACCP system include

- the HACCP Plan
—listing of the HACCP team members and assigned responsibilities
—description of the product and its intended use
—flow diagram for the manufacturing process indicating CCPs
—hazards associated with each CCP and preventive measures
—critical limits
—monitoring system
—corrective action plans for deviations from critical limits
—recordkeeping procedures
—procedures for verification of the HACCP system

In addition, other information can be tabulated as in the HACCP Master Sheet:

- records obtained during the operation of the plan, especially those records of monitoring and verification activities

Principle No. 7: Establish Procedures for Verification That the HACCP System Is Functioning Properly

Four processes are involved in the verification:

- scientific or technical processes to verify that critical limits at CCPs are satisfactory, i.e., validation of the HACCP plan

- verification that the HACCP plan is functioning effectively
- documented periodic validations, independent of audits or other verification procedures
- government's regulatory responsibility and actions to ensure that the establishment's HACCP system is functioning satisfactorily

CHEMICAL ADDITIVES

Another tool for a product development scientist in inducing desirable changes or inhibiting undesirable changes is the use of chemical additives. Chemical additives may be classified as preservatives, processing aids, flavorants, colorants, and bulking agents. When using an additive, it is important that the additive performs a useful function, and that it must be safe under the conditions of effective functionality. The United States Food and Drug Administration (FDA) has a list of additives "Generally Recognized as Safe" (GRAS) under specified conditions of use. When the GRAS compound is used as specified, there is no need for FDA clearance. Other conditions of use may require GRAS affirmation from the FDA. New additives will require extensive testing for safety before FDA will clear it for use in commercially processed foods.

FOOD PROCESSING

Foods are processed to help preserve them, i.e., reduce microbiological safety issues and quality changes and to enhance their sensory qualities. Among the processes employed are heating, cooling, water removal, concentration, mixing, extrusion and filtration, and combinations.

APPLICATION OF THERMAL ENERGY

Heating of food can usually inactivate microorganisms and enzymes and may produce desirable chemical or physical changes in foods. Food is, however, highly susceptible to thermal degradation, and so to produce nutritionally sound and microbiologically safe foods, heat treatment must be carefully controlled.

Heat Treatment

Heat treatment may involve application of heat indirectly, as in a heat exchanger, or through direct contact of the heating medium with the food, as in bread baking in a hot-air oven. The main operations involving heat treatment of foods are blanching, preheating, pasteurization, sterilization, cooking, evaporation, and dehydration.

Blanching is a low-temperature treatment of raw foodstuffs generally to inactivate enzymes. Most vegetables and some fruits are blanched before canning, freezing, or drying. The commonly used types of blanching equipment are rotary drums, screw conveyors, and flume blanchers. Water is the dominant heating medium in drum and flume blanchers; steam is used in the screw-type blancher. Temperature is usually 212°F (100°C) or slightly lower.

Preheating the food before canning ensures the production of a vacuum in the sealed container and uniform initial food temperature before further thermal exposure for pasteurization or sterilization. When the product cools, the headspace water vapor condenses, generating a vacuum. Liquid or slurry foods are usually preheated in tubular heat exchangers. Particulate foods submerged in a brine or syrup in an open-top can are passed through a heated chamber, known as an exhaust box, in which steam is the usual heating medium.

Pasteurization is relatively mild heat treatment and involves the application of sufficient thermal energy to inactivate the vegetative cells of bacteria, molds, and yeasts and enzymes. Typical heating times and temperatures for pasteurization may be 30 min. at 149°F (65°C) or 15 seconds at 160°F (72°C).

Thermal sterilization heat is usually accomplished by heating food which is packed and hermetically sealed in containers. During the heating phase, the food is heated for the proper time by applying the heating medium to the exterior of the container. The cooling phase begins immediately after the heating phase. Because the temperature is at its maximum at the end of the heating phase, it should be reduced as quickly as possible to avoid further thermal destruction of the food quality.

Both batch and continuous heat-sterilization processes are used commercially. A still retort permits thermal processing without product agitation. The heating media are normally steam for food in cans or glasses, and high-temperature water with air overriding pressure or steam-air mixture for retort pouches. Continuous processes are less costly in terms of energy, labor, and time than batch processes, but the cost and complexity of equipment for continuous processing are greater.

The temperatures to which food is heated in conventional sterilization processes depend on the pH of the food. A normal temperature range for the heat sterilization of low-acid food (food with pH 4.6 or above and with water activity equal to or above 0.80) is 221 to 248°F (105 to 120°C); and a range for high-acid food (pH below 4.6, for example, fruit) is 180 to 212°F (82 to 100°C).

Some heat sterilization methods are called high-temperature short-time (HTST) processes. An HTST process usually consists of two separate heat treatments. Liquid food is preheated and then heated to temperatures of up to 280°F (140°C) in less than a few seconds. In another method, liquid food with low viscosity is heated by a tubular or plate-and-frame heat exchanger. Scraped-surface heat exchangers are used to process highly viscous or particulate-containing liquid foods. The heated food then passes through a holding tube before cooling in a tube or a chamber.

Cooling

The rate of spoilage and quality deterioration of fresh food is reduced exponentially as temperature is lowered. Raw meat is generally cooled in still air with high relative humidity to reduce moisture loss. Postharvest cooling methods for fresh fruit and vegetables depend on the type and volume of produce. With hydrocooling, produce is cooled by spraying or immersing in chilled water. Forced-air cooling is accomplished by forced flow of chilled air through the produce. In vacuum cooling, water is evaporated from the surfaces of vegetables or fruits by a vacuum created around the product. This system is especially good for products such as lettuce and spinach which have large surfaces in relation to volume.

Freezing

Many food products, if properly frozen and handled, maintain an acceptable condition for prolonged periods. The shelf life of frozen foods is extended beyond that of the fresh product because the lower temperature decreases the rate of deterioration. and, as the liquid water is changed to a solid, the solutes are immobilized if the temperature is reduced below the glass transition temperature.

Freezing Systems

Commercial freezing methods include forced air and immersion. In forced air freezing, the heat is removed from the product by cold air. In many freezers of this type, the food is stationary during freezing, but in large installations the food may move through the freezer or an intermittent or slow-speed conveyor.

Particulate foods such as peas, strawberries, or French fried potatoes may be individually quick frozen (IQF) by air being forced up through the product. The velocity is sufficiently high to create a fluidized bed, lift the product from the conveyor belt, and keep it agitated so that each particles is individually frozen.

Because the heat transfer rate between air and a solid is relatively low, many freezers are designed to contact cold liquids or solids with the food products. Salt brines and sugar solutions have been used.

Novelty desserts are frozen in molds that are immersed in a refrigerated solution.

In cryogenic freezing, the product is immersed in liquid nitrogen, or either liquid nitrogen or liquid carbon dioxide is sprayed on food products for very rapid freezing without the capital expense of a large mechanical refrigeration system. At atmospheric pressure, liquid nitrogen boils at $-320°F (-196°C)$. Both the vaporization of the liquid nitrogen and the sensible heat required to raise the temperature of the vapor are used to remove heat from the food product.

Concentration

Water-rich liquids are concentrated to remove water and thus

- provide storage stability
- produce saturated solutions
- produce supersaturated solutions which will form glassy or amorphous solids, e.g., sugar candies
- reduce storage, shipment, and packaging volumes and costs
- induce flavor and texture changes

Liquid foods are usually concentrated by evaporation.

Food evaporators usually contain heat-transfer tubes surrounded by a steam- or vapor-filled shell. The liquid to be concentrated flows through the tubes. Part of the water in the liquid is vaporized by the heat provided by the steam or vapor. The water vapor and remaining liquid separate and leave the evaporator through separate lines.

Dehydration

Although some food can be dried beyond its capability of reconstitution, a dehydrated product should be (but not always is) essentially in its original state after adding water. In drying, the water is usually removed below the boiling point as vapor. The basis of dehydration is that energy is transferred into the product to vaporize the water, and moisture is transferred out. The energy supplies the necessary heat, usually the latent heat of vaporization, and is responsible for the water migration.

Many foods are dried to a final moisture content of less than 5%, but this varies with the product and its eventual use. Dehydrated foods lower distribution costs due to the reduction in product volume and weight. Drying typically results in a number of changes including bulk density change; case hardening and toughening; heat damage, which may include browning; loss of ability to rehydrate; loss of nutritive value; loss of volatiles; and shrinkage. These changes are not necessarily desirable.

The dryer selected depends on food product characteristics, desired quality, costs, and volume throughput. Direct dryers include bed, belt, pneumatic, rotary, sheet, spray, through-circulation, tray, and tunnel. Indirect dryers include agitated-pan, drum, conveyor, steam-tube, vacuum, and vibrating-tray.

Drum dryers apply heat on the inside of a rotating drum surface to dry material contacting the outside surface. Heat transfer is by conduction and provides the necessary latent heat of vaporization. The drum rotates, and a thin layer of wet product is applied. Drying rates depend on film thickness, drum speed, drum temperature, feed temperature, etc. The speed of rotation is adjusted so that the desired moisture content of the scraped, dry product is obtained. Quality reduction is minimized by assuring that the product film has uniform thickness and that the dry film is removed completely. An advantage of drum dryers is their capability of handling slurries and pastes of high moisture and viscosity. Products such as mashed potatoes and powdered milk are typical applications.

Spray drying is a method of producing a dried powder out of feed in the form of pastes or slurries, or other liquids containing dissolved solids. Spray drying consists of atomization and moisture removal. Atomization breaks the feed into a spray of liquid droplets. This provides a large surface area for moisture evaporation. The spray is contacted with hot air, and evaporation and drying produce the desired solid product particles. Spray drying is an important preservation method since it removes moisture quickly and continuously without causing much heat degradation. The disadvantages of spray drying are the high energy needed, the requirement that the product be capable of being atomized, and the potential loss of desirable volatiles, as with instant coffee.

Filtration

Filtration processes remove solids from a fluid by using a physical barrier containing openings or pores of the appropriate shape and size. The major force required is pressure to pump the fluid through the filter medium.

Filtration processes cover a wide range of particle sizes. The sieving or screening operations use fairly coarse barriers. The screens may be vibrated or rotated to improve filtration rates.

Equipment for solid-liquid filtration can be classified in a number of ways:

- whether the retained particles (the cake) or the filtrate (the clarified liquid) is the desired product
- batch or continuous operation
- driving force
- mechanical arrangement of the filter medium

Mixing

Mixing provides more-or-less homogeneous compositions and physical properties for food ingredients and combinations of ingredients. Mixing is also used to facilitate heat and mass transfer, create dispersions, produce emulsions and foams, suspend solids, facilitate reactions, and produce textural and structural changes.

Many different mixing methods are used in order to produce the liquid, solid, dough-like, and foamy mixtures and dispersions in food processing. In general, these methods involve repeatedly subdividing the mixture into discrete domains which are then transported relative to each other so that the domains repeatedly encounter and exchange matter with new neighboring domains.

Liquid mixing is usually performed by rapidly turning screw propellers or radially flowing turbines to circulate liquid in baffled tanks.

SUMMARY

Numerous tools are available to a product development scientist/technologist for effective formulation of food products for commercial processing and distribution. The scientist must know the chemistry and functional properties of food constituents and ingredients and apply this knowledge to create commercially successful convenient food products for the mass market.

BIBLIOGRAPHY

Doyle, Michael P., L. R. Beuchat, and T. J. Montville. 1997. *Food Microbiology Fundamentals and Frontiers.* Washington, DC: ASM Press.

Erickson, M. C. and Y. C. Hung. 1997. *Quality in Frozen Food.* New York: Chapman and Hall.

Fennema, O., ed. 1985. *Food Chemistry,* 2nd Ed. New York: Marcel Dekker.

Glicksman, M., ed. 1982. *Food Hydrocolloids.* Boca Raton, FL: CRC Press.

Lopez, A. 1987. *A Complete Course in Canning,* 2nd Ed. Volumes 1, 2, and 3. Baltimore, MD: CTI, Inc.

Potter, N. N. and J. H. Hotchkiss. 1995. *Food Science,* 5th Ed. New York: Chapman and Hall.

Whistler, R. L., J. N. BeMiller, and E. F. Paschall, eds. 1984. *Starch, Its Chemistry and Manufacture,* 2nd Ed. New York: Academic Press.

CHAPTER 8

Development of Packaging for Food Products

AARON L. BRODY

Virtually all food products must be encased in packaging to protect them from the natural environment. With packaging now a representation of the product, the development of effective and attractive packaging is critical to the ability to distribute and deliver food that satisfies consumers. Packaging is not just materials such as paper, metal, glass, or plastic, or structures such as cans, bottles, cartons, or pouches. Rather, packaging is the integration of product content protection requirements with process, and the selection of alternative material/structure combinations with equipment and distribution.

THE role of packaging is to protect the contained food product. Packaging is always corollary to the function of the food contained.

Packaging is one system whose objective is to protect the contained product against an always-hostile environment of water, water vapor, air and its oxygen, microorganisms, insects, other intruders, dirt, pilferage, and so on—because a constant competition exists between humans and their surroundings. Packaging is designed to facilitate the movement of a product from its point of production to its ultimate consumption.

If there is no product, there is no need for a package.

Packaging is arguably the single most important link in the distribution chain that places a product into the hands of the consumer. In a very real sense, in today's society, packaging might be regarded as an integral component of the product contained.

The words *package* or *packaging* have different meanings, intended to convey different images. The *package* is the physical entity that actually contains the product. *Packaging* is the integration of the physical elements

153

through technology to generate the package. *Packaging* is a discipline. The *package* is what the consumer must open to obtain the food.

All definitions of packaging center about a single concept: the protection of the packaged product for the purpose of facilitating its journey to the marketplace and use by the consumer. Packaging is that combination of materials, machinery, people, and economics that together provides protection, unification, and communication.

FUNCTIONS OF PACKAGING

Packaging's roles depend mostly, but not totally, on the food product contained. The main functions of packaging are protection, containment, communication, unitization, sanitation, dispensing, product use, convenience, deterrence of pilfering, and deterrence from other human intrusions such as tampering.

PROTECTION/PRESERVATION

As stated above, product protection is the most important function of packaging. Protection means the establishment of a barrier between the contained product and the environment that competes with man for the product.

For example, the product must be protected to control its moisture content. Most dry products are susceptible to consequences of exposure to moisture or liquid water; hygroscopic foods absorb water and deteriorate. Conversely, most wet products are susceptible to loss of their water content.

Oxygen present in the air reacts with most food products. By establishing a barrier between the air and the product, packaging can retard the oxidation of fat in foods—provided oxygen in the package is first removed. Packaging retards the oxidation of foods and the related product deterioration. Flavoring ingredients of foods may be adversely affected by exposure to oxygen. Carbon dioxide must be contained in products such as beer, champagne, sparkling wine, and carbonated beverages to ensure that the flavor quality of these products is maintained throughout distribution.

Volatile essences and aromatic principles would be lost due to volatilization were it not for gas-barrier packaging. Simultaneously, the absorption of foreign odors and flavors from the environment is deterred by the presence of impermeable packaging.

In today's complex distribution system, the product is manufactured, fabricated, grown, or transformed in one geographic region, and its consumption is in a geographic area far removed from its origin. Packaging is required to ensure the product's integrity during transit and in storage to extend the time and geographic span that is required between origination and final use. Time required can be measured in months or even years, although the distances can, of course, be in thousands of miles.

CONTAINMENT

Food packages hold or contain food products. Packaging permits holding or carrying not only what can be grasped in a person's hands or arms, but also products such as liquids or granular flowable powders that simply cannot be held or transported in industrial- or consumer-sized units if they are not contained. Products such as carbonated beverages and beer could not be consumed at any distance or time from the manufacturing site without packaging. Further, the aging of wines and cheeses require packaging if these processes are to occur without spoilage.

SANITATION

Packaging helps to maintain the sanitary, health, and safety integrity of contained products. Processing and packaging are intended to stabilize food products against degradation during distribution. One purpose of packaging is to reduce food spoilage and minimize the environmental losses of nutritional or functional value of the product.

The presence of debris, foreign matter, microorganisms, the debris of insects, and the droppings of rodents in food products is intolerable. They are especially undesirable if these contaminants have entered the product after it has been properly processed and stabilized. Packaging acts as a barrier to prevent the entry of environmental contaminants. Packaging also minimizes the contamination of contained products by intentional or casual human contact and its potential for infecting the product.

One of the significant objectives of packaging is prevention of reentry of microorganisms after they have been successfully removed from, stabilized, or destroyed within the product. For example, food products are thermally processed in hermetically sealed metal cans or glass jars in order to destroy ubiquitous microorganisms and to ensure against entry of other microorganisms after sterilization. By deterring recontamination, packaging minimizes the probability of disease and infection from food and reduces spoilage that could lead to toxin production or economic losses.

Packaging also retards losses of nutrients from temperature-driven biochemical reactions and prevents nutrient losses that could arise from the adverse actions of microorganisms. Microorganisms grow by consuming the food product and consequently impairing the nutritive value of food products, provided the product is edible after the microbiological activities.

UNITIZATION

Unitization is assembly or grouping of a number of individual items of products or packages into a single entity that can be more easily distributed, marketed, or purchased as a single unit. For example, a plastic ring or shrink film carrier for six cans of beer, or a corrugated fiberboard shipping case containing 12 bottles of salad dressing is far easier to move than attempting to carry six individual cans or 12 bottles. Forty-eight individual jars of baby food would be impossible for a person to carry. A pallet load of 360 cases is far easier to move as a single unit rather than attempting to load and unload a truck with 360 individual cases.

A paperboard folding carton containing three flexible material pouches of seasoning or soup mix delivers more product to a consumer than does a single pouch. A paperboard carton wrapped around 12 beer bottles provides more desired liquid refreshment for home entertainment than does an attempt to carry individual bottles in one's hands.

Unitization reduces the number of handlings required in physical distribution and, thus, reduces the potential for damage. Because losses in physical distribution are significantly reduced with unitization, significant reductions in distribution costs are effected.

COMMUNICATION

Packaging is one of our major communications media. Usually overlooked in the measured media criteria, packaging is the main communications link between the consumer or user and the manufacturer, at both the point of purchase and the point of use.

Mass market self-service retailing that reduces the cost of distributing products from the manufacturer or grower to the consumer virtually could not exist without the graphic communications message on the surface of the package. The task of communicating identity, brand, price, instructions, warnings, warranties, etc., is the responsibility of the package and/or its label. Considerable information is required by both law and regulation for food products. This information is designed to assist the consumer, particularly those consumers who are partially cognizant of the need to be fully informed of the contents.

Self-service retailing requires that the package surfaces bear clear, easily seen messages on the identity of the contents. The package on the shelf is the main link between the producer and the purchaser. At this point in its cycle, the *package* and not the *product* is being purchased. The package is the promise to the consumer of what is inside that package. Thus, recognizable packaging is extremely important, if not essential. Graphic designs that hide, obscure, or otherwise deceive are self-defeating because consumer frustration is perhaps best expressed by rejection during the purchase-decision process. The absence of a sale is telegraphed to the producer by a reduction in purchase by the retail outlet. On the other hand, probably only about half of all products retailed by today's typical supermarkets receive significant media advertising. Consequently, the package itself is an advertising medium communicating the benefits to be received from the investment of money by the prospective purchaser.

Despite this key role of communications, approximately a quarter of all items are distributed with minimal or no packaging. Fresh red meat, delicatessen items, on-premise-baked products, home meal replacement prepared foods, and fresh produce frequently are distributed with almost no graphics on the packaging surface.

DISPERSING AND DISPENSING

The user or consumer often dispenses a product into readily used quantities. Packaging often facilities the safe and convenient use of the product. Thus, bottles may have push-pull or no-drip tops, cartons may have pouring spouts, salt and pepper shakers and spice containers may have openings through which the product may be shaken.

To facilitate opening, the container, such as a carbonated beverage or beer bottle or can, almost invariably have a "finger-friendly," easy-opening device to expose a pouring hole. Many packages, such as coffee or shortening cans and syrup bottles, have reclosure devices that permit the user to effectively reseal the package and protect it during reuse (Figure 8.1).

FIGURE 8.1 An important function of packaging is ease of opening, access to contents, and reclosure if the contents are not fully consumed. For this salty snack composite paperboard can, a metal ring with an adhesive-adhered peel-off flexible membrane is double-seamed to the can body. An injection-molded polyethylene snap-cap serves as a reclosure.

PILFERAGE DETERRENCE

The cost for shoplifting, intentional switching of price markers by consumers, and so on in self-service retail stores is much too high. Despite increasing vigilance by security people (which increases costs), plus numerous attempts made to deter the problem through packaging, this staggering amount unfortunately has not been declining. Nevertheless, packaging helps to keep this figure from reaching astronomical heights.

TAMPERING

The tragic headline events of 1982 and 1986, affecting McNeil Laboratories' Tylenol analgesic capsule products, were this country's most dramatic manifestations of intentional tampering with products. Intentional tampering of fresh fruit and vegetables by consumers has occurred for many years. Intentional opening of packages to taste-test, smell, or examine the contents is not uncommon and is obviously unsanitary. Intentional opening of packages to cause some undesirable event such as the Tylenol poisoning incident is rare, but nevertheless was highlighted by the 1980s incidents. Packaging per se is a deterrent to tampering as has been evidenced by the infinitesimally small number of intentional tamperings with products in the United States.

The laws and regulations promulgated as a result of the Tylenol tragedy have led to overt methods of visibly signaling or further deterring tampering of proprietary drugs. Inspection for evidence of tampering and production of tamper-resistant packages have now become essential functions of over-the-counter drug packaging.

Tamper-evident-resistant packages are not required by law or regulation for foods, but many food packages nevertheless incorporate such features to deter intentional and unintentional tampering.

OTHER FUNCTIONS OF PACKAGING

Other functions of packaging include apportionment of the product into standard units of weight, measure, or quantity prior to purchase. Yet another objective is to facilitate product use by the consumer with devices such as spouts, squeeze bottles, and spray cans. Aerosols not only serve as dispensers, but also prepare the product for use, such as aerating the contained whip toppings. Still other forms of packaging are used in further preparation of the product by the consumer, e.g., tea bags that are plastic-coated, porous paper pouches, or frozen dinner trays, which were originally aluminum and now are fabricated from other materials such as crystallized polyester and polyester-coated paperboard (Figure 8.2).

FIGURE 8.2 Unconventional packaging can sometimes be an effective means to persuade consumers to pick the packaged product off the shelf. Dry pasta mixes are generally packaged in barrier flexible pouches or in barrier flexible pouches in paperboard cartons. Packaging dry pasta mixes in a corked glass bottle certainly qualifies as an unusual package—especially when the retail price is $8 per unit—about five times that of more traditionally packaged product.

PACKAGING: ADVANTAGES

Packaging is a low-cost means of protecting products, reducing waste, and reducing the cost the consumer pays for products. Because of the efficiencies in the American food distribution system, its infrastructure, and its packaging systems, losses of agriculture commodities in the United States between the field and the consumer are less than 20%. In other parts of the world, where little or no packaging is used, over 50% of the food grown in the field never reaches consumers because of spoilage, infestation, weight losses, and theft.

Packaging permits a larger number of different products to be made available to the consumer. For example, the American food industry alone produces more than 50,000 different products. Although some might regard this number as frivolous waste, the food industry, like all other industries in a free-market economy, responds to consumer wants and needs. If there were no market for so many different items, they would not exist, particularly in this period of high recognition of economic value of product inventory. Each year, more than 12,000 new food products are introduced—each in response to some processor/marketer's perception of a consumer desire.

Probably the most significant benefit of packaging to the American consumer is the safety of the products contained. In contrast to the disease, infection, vectors, foodborne toxins and adulterations prevalent in food products in other countries of the world, the American food industries produce products that are overwhelmingly safe and beneficial to consumers. Despite the infamous Tylenol tampering incidents, the number of deaths attributed to packaged products in the United States, due to some error in processing or packaging, stands at fewer than a dozen in a decade. Fewer than a dozen persons have died from toxins produced in commercially canned foods in the last half-century, i.e., the death rate from canned foods is less than six per trillion packages. This is a better safety record than any other entity that we employ in the course of our lifetime (Figure 8.3).

The use of packaging significantly reduces the garbage, waste, and litter in our streets and solid waste streams. The number of apple cores, orange peels and seeds, bones from meat and fish, feathers from poultry, and outer leaves from lettuce is very significantly reduced by the use of processed packaged meats, frozen concentrated orange juice, prebreaded frozen chicken, etc. Thus, the solid waste stream and the unsanitary practices not uncommon in solid waste disposal are reduced significantly by the use of packaging. Packaging generates less solid waste than materials it replaces. Further, the excess materials such as orange peels, potato peels, bones, and trimmings, when in central factory locations, can be employed for beneficial and/or profitable results, e.g., cattle feed, fertilizer fermentation media.

FIGURE 8.3 The food processor/marketer of these pie fillings determined from consumers that their preference was for an all-white (i.e., "clean") barrier plastic can with an easy-open metal end plus a snap-on reclosure, as contrasted to an all-steel can.

In several studies, the U.S. Department of Agriculture has demonstrated that the cost of a food product prepared from a factory-processed convenience food is significantly less than the same food product prepared from scratch raw ingredients purchasable by the consumer. Thus, a cake prepared from a cake mix when this is done, an entree reheated in a microwave oven from a home meal replacement tray, and low-fat ice cream are all significantly less expensive to consumers than the equivalents prepared from basic raw materials. Further, the uniformity of convenience foods is much greater than could be possible with the variability of homemakers and their kitchen facilities.

The types of products available to consumers have multiplied many times in the past three decades. Although some social scientists do not regard this magnification as a measure of progress, consumers apparently feel otherwise. Consumers in the United States today are able to obtain products not available to their ancestors and certainly not available to wide geographic regions just a generation ago. Thus, shrimp, hearts of palm, guacamole, enchiladas, sushi, nan—and even ice cream—are products that the consumer has come to expect to be available at his or her retail shop, that were not available in days past. This availability, of course, has in part been made possible by packaging.

PACKAGING AND THE QUALITY OF LIFE

Efficient food production processing, coupled with packaging in a distributional infrastructure, has been a key element in the contemporary American industrialized society and its standard of living. Until the 20th century, agricultural commodities were produced and consumed only locally. Foods were dispensed almost exclusively from bulk sources. As the population departed from the fields for the factories in the cities in the early years of the 20th century, proportionally fewer products were consumed at the site of production. Thus, food products had to be shipped long distances. The time span between origination and consumption increased significantly and fostered the need for packaging to protect products during their journeys. As the economy developed further, mass production of products and requisite packaging materials also grew. The numbers of factories packaging branded products increased and merchandising and marketing became major tools. Because of the efficiencies and economies of mass production, more people and capital could be diverted from agriculture to the manufacture of a wide variety of industrial and consumer products, better housing, and more rapid and efficient transportation and recreational products. As consumer income expanded, more income could be dedicated to leisure, education, travel, and cultural aspects that contribute to our quality of life. The increase in consumer income expanded the requirements for product safety, reliability, and quality. Consequently, this was translated into a need for more and better packaging that could reduce the investment in time, risks, and permit the increase in the selection of products available.

Self-service, an innovation directed at significantly increasing the productivity of retail labor and reducing prices, is a key contributor to the American food marketing system. Packaging is an essential element of self-service food retailing.

As our population diverted from an agrarian society, it increased in size and income. The way we were as individuals, families, and communities has changed. The automobile and transportation distances, interdependence upon others, entertainment, both active and passive recreation, and the need for self-contained packaged items that could be used without further thought, all are more important than they ever were in the past.

The nature of our work force has changed. Women, first the central focus of agrarian society and homemakers in the urban context, are now employees and entrepreneurs. The pressures of sharing homemaking, motherhood, and parenthood have all led to significant changes in the requirements for products that are safe, convenient, easy to use, and reliable.

DISADVANTAGES OF PACKAGING

Beginning in the late 1960s and persisting through today, counter-packaging elements have been vocal around the world. During the early 1970s, some individuals (many of whom were elected legislators) saw, in the truncation or abolishment of packaging, a true yellow brick road to Utopia. Through their efforts, mountains of legislation and regulation were introduced and argued—with some now the law of the nation, state, or locality. As with all similar national debates, most of the points on both sides have been oversimplified.

Nevertheless, it is wise to enumerate some of the often-voiced concerns and criticisms about packaging.

COST

Obviously, because it is comprised of materials, equipment, people, and thought, packaging has an associated cost. At the outset, the cost of packaging may be greater than the cost of the product without packaging.

RESOURCE UTILIZATION

Because packaging is composed of materials ultimately derived from Earth, it is a user of resources. Most packaging is used only once and discarded, transferring the Earth's natural resources, often to less available or useful applications.

ENERGY

Because energy is required both to make packaging materials and to package, packaging is a net user of energy. Further, because plastics are derived from petrochemicals, they are believed to be more wasteful of energy.

TOXICITY

Packaging is composed of chemicals some of which could be hazardous to humans. Since packaging and product are in proximity with each other, the chemicals of packaging migrate into the product and could be harmful under special circumstances.

LITTER

After use, because they have no further use, packages are discarded in greenlands, streets, and waters to become temporary or permanent eyesores on the landscape. Further, the cost of removing litter from the streets and highways is usually borne by the taxpayer.

SOLID WASTE

Used packaging that has not become litter fills our dumps which reportedly are in short supply. Further, the cost of transporting this used packaging from the home, office, factory, store, or restaurant to the solid waste disposal site must be carried by the consumer. In the past, these locations were unsanitary havens for insects, scavengers, rodents, and other undesirables.

WASTE

Packages such as plastic or paperboard hanging at checkout counters are much larger than the products contained and, therefore, represent a major waste of packaging resources.

DECEPTION

Marketers and packagers employ packaging primarily to hide defects or to deceive consumers into believing the advertising claims on the packaging surface. Further, packaging is designed to convey the impression that the quantity of contents contained is significantly greater than really present.

CONVENIENCE

Because the packagers allegedly care little about their customers, they are careless about their package design and sometimes have failed to incorporate means to open, close, dispense, apportion, etc.

ENCLOSURE

Because packaging encloses the product, the consumer cannot directly touch, test, feel, smell, or taste the product. Thus, the consumer cannot obtain sufficient information on the nature of the product to make an intelligent purchasing decision.

MULTIPLES

Because six or twelve cans or bottles are linked together, the consumer is compelled to buy more units than are desired in any single purchase decision.

SANITATION

The package introduces microorganisms to contaminate the contained product.

QUALITY

The package extracts the natural quality of the contained product and reduces it. By definition, a processed and packaged product is inferior to that original removed from the ground, plucked from the tree, or fashioned by human hand.

The word "plastic" is often used in contemporary conversation to convey the notion of synthetic, unvarying, not real; "packaged goods" are frequently under the same accusation umbrella.

The preceding list is but a sampling of the views of those who would resolve major national and humanistic issues by significantly reducing or putting an end to packaging. Most of the arguments against packaging have a shred of basis in fact, but many are wholly imaginary. In the face of these apparent disadvantages, consumers continue to use packaged foods. To paraphrase Sir Winston Churchill, "Packaging is the worst of solutions, except for all the others."

It has become increasingly important to integrate packaging into the total product system if the objectives of delivering safe and high quality food are to be achieved. This chapter addresses various food product categories and their preservation needs. The potential beneficial effects together with the issues of packaging are introduced. Package materials and structures are described from the perspective of their food preservation characteristics. Finally, the development steps are enumerated in the context of new food product development.

PRESERVATION REQUIREMENTS OF COMMON FOOD CATEGORIES

MEATS

Fresh Meat

Most meat is offered to consumers in a freshly or recently cut form, with little further processing to suppress the normal microbiological flora present from the contamination received during the killing and breaking operations required to reduce carcass meat to edible cuts. Fresh meat is vulnerable to microbiological deterioration from microorganisms. These microorganisms can be as benign as slime formers to stink producers to pathogens such as *E. coli* O157:H7. The major mechanisms to retard fresh meat spoilage are temperature reduction, often coupled with reduced oxygen during distribution, to retard normal spoilage microbial growth. Reduced oxygen also leads to fresh meat color being the purple of myoglobin, a condition changed upon exposure to air which converts the natural meat pigment to bright cherry red oxymyoglobin characteristic of most fresh meat offered to and accepted by consumers. Reduced oxygen packaging is achieved through mechanical removal of air from the interiors of gas barrier multilayer flexible material pouches closed by heat sealing the end after filling (Figure 8.4).

Ground Meat

About 40% of fresh beef is offered in ground form to enable the preparation of hamburger sandwiches and related foods. Ground beef was originally a by-product, that is, the trimmings from reducing muscle to edible portion size. The demand for ground beef is so great that some muscle cuts are specifically ground to meet the demand. Grinding the beef always further distributes the surface and below surface flora and thus provides a rich substrate for microbial growth even under refrigerated conditions. Relatively little pork is reduced to ground fresh form, but increasingly during the 1990s, significant quantities of poultry meat is being comminuted and offered fresh to consumers, both on its own and as a relatively low cost substitute for ground beef. The major portion of ground beef is coarsely ground at abattoir level and packaged under reduced oxygen for distribution under refrigerated temperatures to help retard microbiological growth. The most common packaging technique is pressure

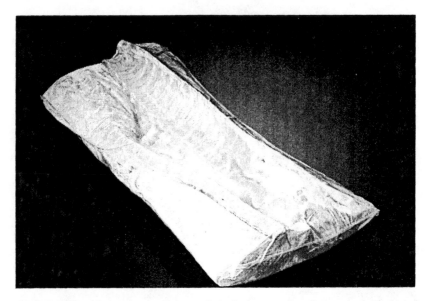

FIGURE 8.4 Beef and pork primal cuts reach the backrooms of retail groceries in the United States in vacuum barrier shrink film bags. Upon opening, the desired red color is restored by exposure to air. Butchers reduce the primal cuts into consumer-size cuts and package them in trays with film overwraps for retail display.

stuffing into chubs which are tubes of flexible gas barrier materials closed at each end by metal clips. At retail level the coarsely ground beef is finely ground to restore the desirable red color and to provide the consumer with the desired product.

In almost all instances, the retail cuts and portions are placed in expanded polystyrene (EPS) trays which are overwrapped with plasticized polyvinyl chloride (PVC) film. The tray materials are fat and moisture resistant only to the extent that many trays are internally lined with absorbent pads to absorb the purge from the meat as it ages and/or deteriorates in the retail packages. The PVC materials are not sealed but rather tacked so that the modest water vapor barrier structure does not permit loss of moisture during distribution. Being a poor gas barrier, PVC film permits access air and hence the oxymyoglobin red color is retained for the short duration of retail distribution.

Case-Ready Meat

For many years, attempts have been made to shift the retail cutting of the beef and pork away from the retailer's back room and into centralized factories. This movement has been stronger in Europe than in the United States but some action has been detected in the United States in the wake of the pathogenic microbiological incidents. Case-ready retail packaging involves the cutting and packaging under hygienic conditions to reduce the probability of microbiological contamination. The package is usually in a gas barrier structure, typically gas/moisture barrier expanded polystyrene trays heat seal closed with polyester/gas barrier film. The internal gas is altered to a high oxygen/high carbon dioxide internal atmosphere. The high oxygen concentration fosters the retention of the consumer desired red color while the elevated carbon dioxide suppresses the growth of most spoilage microorganisms. Using this or similar technologies, refrigerated microbiological shelf lives of retail cuts may be extended from a few days to as much as a few weeks permitting long distance distribution, e.g., from a central factory to a multiplicity of retail establishments. One thesis with central packaging of ground beef is that the probability of the presence of the *E. coli* O157:H7 pathogen is reduced.

Alternative packaging systems for case-ready beef and pork include the master bag system used widely for cut poultry in which retail cuts are placed in printed polyolefin film overwrapped EPS trays and the trays are multipacked in gas barrier pouches whose internal atmospheres are carbon dioxide to retard the growth of aerobic spoilage microorganisms. Another popular system involves the use of gas barrier trays with heat-seal closure by means of flexible gas and moisture barrier. Conventional non–gas barrier trays such as EPS may be overwrapped with gas/moisture barrier flexible films subsequently shrunk tightly around the tray to impart an attractive appearance. Other systems, all of which involve removal of oxygen, include vacuum skin packaging (VSP) in which a film is heated and draped over the meat on a gas/moisture barrier tray. The film clings to the meat so that no headspace remains meaning that the color of the contained meat is purple myoglobin. In one such system the drape film is multilayer with an outer gas barrier layer which may be removed by the retailer, exposing a gas permeable film that permits the entry of air that reblooms the pigment and restores the desired color. Variations on this double film system include one or more in which the upper film is not multilayer but rather independent flexible materials one of which is gas/moisture barrier and the other of which is gas permeable to permit air entry to restore the red color (Figures 8.5, 8.6, and 8.7).

Processed Meat

Longer-term preservation of meats is achieved by curing using agents such as salt, sodium nitrite, sugar, seasonings, spices, and smoke, and processing methods such as cooking and drying to alter the water activity, add antimicrobials, provide a more stable red color, and generally enhance the flavor and mouthfeel of the cured meats. Cured meats such as frankfurters and ham are often offered in tubular or sausage form which means that the shape is dictated by the traditional process and consumer preference. Because of the preservatives, refrigerated shelf life is generally several times longer than for fresh meats. Because the cured meats are not nearly as sensitive to oxygen variations as fresh, the use of reduced oxygen to enhance the refrigerated shelf life is quite common.

Packaging for reduced oxygen packaging of cured meats is drawn from a multiplicity of materials and structures depending on the protection and the marketing needs; frankfurters are generally in twin web vacuum packages in which the base tray is an in-line thermoformed nylon/polyvinylidene chloride (PVDC) web and the closure is a heat-sealed polyester (PET)/PVDC flexible material. Sliced luncheon meats and their analogs are in thermoformed unplasticized PVC or polyacrylonitrile (PAN) trays heat-seal closed with PET/PVDC film. Sliced bacon packaging employs one of several variations of PVDC skin packaging (in contact with the surface of the product) to achieve the oxygen barrier. Ham may be fresh cured or cooked, with the cooking often performed in the package. The oxygen barrier materials employed are usually a variation of nylon/PVDC in pouch form.

Poultry

Poultry is largely chicken, but turkey has become a much more significant category of protein. Further, chicken is increasingly penetrating the cured

FIGURE 8.5 Centralized packaging of red meat has been a vision of packers and packagers for four decades. Although a factor in the United Kingdom, case-ready packaged retail cuts of beef have yet to make an impact in the United States. Among the many technologies offered is a high gas barrier foamed polystyrene tray, heat-seal closed after evacuation and replacement with a high oxygen/carbon dioxide gas mixture. The gas mix retains the red color while retarding microbiological spoilage.

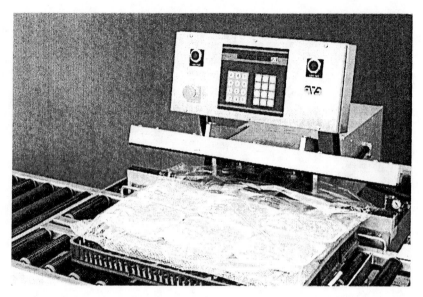

FIGURE 8.6 Centralized packaging of fresh red meat may be achieved by master packing. The retail meat cuts are conventionally packaged and then placed in a large bag from which the air is removed and replaced with a modified atmosphere to retain quality.

meat market as a less expensive and nutritionally and functionally similar meat to beef and pork. Since the mid-1970s, poultry processing has shifted into large size almost automated killing and dressing operations. In such facilities the dressed birds are chilled in water close to their freezing points after which they are usually cut into retail parts and packaged in case-ready form: expanded polystyrene trays overwrapped with printed PVC or polyethylene film. The package is intended to appear as if it has been prepared in the retailer's back room, but in reality, it is only a moisture and microbial barrier. Individual retail packages, however, may be multipacked in gas barrier flexible package materials to permit gas flush packaging and thus extending the refrigerated shelf life of the fresh poultry products.

All meat products may be preserved by thermal sterilization in metal cans. Product is inserted and the container is hermetically sealed usually by double seam metal end closure. After sealing, the cans are retorted to destroy all microorganisms present and cooled to arrest further cooking. The metal serves as a gas, moisture, microbial, etc., barrier to ensure indefinite microbiological preservation. Cans or jars do not, however, ensure against further biochemical deterioration of the contents.

FISH

Varieties of fish are among the most difficult of all foods to preserve

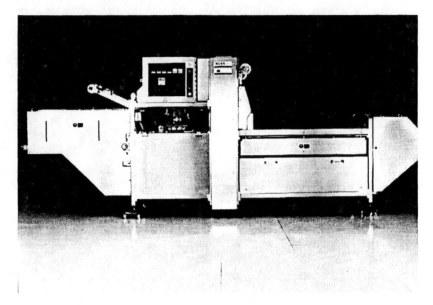

FIGURE 8.7 Just as important to packaging as the package material and structure is the equipment to marry the product contents with the package at economical output rates. This is a Ross-Reiser Inpack machine, which uses preformed plastic trays and flexible material in roll form. Product such as fresh red meat is placed in the tray which is evacuated and gas-flushed prior to being hermetically heat sealed.

in their fresh state because of their inherent microbiological populations many of which are psychrophilic, i.e., capable of growth at refrigerated temperatures. Further, seafood may harbor a nonproteolytic anaerobic pathogen, *Clostridium botulinum* type E, capable of toxin production without signaling spoilage.

Packaging for fresh seafood is generally moisture resistant but not necessarily against microbial contamination. Simple polyethylene film is employed often as a liner in corrugated fiberboard cases. The polyethylene serves not only to retain product moisture but also to protect the structural case against internal moisture.

Seafood may be frozen in which case the packaging is usually a form of moisture resistant material plus structure such as polyethylene pouches or polyethylene coated paperboard cartons.

Canning of seafood is much like that for meats because all seafoods are low acid and so require high pressure cooking or retorting to effect sterility in hermetically sealed metal cans.

One variation unique to seafood is thermal pasteurization in which the product is packed into plastic cans, under the reasonably clean conditions achievable in contemporary commercial seafood factories. The filled and hermetically sealed cans are heated to temperatures of up to 80°C to effect pasteurization, which allows several weeks of refrigerated shelf life.

DAIRY PRODUCTS

Milk

Milk and its derivatives are generally excellent microbiological growth substrates and therefore potential sources for pathogens. For this reason, almost all milk is thermally pasteurized or heated short of sterility as an integral element of processing. Refrigerated distribution is generally dictated for all products that are pasteurized, to minimize the probability of spoilage.

Milk is generally pasteurized and packaged in relatively simple polyethylene-coated paperboard gable top cartons or extrusion blow-molded polyethylene bottles for refrigerated short-term (several days to 2 weeks) distribution. Such packages offer little beyond containment and avoidance of contamination as protection benefits. Obviously, such packages retard the loss of moisture and resist fat intrusion. In recent years, milk packaging has been upgraded to incorporate reclosure, a feature that has been missing from gable top polyethylene-coated paperboard cartons. Further, in recent years, the packaging environmental conditions have been upgraded microbiologically to enhance refrigerated shelf life.

An alternative, popular in Canada, employs polyethylene pouches. This variant has been enhanced by reengineering into aseptic format, a system that has not become widely accepted. Pouch systems are generally less expensive than paperboard and semirigid bottles, but are less convenient for consumers.

In other regions, aseptic packaging is employed to deliver ambient temperature shelf stable fluid dairy products. The most common processing technology is ultra high temperature short time thermal treatment to sterilize the product, followed by aseptic transfer into the packaging equipment. Three general types of aseptic packaging equipment are employed commercially: vertical form/fill/seal in which the paperboard composite material is sterilized by hydrogen peroxide, erected preformed paperboard composite cartons that are sterilized by hydrogen peroxide spray, and bag-in-box in which the plastic pouch is presterilized by ionizing radiation. The former two are generally employed for consumer sizes while the last is applied to hotel/restaurant/institutional sizes, largely for ice cream mixes. Fluid milk is generally pasteurized, cooled, and filled into bag-in-box pouches for refrigerated distribution.

Cheese

Fresh cheeses such as cottage cheese, fabricated from pasteurized milk, are generally packaged in polystyrene or polypropylene tubs or polyethylene pouches for refrigerated distribution. These package forms do not afford significant protection beyond barrier against recontamination, i.e., they are little more than rudimentary moisture loss and dust protectors because the refrigerated distribution time is so short.

Fermented Milks

Fermented milks such as yogurts fall into the category of fresh cheeses from a packaging perspective, i.e., they are packaged in either polystyrene or polypropylene cups or tubs to contain and to protect minimally against moisture loss and microbial recontamination. Because the refrigerated shelf life is relatively short, however, few measures are taken from a packaging standpoint to lengthen the shelf life. Aseptic packaging of such desserts is occasionally performed to achieve extended ambient temperature shelf life. Two basic systems are employed, one with preformed cups and the other, thermoform/fill/seal (Figure 8.8).

Recently, aseptic packaging of dairy products has been complemented by ultra clean packaging on both preformed cup deposit/fill/seal and thermoform/fill/seal systems. In these systems intended to offer extended refrigerated shelf

FIGURE 8.8 Compartmented packages maintain separation of food components that are intended to be mixed prior to consumption. This thermoform/fill/seal package contains two sealed cavities, one of which has the pudding-like base and the other, the fruit topping. These two foods would be incompatible together, even in chilled distribution, but the quality of each may be retained if kept apart. When the consumer is ready to eat this dessert, the flexible closure is removed, the fruit topping compartment is broken off, and its contents are poured on top of the food in the larger cavity.

FIGURE 8.9 Although reclosable zippers have been available to packagers for decades, only when one of the smaller shredded cheese packagers introduced it as a means to offer the consumer more convenience—and to differentiate itself—did the zipper become an integral component of increasing numbers of flexible pouches, and probably influence the number-one shredded cheese packager to use zipper pouches.

life for low-acid dairy products, the microbicidal treatment is with hot water to clean the packaging material surfaces. The same systems may be employed to achieve ambient temperature shelf stability for high-acid products such as juices and related beverages.

Cured cheeses are subject to surface mold spoilage as well as to further fermentation by the natural microflora. These microbiological growths may be retarded by packaging under reduced oxygen and/or elevated carbon dioxide. Commercially, gas barrier packaging is used to retain the internal environmental condition. Generally, flexible barrier materials such as nylon plus polyvinylidene chloride or polyester/polyvinylidene chloride are employed.

In recent years, shredded cheeses have been popularized. Shredded cheeses have increased surface areas, which thereby increase the probability of microbiological growth. Gas packaging under carbon dioxide in gas barrier material pouches is mandatory. One feature of almost all shredded cheese packages today is the zipper reclosure (Figure 8.9).

Ice Cream

Ice cream and related frozen desserts are distributed under frozen conditions. The product must be pasteurized prior to freezing and packaging. The packaging is basically moisture resistant because of the presence of liquid water prior to freezing and sometimes during removal for consumption. Water-resistant paperboard, polyethylene-coated paperboard, and polyethylene structures are usually sufficient for containment of frozen desserts (Figure 8.10).

FRUITS AND VEGETABLES

In the commercial context, fruit are generally high acid and vegetables are generally low acid. Major exceptions are tomatoes, which commercially are regarded as vegetables, and melons and avocados, which are low acid.

The most popular produce form is fresh and increasingly fresh-cut or minimally processed. Fresh produce is a living, "breathing" entity fostering the physiological consumption of oxygen and production of carbon dioxide and water vapor. From a spoilage standpoint, fresh produce is more subject to physiological than to microbiological spoilage, and measures to extend the shelf life are designed to retard such reactions and water loss.

The most fundamental means of retarding fresh produce deterioration is temperature reduction, ideally to near the freezing point, but more commonly to

FIGURE 8.10 Ice cream has reigned as a leading dessert for most of the 20th century. Previously obtainable solely from hotel/restaurant/institutional outlets, ice cream became a packaged product during the 1960s. Although numerous packaging structures ranging from paperboard tubs to plastic cups have been used, the economic paperboard folding carton is favored despite problems with opening and reclosure. The structure shown includes a two-piece rectangular paperboard "can" fabricated in-line from two flat-printed paperboard blanks. The closure is an insert-injection-molded paperboard plastic snap-on device.

the 40°F range. Temperature reduction also reduces the rate of microbiological growth which is usually secondary to physiological deterioration and follows it.

Since the 1960s, alteration of the atmospheric environment in the form of modified or controlled atmosphere preservation and packaging have been used commercially to extend the refrigerated shelf life of fresh produce items such as apples, pears, strawberries, lettuce, and now fresh-cut vegetables. Controlled atmosphere has been largely confined to warehouse and transportation vehicles such as trucks and seaboard containers. In controlled atmosphere preservation, the oxygen, carbon dioxide, ethylene, and water vapor levels are under constant control to optimize refrigerated shelf life. For each class of produce a separate set of environmental conditions is required for optimum preservation effect. In modified atmosphere packaging, the produce is placed in a package structure and an initial atmosphere is introduced. The normal produce respiration plus the permeation of gas and water vapor through the package material and structure drive the interior environment toward an equilibrium gas environment that extends the produce quality retention under refrigeration.

The target internal atmosphere is to retard respiration rate, but reduced oxygen and elevated carbon dioxide independently or together retard the usual microbiological growth on fruit and vegetable surfaces.

One major problem is that produce may enter into respiratory anaerobiosis if the oxygen concentration is reduced to near extinction. In respiratory anaerobiosis, the pathways produce undesirable flavor compounds. To minimize the production of these undesirable end products, elaborate packaging systems have been and continue to be developed. Most of these involve mechanisms to permit air into the package to compensate for the oxygen consumed by the respiring produce. High gas permeability plastic films, microperforated plastic films, plastic films disrupted with mineral fill, and films fabricated from temperature-sensitive polymers have all been proposed or used commercially.

The need for reduced temperature is emphasized in modified atmosphere packaging because the dissolution rate of carbon dioxide in water is greater at lower temperatures than at higher temperatures.

Since the late 1980s, fresh-cut vegetables, especially lettuce, cabbage, and carrots have been a major product in both the retail and the hotel/restaurant/institutional market (Figure 8.11). Cleaning, trimming and size reduction lead to greater surface to volume of the produce and to the expression of fluids from the interior to increase the respiration and microbiological growth rate. On the other hand, commercial fresh-cutting operations generally are far superior to mainstream fresh produce handling in cleanliness, speed through the operations, temperature reduction, and judicious application of microbicides such as chlorine. Some would argue, on the basis of microbial counts found in fresh-cut produce in distribution channels, that fresh-cut produce is less safe than uncut produce. The paucity of cleaning undergone by whole produce, coupled with the relative absence of reported adverse public health

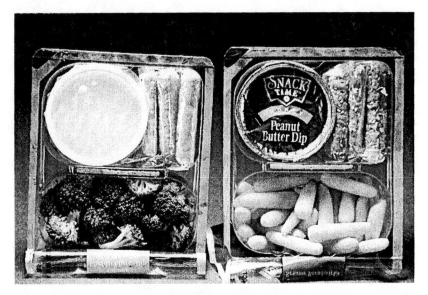

FIGURE 8.11 Fresh-cut vegetables coupled with bread sticks and dip constitute a reasonably flavorful mouthful and nutritious snack for adults and younger persons.

incidents, leads to the opposite conclusion, that fresh-cut is significantly safer microbiologically.

Uncut produce packaging is really a multitude of materials, structures, and forms that range from the old and traditional, such as wood crates, to inexpensive, such as injection-molded polypropylene baskets, to polyethylene liners within waxed corrugated fiberboard cases. Much of the packaging is designed to help retard moisture loss from the fresh produce or to resist the moisture evaporating or dripping from the produce (or, occasionally, its associated ice) to ensure the maintenance of the structure throughout distribution. Some packaging recognizes the issue of anaerobic respiration and incorporates deliberate openings to ensure passage of air into the package, as, for example, perforated polyethylene pouches for apples or potatoes.

For freezing, vegetables are cleaned, trimmed, cut, and blanched prior to freezing and then packaging, or prior to packaging and then freezing. Blanching and the other processing operations reduce the number of microorganisms. Fruit may be treated with sugar to help retard enzymatic browning and other undesirable oxidations. Produce may be individually quick frozen (IQF) using cold air or cryogenic liquids prior to packaging, or frozen after packaging as in folding paperboard cartons. Frozen food packages are generally relatively simple monolayer polyethylene pouches or polyethylene-coated paperboard to retard moisture loss.

Canning of low-acid vegetables to achieve long-time ambient-temperature microbiological stability is conventional for low-acid foods, with blanching prior to placement in steel cans. Today food cans are almost all welded side seam tin-free steel in the United States, with some two-piece cans replacing the traditional three-piece ones. Hermetic sealing by double seaming, retorting, and cooling follows fill up. Canned fruit is generally placed into lined three-piece steel cans using hot filling coupled with post-fill thermal treatment. Increasingly one end is easy-open for consumer convenience. In recent years, fruit is being placed hot into multilayer gas/moisture barrier tubs and cups prior to heat sealing with flexible barrier materials and subsequent thermal processing to achieve ambient-temperature shelf stability or extended refrigerated-temperature shelf life. These plastic packages are intended to provide greater convenience for the consumer as well as to communicate that the contained product is not "overprocessed canned food."

Tomato Products

The highly popular tomato-based sauces, pizza toppings, etc., must be treated as if they were low acid if they contain meat as so many do. For marketing purposes, tomato-based products for retail sale are more commonly packaged in glass jars with reclosable metal closures. The glass jars are often retorted after filling and hermetic sealing (Figure 8.12).

Juices and Juice Drinks

Juices and analogous fruit beverages may be hot filled or aseptically packaged. Traditional packaging has been hot filling into steel cans and glass bottles and jars. Aseptic packaging, described previously for paperboard composite cartons, is being applied to polyester bottles using various chemical sterilants to effect the sterility of the package and closure interiors. Much fruit beverage is currently hot filled into heat-set polyester bottles capable of resisting temperatures of up to 80°C without distortion. Hermetic sealing of the bottles provides microbiological barriers but the polyester is a modest oxygen barrier and so the ambient temperature shelf life from a biochemical perspective is somewhat limited (Figure 8.13).

For more than 20 years high-acid fluid foods such as tomato pastes and nonmeat containing sauces have been hot filled into flexible pouches. The hot filling generates an internal vacuum within the pouch after cooling so that the contents are generally ambient temperature shelf stable. The package materials used are generally laminations of polyester and aluminum foil with a linear low density polyethylene (LLDPE) internal sealant to achieve an hermetic heat seal. Some efforts have been made to employ transparent gas/water vapor barrier films in the structures: polyester/ethylene vinyl alcohol

FIGURE 8.12 For more than a century, Campbell's® Condensed Tomato Soup has been a reference standard for ambient-temperature shelf-stable canned foods—and perhaps all foods. With the coming of plastic packaging and home meal replacement foods, Campbell's was faced with the dual challenge of penetrating the market—perhaps using plastic bottle technology. Their answer was in combination technologies which deviated from traditional but retained the quality and quality image. The product was prepared in single-strength/ready-to-heat-and-eat, slightly acidified to remove the need for postfill retorting, and reduced in distribution time to permit the use of modest oxygen barrier transparent polyester plastic jars.

laminations with the same LLDPE sealant. Transparent flexible pouches offer the opportunity for the retail consumer to see the contents or the hotel/restaurant/institutional worker to identify the contents even without being able to read the label.

OTHER PRODUCTS

A variety of food products that do not fall clearly into the meat, dairy, fruit or vegetable categories may be described as prepared foods, a rapidly increasing segment of the industrialized society food market during the 1990s. Prepared foods are those that combine several different ingredient components into dishes that are ready to eat or nearly ready to (heat and) eat. If packaged in a can, the thermal process must be adequate for the slowest heating component, meaning that much of the product is overcooked to ensure microbiological stability. If frozen, the components are separate but the freezing process reduces the eating quality. The preferred preservation technology from a quality retention or consumer preference perspective is refrigeration.

Incorporation of multiple ingredients from a variety of sources correctly implies many sources for microorganisms. With refrigeration as the major spoilage barrier, problems are obviated mostly by reducing the time between preparation and consumption to less than one day under refrigeration above freezing, plus cleanliness during preparation. As commercial operations have attempted to prolong the quality retention period beyond same- or next-day consumption, enhanced preservation or "hurdles" have been introduced.

Packaging for air-packaged prepared dish products is generally thermoformed oriented polystyrene trays with oriented polystyrene dome closures snap-locked into position, i.e., no gas, moisture, or microbiological barriers of consequence (Figure 8.14). Refrigerated shelf life is measured in days. When the product is intended to be heated for consumption, the base tray packaging may be thermoformed polypropylene or crystallized polyester with no particular barrier closure. For modified atmosphere packaging the tray material is a thermoformed coextruded polypropylene/ethylene vinyl alcohol with a flexible gas/moisture barrier lamination closure heat sealed to the tray flanges. Refrigerated shelf life for such products may be measured in weeks (Figure 8.15).

For several years, concepts of pasteurizing the contents, vacuum packaging, and distribution under refrigeration have been both debated and commercial

FIGURE 8.13 Fruit beverages traditionally were in metal cans and glass bottles, both heavy, with the latter breakable. Metal cans are difficult to open and reclose, as well as not easy for dispensing contents. The advent of oxygen barrier plastics, coupled with reduction in the previously stringent and and lengthy distribution requirements, has sparked the rapid ascension and acceptance of plastic bottles for most fruit beverages in larger sizes. Technology may be either aseptic or hot fill.

FIGURE 8.14 Rather than the tedious and skilled effort to manually prepare elegant food dishes for guests, alternatives include acquisition from a professional chef—or prepackaged by a food processor from a grocery store. Chilled shrimp cocktail in a decorative package includes the cocktail sauce ready for dipping—a meal solution.

FIGURE 8.15 Home meal replacement food products have been a major growth category in the final years of the 20th century. Such products are dishes prepared to be ready-to-eat or ready-to-heat-and-eat. Quality retention and most microbiological spoilage retardation are achieved by refrigerated distribution coupled with modified-atmosphere packaging. Roasted chicken is placed in an oxygen barrier plastic tray which is closed with a flexible oxygen barrier film after removing the air and displacing with a carbon dioxide/nitrogen gas blend.

in the United States and Europe. Sous vide is the most publicized pouch of this type. In sous vide, the product is packaged under vacuum and heat sealed in an appropriate gas/water vapor barrier flexible package structure such as aluminum foil lamination. The packaged product is thermally processed at less than 100°C to destroy spoilage microorganisms and then chilled for distribution under refrigerated or, in the United States, frozen conditions. The United States option is to ensure against the growth of pathogenic anaerobic microorganisms. A very similar technology is cook/chill in which pumpable products such as chili, chicken a la king, and cheese sauce are hot filled at 80+°C into nylon-based pouches which are immediately chilled (in cold water) to 2°C and then distributed at temperatures of 1°C in hotel/restaurant/institutional channels. The hot filling generates a partial vacuum within the package to virtually eliminate the growth of any spoilage microorganisms that might be present.

Dry Foods

Removing water from food products markedly reduces water activity and its subsequent biochemical activity, and thus also significantly reduces the potential for microbiological growth.

Dry products include those directly dried from liquid form such as instant coffee, tea, milk, etc. The liquid is spray-, drum-, or air-dried, or even freeze-dried, to remove water.

Moisture can change physical and biological properties. Engineered dry products include beverage mixes such as blends of dry sugars, citric acid, color, flavor, etc.; and soup mixes, which include dehydrated meat stock plus noodles, vegetables, meats, etc., that become particulate-containing liquids on rehydration with hot water. Such products must be packaged in moisture-resistant structures to ensure against water vapor entry which can damage the contents.

Products containing relatively high fat such as bakery goods or some soup mixes also must be packaged so that the fat does not interact with the packaging materials. Flavoring mixes that contain seasonings and volatile flavoring components can unfavorably interact with interior polyolefin packaging materials to scalp or remove flavor from the product if improperly packaged. Packages for dry products must be well sealed, for example, to provide a total barrier against access by water vapor, and, for products susceptible to oxidation, also to exclude oxygen after removal of air from the interior of the package.

Fats and Oils

Fats and oils may be classified as those with and those without water. Cooking oils such as corn or canola oil and hydrogenated vegetable shortenings contain no water and so are stable at ambient temperatures if treated to preclude rancidity.

Unsaturated lipids are susceptible to oxidative rancidity. Oils are more subject to oxidative rancidity than fats, but both are usually sparged with and packaged under nitrogen to reduce oxygen. Hydrogenated vegetable shortenings generally are packaged under nitrogen in spiral-wound composite paperboard cans to ensure against oxidative rancidity. Edible liquid oils are packaged in injection blow-molded polyester bottles usually under nitrogen.

Margarine and butter and analogous bread spread products consist of fat plus water and water-soluble ingredients which contribute flavor and color to the product. Often, these products are distributed at refrigerated temperatures to assist in quality retention. Fat-resistant packaging such as polyethylene-coated paperboard, aluminum foil/paper laminations and parchment paper wraps, and polypropylene tubs are used to package butter, margarine, and similar bread spreads.

Cereal Products

Dry breakfast cereals generally are sufficiently low in water content to be susceptible to water vapor absorption and so require good moisture- as well as fat-barrier packaging. Further, packaging should retain the product flavors. Breakfast cereals are usually packaged in coextruded polyolefin films fabricated into pouches or bags inserted into or contained within printed paperboard carton outer shells. Sweetened cereals may be packaged in aluminum foil, metallized plastic, or gas barrier plastic films or laminations to retard water vapor and flavor transmission (Figure 8.16).

Soft bakery goods such as breads, cakes, and muffins are highly aerated structures subject to dehydration and staling. In moist environments, baked goods are often subject to microbiological deterioration as a result of surface growth of mold and other microorganisms. To retard water loss, good moisture barriers such as coextruded polyethylene film bags or polyethylene extrusion-coated paperboard cartons are used for packaging.

Hard baked goods such as cookies and crackers generally have low water and often high fat contents. Water, however, can be absorbed, so that the products can lose their desirable texture properties and become subject to oxidative or even hydrolytic rancidity. Package structures for cookies and crackers include fat- and moisture-resistant coextruded polyolefin film pouches within paperboard carton shells and thermoformed polystyrene trays overwrapped with polyethylene or oriented polypropylene film. Soft chewy cookies

FIGURE 8.16 Introduced first during the 1960s, stand-up flexible pouches generated enormous interest in the 1990s and have been increasingly popular for dry foods. Dry breakfast cereal manufacturers have been seriously considering stand-up flexible pouches as substitutes for traditional bag-in-carton or lined paperboard folding cartons.

are packaged in high moisture-barrier laminations containing metallized film to improve the barrier.

Salty Snacks

Snacks include dry cereal or potato products such as potato and corn and tortilla chips, and pretzels, and include roasted nuts, all of which except pretzels have low water and high fat contents. Snack food packaging problems are often compounded by the presence of flavorings such as salt, a catalyst for fat oxidation. Snacks are usually packaged in flexible pouches made from oriented polypropylene or metallized oriented polypropylene to provide low moisture and gas transmission. Snack food producers depend on rapid and controlled product distribution to minimize fat oxidation. Many salty snacks are packaged under nitrogen both in pouches and in rigid containers such as spiral-wound paperboard composite cans to extend shelf life. Generally, light, which catalyzes fat oxidation, harms snack products, and so opaque packaging is often employed.

Candy

Chocolate, a mixture of fat and nonfat components such as sugar, is subject to slow flavor change. Ingredients such as nuts and caramel are susceptible to water content variation. Chocolates, which are generally shelf stable at ambient temperatures, are packaged in fat-resistant papers and moisture/fat barrier such as pearlized polypropylene film.

Hard sugar candies are flavored amorphous sugars which are very hygroscopic because of their extremely low moisture contents. Sugar candies are packaged in low-moisture-transmission packaging such as unmounted aluminum foil, oriented polypropylene film, or metallized oriented polypropylene film.

Obviously, this listing is only a sampling of the many alternative packaging forms offered and employed commercially for foods subject to immediate microbiological deterioration. An entire encyclopedia is required to enumerate all of the known options available to the food packaging technologist with the advantages and issues associated with each.

PACKAGE MATERIALS AND STRUCTURES

PACKAGE MATERIALS

In describing package materials, different conventions are employed depending on the materials and their origins. These terminologies are standard for each material in the United States. In other countries, different conventions are employed.

Paper and Paperboard

The most widely used package material in the world is paper and paperboard derived from cellulose sources such as trees. Paper is less used in packaging because its protective properties are almost nonexistent and its usefulness is almost solely as decoration and dust cover. Paper is a cellulose fiber mat in gauges of less than 0.010 inches. When the gauge is 0.010 to perhaps as much as 0.040 inches, the material is paperboard which, in various forms, can be an effective structural material to protect contents against impact, compression, vibration, etc. Only when coated with plastic does paper or paperboard possess any sort of protection against other environmental variables such as moisture. Only when coated and laminated does a paper or paperboard structure such as a carton or case become a good moisture and gas barrier. For this reason, despite its long history as a package material, paper and paperboard are only infrequently used as protective packaging against moisture, gas, odors, or microorganisms.

Paper and paperboard may be manufactured from either trees or recycled paper and paperboard which are present in large volumes in commerce. Virgin paper and paperboard, derived from trees, has greater strength than recycled materials whose fibers have been reduced in length due to the multiple processing. Therefore, increased gauges or calipers of recycled paper or paperboard are required to achieve the same structural properties. On the other hand, because of the short fiber lengths, the printing and coating surfaces are smoother. Paper and paperboard are moisture sensitive, changing their properties significantly and thus often requiring internal and external treatments to ensure performance.

Metals

Two metals are commonly employed for package materials: steel and aluminum. Steel is traditional for cans and glass bottle closures, but is subject to corrosion in the presence of air and moisture and so is almost always protected by other materials. Until the 1980s, the most widely used steel protection was tin which also acted as a base for lead soldering of the side seams of "tin" cans. When lead was declared toxic and removed from cans during the 1980s in the United States, tin was also found to be superfluous and declined as a steel can liner. Replacing tin in "tin-free" cans was chrome and chrome oxide.

In almost every instance the coated steel is further protected by organic coatings such as vinyls and epoxies which are really the principal protection.

Steel is rigid, an ideal microbial, gas and water vapor barrier and resistant to every temperature to which a food may be subjected. Because steel to steel or steel to glass interfaces are not necessarily perfect, the metal is complemented by resilient plastic to compensate for the minute irregularities in closures (Figure 8.17).

Aluminum is lighter in weight than steel and easier to fabricate. Therefore aluminum has become the metal of choice for beverages in the United States and favored in other countries. As with steel, the aluminum must be coated to protect it from corrosion. Although it is the most used material for can making in the United States, all such cans must have internal pressure from carbon dioxide or nitrogen to maintain structure, and so aluminum is not widely used for food can applications in which internal vacuums and pressures change as a result of retorting (Figure 8.18).

Aluminum may be rolled to very thin gauges to produce foil, a flexible material with excellent microbial, gas and water vapor barrier properties when it is protected by plastic film. Aluminum foil is generally regarded as the only "perfect" barrier flexible package material. Its deficiencies include a tendency to pin holing especially in thinner gauges and to cracking when flexed (Figure 8.19).

FIGURE 8.17 Steel cans may be fabricated in shapes other than traditional cylindrical. Often the shapes cannot be effectively described in words as can be seen in this array of cans from around the world produced on equipment from Oberburg in Germany.

FIGURE 8.18 In a major thrust to improve the market performance of cans, both steel and aluminum can makers have been changing the shape from traditional cylindrical to ''contoured.''

In recent years, some applications of aluminum foil have been replaced by vacuum metallization of plastic films such as polyester or polypropylene.

Glass

The oldest and least expensive package material per se is glass, derived from sand. Further, glass, by itself, is well known as a perfect barrier material against gas. water vapor, microorganisms, odors, etc. Closure may, however, be less than perfect. Further, the transparency of glass is often regarded by marketers and consumers as a desirable property. Technologists may view the transparency as less than desirable because visible and ultraviolet radiation accelerates biochemical and particularly oxidative reactions.

Glass is energy intensive to produce, heavy, and, of course, vulnerable to impact and vibration even though it has excellent vertical compressive strength. For these reasons, glass is being displaced by plastic materials in industrialized societies (Figure 8.20).

Plastics

Plastics is a term describing a number of families of polymeric materials each with different properties. Most plastics are not suitable as package materials because they are too expensive or toxic in contact with food, or do not possess properties desired in packaging applications. The most commonly

FIGURE 8.19 Roasted and ground coffee may be packaged in several different formats to achieve the technical objectives of protecting against oxidation, moisture gain, and aroma loss. Although vacuum steel cans have been traditional, vacuum bricks from flexible barrier laminations have gained some favor due to better economics. The disadvantages of flexible material vacuum bricks—appearance, absence of reclosure, etc.—may be overcome by this package from Italy which combines a paperboard folding carton and a reclosable injection-molded polypropylene lid to offer a ''splendid'' reclosable canister.

used plastic package materials are polyethylene, polypropylene, polyester, polystyrene, and nylon. Each has quite different properties. Plastics often may be combined with each other and with other materials to deliver desired properties.

Polyethylene

Polyethylene is the most used plastic in the world for both packaging and nonpackaging applications. Polyethylene is manufactured in a variety of densities ranging from 0.89 grams/cc or very low density to 0.96 or high density. It is light weight, inexpensive, impact resistant, relatively easily fabricated, and forgiving. Polyethylene is not a good gas barrier and is generally not transparent, but rather translucent. Polyethylene may be extruded into film with excellent water vapor and liquid containment properties. Low-density polyethylene film is more commonly used as a flexible package material. Low-density polyethylene is also extrusion coated onto other substrates such as paper, paperboard, plastic, or even metal to impart water and water vapor resistance or heat sealability.

Although used for flexible packaging, high-density polyethylene is more often seen in the form of extrusion blow-molded (squeezable) bottles with impact resistance, good water and water vapor barrier, but poor gas barrier properties.

Polypropylene

Polypropylene is another polyolefin but with better water vapor barrier properties and greater transparency and stiffness than polyethylene. Although more difficult to fabricate, polypropylene may be extruded and oriented into films that are widely used for making pouches particularly on vertical form/fill/seal machines. In cast film form, polypropylene is a good twistable candy wrapper.

Polypropylene's heat resistance up to about 133°C permits it to be employed for microwave-only heating trays.

Polyester

A relatively difficult-to-fabricate polymer, polyethylene terephthalate polyester (PET) is increasingly the plastic of choice as a glass replacement in

FIGURE 8.20 Fabricating conventional package materials into structures that have special appeal to targeted groups has been a marketing strategy for centuries. A major American brewer/marketer offers its product in a glass bottle shaped and surface-embossed like an old-fashioned wood baseball bat, and signed (in decoration) by the fan's favorite major league hitter (in this example, Hall-of-Famer Reggie Jackson).

making food and beverage bottles (Figure 8.21). Polyester plastic is a fairly good gas and moisture barrier, is difficult to fabricate, but in bottle, tray or film form is dimensionally quite stable and strong. Heat resistance in amorphous form is sufficient to permit its use in hot fillable bottles such as for juices. When partially crystallized the heat resistance increases to the level of being able to resist conventional oven heating temperatures. For this reason crystallized polyester is employed to manufacture dual ovenable trays for heat-and-eat foods. Dual ovenable means that the plastic is capable of being heated in either conventional or microwave ovens.

The transparency of polyester makes it highly desirable from a marketing standpoint after taking into account the issues of light sensitivity of contained foods.

Nylon

Nylon is a family of polymers noted for their very good gas barrier properties. Moisture barrier tends to be lower than polyolefin polymers and nylon is somewhat hygroscopic meaning that gas barrier may be reduced in the presence of moisture. Gas and water vapor barriers are enhanced by the multilayering with polyolefins and high gas barrier polymers. Nylons are

FIGURE 8.21 The end of the millennium appeared to mark the beginning of the elusive plastic beer bottle. Previously a major challenge to brewery packaging managers because of the extreme oxygen sensitivity of the contents, molders have introduced several different mechanisms to reduce the oxidation problem. This United Kingdom beer bottle is a coinjection blow-molded polyester plus ethylene vinyl alcohol, a very high oxygen barrier structure.

thermoformable and both soft and tough, and so are often used for thermo-formed processed meat and cheese package structures in which the oxygen within the package is reduced to extend the refrigerated shelf life.

Polystyrene

Polystyrene is a plastic with little moisture or gas barrier. It is, however, very machineable and usually highly transparent. Structural strength is not good unless the plastic is oriented or admixed with a rubber modifier which reduces the transparency. Polystyrene is often used as the easy and inexpensive tray material for prepared refrigerated foods.

Polyvinyl Chloride

Polyvinyl chloride (PVC) is a polymer capable of being modified by chemi-cal additives into plastics with a wide range of properties. The final materials may be soft films with high gas permeabilities such as those used for overwrap-ping fresh meat in retail stores; stiff films with only modest gas barrier properties; readily blow moldable semi-rigid bottles; or easily thermoformed sheet for trays. Gas and moisture barriers are fairly good but must be enhanced to achieve "barrier" status.

PVC falls into a category of chlorine-containing polymers that are regarded by some environmentalists as less than desirable. For this reason, in Europe and, to a lesser extent in the United States, PVC has been resisted as a package material.

Polyvinylidene Chloride

Polyvinylidene chloride (PVDC) is a high gas, moisture, fat, and flavor barrier plastic which is so difficult to fabricate on its own that it is almost always used as a coating on other substrates to gain the advantages of their properties.

Ethylene Vinyl Alcohol

Ethylene vinyl alcohol (EVOH) is an outstanding gas and flavor barrier polymer which is highly moisture sensitive and so must be combined with polyolefin to render it an effective package material. Often EVOH is sand-wiched between layers of polypropylene, which act as water vapor barriers and thus protect the EVOH from moisture (Figure 8.22).

PACKAGE STRUCTURES

Currently, rigid and semi-rigid forms are the most common commercial

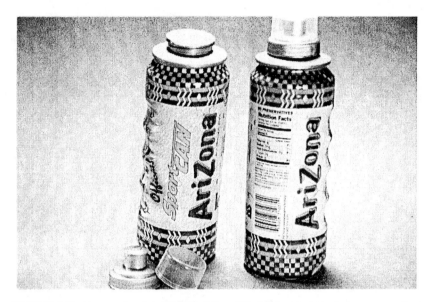

FIGURE 8.22 An example of AriZona Iced Tea differentiating its isotonic beverage product from competitive beverages is in this oversized 24-ounce multilayer coextrusion-blow-molded oxygen barrier bottle with finger grips suggesting greater ease of holding. The closure is a push-pull for convenience of dispensing directly into the mouth of the consumer. A snap overcap reclosure protects the push-pull valve. Decoration is full-body, reverse-print shrink film that conforms to the irregular body shape. Marketer nomenclature for the bottle is ''SportCan.''

structures to contain foods. Paperboard is most common in the form of corrugated fiberboard cases engineered for distribution packaging. In corrugated fiberboard three webs of paperboard are adhered to each other with the central or fluted section imparting the major impact and compression resistance to the structure. Folding cartons constitute the second most significant structure fabricated from paperboard. Folding cartons are generally rectangular solid shape and often are lined with flexible films to impart the desired barrier.

Metal cans have traditionally been cylindrically shaped probably because of the need to minimize problems with heat transfer into the contents during retorting. Recently, metal and particularly aluminum has been fabricated into tray, tub, and cup shapes for greater consumer appeal, with consequential problems with measuring and computing the thermal inputs to achieve sterilization. During the 1990s shaped cylinders have entered the market again in an effort to increase consumer market share. Few have been applied for cans requiring thermal sterilization, but barrel and distorted body cans are not rare in France for retorted low-acid foods. Analogous regular-shaped cans are being used for hot filling of high-acid beverages.

Noted for its formability, glass has traditionally been offered in a very wide range of shapes and sizes including narrow neck bottles and wide mouth jars.

Each has its own singular problems in terms of fabrication, closure, and, when applicable, thermal sterilization.

Plastics are noteworthy for their ability to be relatively easily formed into the widest variety of shapes. Thin films can be extruded for fabrication into flexible package materials. These flexible materials may then be employed as pouch or bag stock, as overwraps on cartons or other structures, or as inner protective liners in cartons, drums, cases, etc. Thicker films, designated sheet, may be thermoformed into cups, tubs, and trays for containment. Plastic resins may be injection or extrusion molded into bottles or jars by melting the thermoplastic material and forcing it, under pressure, into molds that constitute the shape of the hollow object, i.e., the bottle or jar (Figure 8.23).

PACKAGING DEVELOPMENT

The primary packaging development path is the development of the total system from concept to marketplace, and the secondary path is the development of the package itself as an integral part of the whole system. Many organizations call on package materials suppliers and converters to assist in the development of packaging suitable for their product.

FIGURE 8.23 Decoration of plastic and glass jars and bottles, and even of metal cans, has been enhanced by the application of reverse-printed plastic shrink film labels. The film may be printed in roll form on narrow-web rotogravure presses to permit fine line pictorials. By back-side printing, the graphics are viewed through the film. Heat-shrinking the film enables the label to conform to the contour of the package and appear to be an integral element of the container.

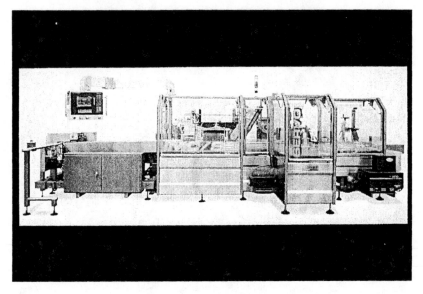

FIGURE 8.24 Package machinery to marry primary to secondary packages is important to unitization, multipacking, and distribution. Today's packaging machinery can be servo motor driven for efficiency and reliability of operation. (Photo courtesy of Klockner Bartelt.)

THE TOTAL SYSTEM PATH

The total system or stage/gate path involves management, marketing, sales, manufacturing, and the packaging development departments or staffs in a coordinated effort, and includes the following steps:

- definition of the goal (initial reason for the development, forecasting of market potential, projection of acceptable developmental cost, and final package cost)
- package development path
- marketing testing (planning, execution, and analysis of results)
- decision whether to proceed, modify, and retest or to drop further effort
- full production (planning and execution of gear-up to full production coupled with communications programs) (Figure 8.24)

THE PACKAGING DEVELOPMENT PATH

The packaging development path involves packaging management and in-house, supplier, etc., staffs. It comprises the following steps:

- definition of the food product properties as they relate to package technical requirements

- definition of package technical and functional requirements
- definition of package marketing and hence design requirements
- identification of legal and regulatory requirements
- selection of potential package designs and materials
- involvement of production and engineering for equipment (Figure 8.25)
- estimate of the probable time, resource requirements, and cost of development
- decision to proceed
- package preparation and testing for technical performance, consumer preference, and economic feasibility
- shelf life testing
- market testing

The initiation for the development may originate from any source. A new food product or technical development may produce a more efficient or less costly process, a new source of supply may develop for a cheaper or higher performance material, a competitor introduces a new package, marketing may want to create a "new image," or, as is common today, consumers signal their desire for a new package.

Marketing research should enter the picture at this point to establish basic economic criteria. What is the predicted market potential for the proposed

FIGURE 8.25 Machinery for packaging foods and beverages can become complex as this equipment for aseptic packaging of beverages in a composite paperboard block suggests. (Photo courtesy of Combibloc.)

product? Does the projected selling price include a sufficient margin to recover the developmental cost in a reasonable time? Will the consumer pay a premium? What special features will the package have to have a command such a premium? Sketches or models of proposed packages can be included in the marketing research element of the investigation. If careful marketing research is not done at this point, but is deferred until the test market stage, considerable development money may be spent only to find that the package meets all requirements except economics.

Definition of Food Properties Affecting Packaging
Technical Properties

Package requirements are dictated by product requirements. The general physical form of the product influences the type of package to be used. It is therefore necessary to know, for example, whether the food unit is large or small; a solid or liquid or a combination, such as a form or an emulsion; massive, chunky, granular, or powdery, if a solid, or watery and thin, or thick and viscous, if a liquid; and soft and light or hard and dense.

It is necessary to comprehend special properties of the product that will require special features in the package. Is the product sensitive to temperature? Must it be protected against extreme heat or cold? Will it be marketed frozen? Will the entrance of moisture or evaporation of the product make it unsaleable?

Definition of Package Technical and Functional Requirements

The functional requirements for a new package must be defined precisely, accurately, and completely if the development of the package is to be accomplished with economy and dispatch. All too often a packaging development nears completion and then the packaging scientist is informed by marketing, management, or product development, "This is not what we want! We forgot to tell you that another property is needed!" Therefore, it is vital to the success of packaging development to be complete in fact gathering so as to define the target accurately. A checklist is helpful to keep track of and properly organize the process (Figure 8.26).

From the data supplied by markets and the information gathered about properties of the product itself, the food packaging technologist enumerates the properties the package must have and at the same time begins to select from among the vast array of materials and package structures for consideration. In addition, unless a radical departure from the mainstream is desired, experience can be drawn on for guidance. For this reason, it is extremely desirable to know what packages have been or are being used for similar products. A study of the advantages or disadvantages of these packages will help in making a selection. Conventional packages will reveal the current market unit sizes

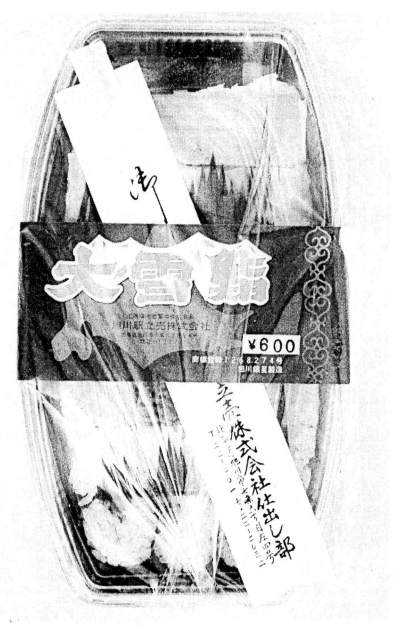

FIGURE 8.26 Japanese packaging, especially for food, portrays extraordinary visual appeal as can be seen in this ready-to-eat package of sushi, complete with ginger root, wasabi, and chopsticks.

or counts, existing prices, whether product visibility is expected, etc., and will offer some evidence on the shelf life that may be anticipated.

In considering the types of primary and secondary packaging required for a product, it is necessary to determine the nature of the handling, storage, and distribution cycle from the point of manufacture to the point of consumption. No packaged product under any circumstance should be introduced into commercial distribution without thorough shelf life testing. As many expert opinions, predictive models, and analogues as have been suggested, the only reasonably reliable method to determine how a packaged product will perform in the marketplace is actual real-time/real-condition testing. Even here, however, testing is sometimes insufficient to determine what will happen in real-life commercial situations. Thus, shelf life testing should be followed up with field observations to confirm the test results or to modify the product, packaging, or distribution conditions.

The effect of climate in a given market area must be considered. High altitudes can cause packages to explode due to reduced atmospheric pressure. Thus the environment within the vehicles of transportation and within warehouses is equal in importance to the type of packaging chosen.

A number of package limitations are imposed on the packaging technology by the very nature of the package manufacturing process. For example, some methods of forming aluminum cans or containers will not permit the depth-to-diameter ratio to exceed specified limits. Other limitations may be imposed by the manufacturing line for the product itself. Filling speeds may require a wider can or jar mouth. Production line speeds may dictate what type of heat-sealing compound or labeling adhesive must be used. Thus the packaging technologist should be cognizant of all phases of packaging and product manufacturing that can affect the package's functional requirements.

Definition of Package Design/Marketing Requirements

Marketing staff should study the market and propose the design requirements for the package. The packaging technologist must be aware of these proposals from the outset, as they may place boundaries on the technical and functional properties of the package.

Conventional packaging will reveal what convenience features may be expected by the consumer in the marketplace. These include such items as easy-opening devices, dispensers, reclosure features, and pilferage protection.

Identification of Legal and Regulatory Requirements

The packaging technologist must be fully aware of all legal and other regulatory restrictions that may influence the choice of packaging material or design (Chapters 16 and 18).

The packaging technologist must also be aware of the nonstatutory regulations that may affect his choice of package. Industry-accepted standards may not include a dimension or size desired. Some religions (particularly the Jewish and Muslim faiths) have strict rules applying to packages that are used for foods.

Selection of Potential Package Designs and Materials

The packaging technologist, working together with graphic design professionals, should prepares a list of optional package designs considered to be technically feasible, including specifications on materials, methods of manufacture, and estimated costs. This list will be subjected to a preliminary screening, and the preferred few possibilities will then be rendered into artists' computer drawings or actual package mockups so that some idea of consumer preference can be obtained by marketing research.

Estimation of the Probable Cost of Development

The packaging technologist should have a grasp of the magnitude of the problem. For each proposed package design he/she should be able to estimate the developmental cost needed to bring it to the marketplace and the probability of success. He/she should also be able to point out whether new or modified packaging equipment might be required. Engineering should be involved in the process.

Testing for Technical Performance, Marketing, and Economic Feasibility

The packaging technologist should obtain a number of sample packages for each design concept so that evaluation tests can be conducted. He/she may purchase the packages or fabricate them by hand, by pilot machinery, or on a regular manufacturing line. He/she may choose several alternative materials or fabrication methods for each concept. During this procedure, he/she eliminates only those items which are probably extremely difficult to manufacture or too costly. When the packages are in hand, he/she subjects them to product compatibility tests, design fulfillment tests, and distribution evaluation.

Product compatibility tests determine whether the package tends to adversely affect the product and, almost as important, whether the product tends to adversely affect the package. Design fulfillment tests determine whether the package meets the pre-established design and performance criteria. Will it hold the desired quantity, will it dispense, does it provide critical protection needs, does it provide minimum shelf-life requirements? Distribution tests determine whether the package will survive the normal handling to be expected in passing through distribution channels. Abuse testing determines the margin of safety built into the package.

Finally, it may be desirable to submit the best concepts to consumer testing. This helps gauge the consumer's reaction to the design and the probability of its being a commercial success. Unexpected consumer dislikes may eliminate some concepts or require some design modifications.

The packaging technologist must be familiar with the costs of materials and of the final package forms so that in selecting a package structure and design he/she will not choose one that is too expensive for the product need. He/she walks a fine line between inadequate packaging, which may lead to complaints or loss of sales, and overpackaging, which gives unnecessary protection at too high a cost. In the beginning it is best to err toward overpackaging, as later process refinements or new technological developments frequently permit the package costs to be reduced without loss of performance in the marketplace.

It is also difficult to judge how much the consumer will pay for the added value provided by a new product-package combination.

Decision to Proceed to Marketing Test

Management must make a decision. Now one or two package concepts have been tested. Limited production runs have indicated that the package can be made and used. Distribution tests and shelf life tests have indicated that it is functional. Estimated costs are reasonable. All that remains is to find out whether the consumer will buy the product and at what price. Management must decide whether the cost of a marketing test is justified.

The decision is one of the most critical management decisions in the development of a food product, and the package design is a major factor in whether the marketing test will succeed or fail. Up to this point, only development costs have been risked. Beyond this point not only are larger expenses for materials and services risks, but the entire company image is exposed to the adverse reactions that may occur in the marketplace should the package fail to perform. A successful package development may or may not help a product attain marketing success, but a package failure almost certainly will damage the chances of a good product attaining success.

Marketing Testing

The planning, execution, and analysis of results of a marketing test are primarily the responsibility of marketing personnel. The marketing test will indicate whether the consumer will buy the product and at what price, in what sales units, etc. It can also indicate whether sales are sustained, indicating repurchase, or whether they taper off, indicating one-time buying, signaling product deficiency. It will reveal shortcomings in the package design, if any, and manufacturing problems.

After examining the results of the marketing test, management must decide whether to terminate the program, make modifications and retest, or go into full-scale production.

BIBLIOGRAPHY

Brody, Aaron L. 1989. *Controlled/Modified Atmosphere/Vacuum Packaging of Foods.* Trumbull, CT: Food & Nutrition Press.

Brody, Aaron L. 1994. *Modified Atmosphere Food Packaging.* Herndon, VA: Institute of Packaging Professionals.

Brody, Aaron L. and Kenneth S. Marsh. 1997. *The Wiley Encyclopedia of Packaging Technology,* 2nd Ed. New York: John Wiley & Sons.

David, Jairus, Ralph Graves, and V. R. Carlson. 1995. *Aseptic Packaging of Food.* Boca Raton, FL: CRC Press.

Paine, Frank A. and Heather Y. Paine. 1983. *A Handbook of Food Packaging.* London, United Kingdom: Blackie & Son Ltd.

Robertson, Gordon L. 1993. *Food Packaging.* New York: Marcel Dekker, Inc.

Soroka, Walter. 1999. *Fundamentals of Packaging Technology,* 2nd Ed. Herndon, VA: Institute of Packaging Professionals.

Wiley, Robert C. 1994. *Minimally Processed Refrigerated Fruits and Vegetables.* New York: Chapman & Hall.

CHAPTER 9

New Product Organizations: High-Performance Team Management for a Changing Environment

ROBERT E. SMITH
JOHN W. FINLEY

Different organizations have different notions of who drives their new food products. Increasingly, consumers are being recognized as the principal drivers, with marketing and new product sectors as the organization's consumer representatives. Food product development must have organizational support and the enthusiastic cooperation of those who are responsible for the process. Incorporating all contributors as early in the process as possible appears to be the mechanism to optimize and speed progress. Every organization should have a positive attitude towards reaching out to suppliers and consultants to extend resources. New product departments and project venture and cross-functional teams each offer benefits and drawbacks that should be considered in this perspective of organizational objectives and resources.

INTRODUCTION

DURING the 1960s, like many other industries, the food industry benefited from the investments that resulted in the technological boom of the post-Sputnik era. The rapid expansion in technology developed in the universities was adapted and converted into new products by the food industry. Universities rapidly expanded their food science departments, merging many disciplines, such as dairy science, nutrition, microbiology, plant science, animal science, and vegetable crops, to form mega food science departments. These departments produced both undergraduate and postgraduate students who were welcomed with open arms by the rapidly expanding R&D departments of the food industry. The industry embraced these individuals and incorporated them into rapidly growing research programs.

At the time many major food manufacturers had comprehensive research programs that spanned from discovery research to product development programs. Throughout this era universities developed high-quality basic science which found its way to the ''basic'' research in industry. Companies such as General Foods, Ralston Purina, Procter & Gamble, and Swift explored and developed new technologies, which in turn led to families of new products. The results of the technologies were reflected in a rapid expansion of innovative new products ranging from sophisticated frozen foods to multi-layered dessert products.

In the late 1970s and into the 1980s, corporate America began to retrench and emphasize much more short-term financial return versus longer-term development programs. For the corporations to remain competitive, product development evolved to development of ''now'' products. Although product development was cut back relative to sales in the food industry, it survived by applying technologies that were in the pipeline or by developing line extensions. The longer-term work became minimal. The technology pipeline was beginning to run dry because industry was moving away from basic research and many universities were often finding more support in applied areas. This change in emphasis was fueled by the fact that government cutbacks in research funding limited the ability to perform longer-term or higher-risk research projects. It became evident that balance needed to be reestablished between fundamental research and product development for the food industry.

What is the best organizational model for product development? There probably is no correct answer to this question. All organizations have existing cultures that in part will dictate what type of new product organization will work. In the food industry we have research and development organizations ranging in size from just a few person operations to the few mega organizations such as those held by Nestlé, Unilever, or Kraft. Certainly the smaller companies cannot support major longer-term research programs. Essentially these organizations need to get the product out the door. This does not imply, however, that they lack innovation. Frequently small manufacturers are willing to take greater risks than more conservative larger organizations. Currently, in the areas of supplements, ''healthy'' foods, and ''functional'' foods, we are seeing many small companies come to the market with innovative new products. Often the supporting data for claims or promotion are limited. The more traditional companies are taking a lower-risk approach, moving slowly making sure they can support the health claims of the new products.

THE CHANGING FOCUS OF THE WORKPLACE

In the September 1994 issue of *Fortune* magazine William Bridges published an article titled ''The End of the Job.'' This article focuses on the changing environment in the workplace, resulting in the demise of the traditional job as defined by explicit job descriptions. During the age of mass production and huge production lines, well-defined jobs were necessary for production and assembly of goods. Frequently, research and development organizations in the food industry evolved to be somewhat like production lines. The product was developed, handed off to a pilot plant, and later to production. Microbiology and quality assurance were brought in later in the development phase to prevent disasters. The product was passed to packaging and then to distribution to deliver for sales, sometimes after production was initiated. Each contributing person, from product design through production and delivery, had a specific job or set of responsibilities. This job definition usually determined where the individual's next opportunity would be found in the organization. Employees were evaluated on how they did their "job," not on what problems were solved or how the company did at the end of the year. Many employees did their "jobs" in now defunct companies.

The modern worker is presented with unfamiliar risks, but much greater opportunities to develop his or her skills and ultimately achieve greater independence while helping develop added value to the company. Now project teams are replacing many middle management jobs in R&D. Through the appropriate use of research teams middle management roles can be significantly reduced. In traditional organizations the managers fulfilled the role of communicator. They funneled the research goals from management to the researcher and the results back up through the hierarchy and then back down through another function. With the advent of team research, the crossfunctional teams provide the cross communications between groups such as research and production. This is also facilitated by the use advanced information systems, that provide cross communications on a real time basis. Another advantage is that E-mail and other telecommunications allow teams to work together from distant locations. A researcher can "watch" a production run from an R&D location halfway around the world and contribute ideas in real time without travel.

The fast moving organizations operate on projects rather than the traditional structured environments. The "job" as it was known from the onset of the industrial revolution is dead. We are evolving into an environment where work is defined as a series of tasks which need to be completed, but not in the traditional job environment. Instead the tasks will be completed by groups of individuals brought together in teams that have clearly defined goals. In food research and development the goal may be a new ingredient, a new product, or a new process. The team will include members who bring together all of the necessary skills to complete the task. Depending on the specific goal it is likely all aspects from design through distribution and quality assurance will be included. The role of marketing is to help define the target. In some cases marketing will develop subteams which include consumers and health professionals to obtain clear definition of the proposed product. In functional foods the product definition and the target consumers need to be carefully identified. This will significantly improve the team's ability to provide specifications for the product.

WHAT IS A HIGH-PERFORMANCE TEAM?

A team is defined as a group of individuals drawn together to work toward a common goal. Although this is a simple definition, it is anything but trivial. First it should be recognized that every individual on a team is there for a specific reason. World Cup soccer teams and Super Bowl–winning football teams are all composed of individuals who bring special skills to help make the team function as a whole. For example, without a skilled place kicker it is not likely that a football team would make it to the Super Bowl, let alone win it. When players are paid in millions of dollars per season no team carries any "extra" players. Similarly with teams developing new products or processes, a multitude of skills are required, but there are not likely to be any extra members.

In food research, the high-performance team is composed of members who contribute to all aspects of the project from research, marketing, packaging, production, through quality assurance. These teams differ from other teams because they address all aspects of the project. They communicate regularly on an open and frank basis and all success and failure is shared. Each member is responsible for his or her own contribution and for that of the other team members. The teams are built around trust, communication, and commitment to the final success of the project. The elimination of layers of middle management frees individuals to contribute their technical knowledge to the project team. As stated previously, with the disappearance of the "job" concept, the level of the team members within the conventional corporate hierarchy is not an issue. In the conventional hierarchy the team traditionally would be led by the highest-ranking member. In the new vision the team might be coached or sponsored by a higher-ranking person, but the team leader could come from anywhere in the organization. The critical factor is that the areas of expertise are present to accomplish the goal expeditiously. As mentioned for a product development effort, the team might consist of a marketing person and several technical persons with expertise in all of the operations needed. Fundamental scientists with appropriate expertise, in areas such as biochemistry, enzymology, or polymer chemistry would work with other technical individuals experienced in sales, manufacturing, distribution, packaging, analytical services, and quality assurance. Such a team accomplishes "buy in" from every area of the company that will be impacted by the new product.

This vertical integration of resources in the development team facilitates "getting it right the first time." It is important that these members communicate regularly and openly throughout the process and establish clearly defined conditions of satisfaction for the final product. The conditions of satisfaction describe what it will take to make the new product meet the needs of each phase of development. Vertical integration helps assure that all of the product "needs" are met throughout each phase of the development, and prevents products or concepts from being thrown over the transom to the next guy to fix.

HIERARCHY VERSUS TEAM MANAGEMENT

Research organizations have traditionally been set up like manufacturing organizations where each individual had a job in a well-defined hierarchy. The top of a generic R&D organization is shown in Figure 9.1. In this traditional organization a number of directors report to a vice president. The directors all have organizations of differing sizes reporting to them. Typically, group leader, scientist, and technician functions will have similar structures below the directors. In this model the director of product development is likely to have the largest organization reporting to him or her. Each individual in the organization has a defined "job." This job is carefully spelled out in a job description. In most organizations the group leaders would be in some level of competition to eventually inherit the director's job, and the directors, in turn, would compete for the vice president's job. In an organization of this sort, research programs and assignments typically follow the lines on the chart. In this mode the assignments generally come down from the vice president

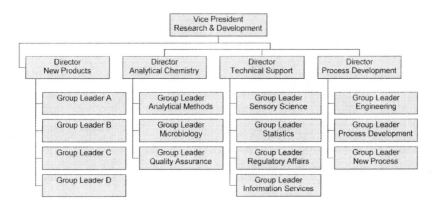

FIGURE 9.1 Generic R&D organization.

to directors, etc. A request for a product might come from a marketing VP to the R&D VP who then assigns the project as appropriate in his organization.

In Table 9.1, the responsibilities of the manager and the scientist in the hierarchical organization are summarized. For the last 50 years, and in certain current situations, this is the way R&D gets done. In this top-down model, the manager is responsible for the connection to the business and the scientist stays in the laboratory making new discoveries. The scientist is judged solely on technical productivity. When we look at the success of the industry over the last 50 years we can hardly say it doesn't work. However, the pressures on contemporary society and resultant changes in the work environment would suggest that another model, such as the team approach, should be considered. In this new paradigm, research efforts can be more sharply focused on a specific goal or concept and conducted by cross-functional teams.

The model for team research, on the other hand, is to have clearly defined goals and teams that include all of the core competencies required to success-fully complete the project. Scientists in the team approach must become much more focused on the business. The team has the responsibility for considering customer needs, business objectives, and technological capabilities. The technological capabilities should be based on core competencies. When the core competencies are not available to the team they must be obtained from the outside. As described previously, the outside capabilities may come from partnerships with suppliers, government laboratories, university centers, or individual consultants. It is important to emphasize that this model for product development includes marketing and sales working on the same team with the technical staff. In the section below on team work there will be greater discussion of the actual team make-up and responsibility.

In Table 9.2, the responsibilities of management and scientists in this new paradigm are summarized. In the team paradigm the management is essentially taken over by the team. Senior management coaches the team to help achieve the goals. Teams will replace the traditional reporting hierarchy where the

TABLE 9.1. **Manager and Scientist Responsibilities in the Hierarchical Organization.**

• Manager
Manager knows best
Plans, leads, organizes, controls
Communicates to clients and head office
Power to commit
Evaluated by superiors
• Scientist
Mainly performs research
New discoveries are generally serendipitous
Technical excellence validated by peer reviewed publications

TABLE 9.2. Responsibilities of Management and Scientists in the Team Research Paradigm.

- Management
  Team members know best
  The team sets goals, management coaches for achievement
  Communications are facilitated because all interested parties are team members
  Staff empowered to make commitments
  Leaders evaluated by staff and superiors
- Scientists
  Research is conducted to support team goals
  Search for breakthroughs to facilitate project success
  Technology evaluated by commercial success

individuals on the team report to each other. With the disappearance of middle managers we will see the emergence of leaders who act as coaches to oversee the process from beginning to end. These coaches will support and nurture employees as senior management has always done. The new employee team member will be provided access to the same information that traditionally was the exclusive domain of senior management. As a result he or she must accept the responsibility of ownership of the company. In other words, the employee must take responsibility for the task at hand and the future of the organization. When the team succeeds everybody succeeds. The employee will participate actively and be held accountable for key decisions related to the project. Ongoing training is essential to support the greater knowledge base of the employee. Research and development staff members must understand the needs of marketing, distribution, and be familiar with how the decisions that their teams make impact the balance sheet of the company.

ROLE OF SENIOR MANAGEMENT

As the role of the team changes, so does the role of senior management in the R&D organization. In product development organizations, as increased levels of teamwork take over, less ''management'' will be required. The workers in the organization will no longer be burdened with the constraints of the hierarchical organization. The elimination of job boundaries will result in a workforce that is much more independent and self-directed than in traditional organizations. Middle management as we have seen it before will erode away. Former managers will go back to doing real work, contributing to teams based on their expertise and experience. There will still be a need for managers who will oversee the product development process from start to finish, thus assuring that the teams remain focused on the desired end products and that the required resources are made available.

A critical aspect of the remaining leadership role is that of employee coaching. The coaching role includes nurturing and developing high-potential employees in the same way senior management does in today's environment. This means that the leader will provide staff with direct access to information that was once only in the domain of the decision makers. The employees will be expected to understand the why and wherefore of the corporations business strategy to a far greater degree than in conventional environments. The leaders must assure that the vision and strategy of the organization is clearly understood by the staff. Thus, to a greater degree than ever the senior management must become messengers, keeping employees informed of the strategic roadmap, and the leaders must become champions of the projects within the organization helping assure adequate resources to conduct the work in a timely way.

WHAT ARE THE SIGNIFICANT ATTRIBUTES OF A HIGH-PERFORMANCE TEAM?

Every activity of mankind is based on some sort of communication. High-performance teams are primarily based on precise communication. Such open and frank communication leads to the second important characteristic of the high-performance team—trust. One of the underlying prerequisites of good communications is critical listening. Each team member needs to feel free to state their views and all must listen. Listening to understand and to clarify will help build the trust and relationships required for success.

An important feature of the high-performance team is its ability to surface breakdowns in order to accelerate projects. When a breakdown is declared the team needs to come together and help find a resolution to the problem. Frequently when challenging breakdowns are solved it is through innovation. Teams typically come up with innovative ways to solve many problems. The team should listen to all approaches to problems and their potential solutions. Breakthroughs should be acknowledged and those who made the breakthroughs recognized as expeditiously as possible. It is simply human nature that we all like to be acknowledged and recognized for jobs that are well done. By making this part of every team meeting, participants feel reinforced and as a result much more motivated.

HIGH-PERFORMANCE TEAMS CAN MEET INDIVIDUAL NEEDS

As mentioned, the way we work is also changing with society. In the future, emphasis will be placed on accomplishing the task, not the location where it must be done. With advanced telecommunications more employees can work at home and in the time frames they choose. The individual will be able to work within a style that fits his or her personal needs. In the old production line job a geographic proximity was essential. Now technology allows co-workers to live thousands of miles apart and still work "together" in real time. The expanded modern workplace will accommodate much more individuality, with the proviso that the individual is part of a team.

CONCEPT OF THE CHAMPION

The leadership and members within teams also need to "champion" ideas and projects to convince those who control resources that it is in their best interest to come on board. The champion should passionately sell the innovation and its commercial potential but always be cautious not to oversell and create unachievable goals. Typically the leaders in the organization or the more experienced team members are the champions. As such, they speak to those who control resources and communicate to them the technical achievements as well as the commercial ramifications. An advantage of the integrated high-performance team is that members come from various disciplines within the company beyond R&D. Thus, when the VP of R&D meets with the VP of Manufacturing to sell the latest innovation in processing, the VP of Manufacturing is aware that members of his staff are already on board with the project. The VP of R&D then is in the position of championing an idea that should already have some support from manufacturing. Selling the concept or innovation must be conducted at all stages in the development of the project, from conception through commercialization. At each stage of product development the champion must agree to the targets and hold the team accountable for meeting the targets.

The champion must believe in the innovation and be willing to expend the effort and take the risk. Occasionally, for a breakthrough project, the champion and/or senior management must be willing to "bet their job" to sell the concept. At the same time he/she must be willing to share in the responsibility if the project fails. For this reason, no project should be oversold or over promised. The champion must support projects that are good for the company always keeping in mind the importance of return on investment. Will the project ultimately improve the bottom line of the company? Clear objective evaluation and explanation of current status at all phases of the project garnished with enthusiasm and belief in the project will make the role of the champion enjoyable and rewarding for the individual as well as the company.

CLEAR GOAL—UNCLEAR PATH

The high-performance team must have a clearly defined goal, but it is generally safe to say that the path to achieve the goal is anything but clear. When the project is started the path is usually not clear. For example, in 1960 John F. Kennedy promised to put a man on the moon by the end of the decade.

In 1960 we had no idea how to accomplish this incredibly complex task. In 1969, we took the "small step for a man," the "giant leap for mankind." The goal was clear, the work to be done was monumental. Teams were established with extraordinary budgets and the task was done. There were, however, a myriad of daily problems that had to be overcome.

On a smaller scale all of these same things happen with product development projects. A successful project is the one where these daily surprises are overcome in a timely and effective way so that the project can keep moving. Planning of the project automatically accounts for the things that do not go wrong. It is the unforeseen adverse occurrences that frequently destroy timelines and projects. In order to overcome this inherent obstacle we define a project as a series of breakdowns. This accomplishes two things. First it gives "permission" to identify breakdowns, and second it surfaces breakdowns before they interfere with project timelines.

The management of breakdowns is one of the most critical factors in the success of a development project. The management team of the project must meet regularly (weekly) and discuss the progress of the project. These meetings should have an agenda that covers all critical issues and minimizes superfluous discussion. As part of those discussions it is critical that any breakdowns be brought to the attention of the group. It is every individual's responsibility to identify breakdowns; it is the team's responsibility to "put there heads together" to solve the breakdown. Solving breakdowns is one of the most important reasons that all team members must try to attend all meetings. During these meetings breakthroughs are surfaced, discussions convert breakdowns into solutions, and then the solutions become opportunities.

Breakdowns are an inevitable part of any project, particularly where research is involved, because, by definition, research is like sailing over uncharted waters. The focus should be on the timely and effective repair of the breakdown. While the new work environment has fewer rules than the "job-based" environment, there is one cardinal rule on the team: "No finger pointing." When a breakdown occurs, it has happened—nothing can be gained by retribution. The goal is to fix it.

In a real high-performance project, breakdowns should be forced. Milestones should be established on time frames that are aggressive enough to find the weak points. For example, if the breakdown is that the analytical chemistry cannot be done in time for the next step, perhaps a new chemical method is required. Traditional "job-based" organizations would immediately say we need more staff. In modern American industry this is not likely to be the solution. What we need to do is work smarter through better use of new technology. The technology may be developed internally or contracted from an outside vender. Whichever, it must be identified and implemented quickly and effectively without "egos" or "turf" issues getting in the way. The way we have always done it no longer applies.

INDIVIDUAL COMMITMENT

The commitment of individuals to the success of the project is the fuel that makes it work. How does this work in the "jobless" environment? The project-driven work environment provides the target and challenge for the individual. Commitment, as always, must come from inside each individual on the team. When individuals are part of a team there is a mutual support system that encourages enthusiasm and commitment. The successful worker in the future will rally around the project or the task and not the job. The end goal of getting the new product into the market is the focus. The committed worker will do what it takes to get it there. The commitment to do what it takes is what makes a breakthrough team work. The new environment allows the individual much more free expression such as working at home, frequently less-traditional business dress, and a more open and supporting, flexible work environment.

The commitment for the goal of the team needs to be established early and unanimously by the team. The aligned commitment to the project goal is a major first step in the initiation of an effective high-performance team. Again it is important to acknowledge that the team may not know how they will reach the goal; it is alliance on the goal that is critical.

The individual working in the "jobless" environment must bring a multitude of contemporary skills to the organization. The role or need for the nontechnical manager in the R&D organizations is rapidly going the way of the "traditional job." The individual must come to the team with technical skills and a willingness to expand their technical knowledge. In addition to technical skills they must be adaptable in the way they apply their skills, remaining flexible to make the project work for the goals of the company and not for the individual job. These skills range from marketing, through basic science, through manufacturing, distribution, and sales. Every team that is assembled is unique and is faced with a unique set of problems. The brilliant leadership of the future should know how to organize these teams and then let the teams manage themselves. The leadership should help the teams acquire the needed resources to effectively complete the desired task. The team must be empowered to make critical decisions and be provided the information and resources to do so.

THE NEED FOR STRATEGIC PARTNERSHIPS

In modern industry, the fixed costs of research and development are under constant scrutiny. As a result, a food company cannot necessarily maintain in-depth experts in all the fields necessary to develop new products. Clearly companies must have core competencies such as chocolate technology at Hershey, baking at Nabisco, and refrigerated dough at Pillsbury. But modern food product development requires technologies that frequently extend beyond a company's core competencies. These should be handled by alliances. For example flavor formulation has become increasingly complex. Low-fat/no-fat products have different physical-chemical properties and consequently major modifications must be made in flavor formulations. The integration of the supplier into a development team is much more cost effective than establishing an extensive flavor development program within the company. Flavor companies can provide the needed technology and apply the flavor delivery technology over a spectrum of products without necessarily betraying the confidentiality of the food companies. In other words high-flavor technology developed for a low-fat baked product by a flavor supplier may also have similar application in low-fat confection developed by a candy manufacturer. When confidentiality is assured, the strategic partnership between suppliers and manufacturers can lead to new business opportunities and innovative technology breakthroughs. The synergistic relationships help limit needless duplication of effort within a corporation and encourage the development and support of core competency focused on the strength and strategic position of the company.

Strategic partnerships are now starting to emerge between food and pharmaceutical companies. Functional foods or nutraceuticals appear to offer unique opportunities for new healthy food products. Pharmaceutical manufacturers have experience identifying bioactive components in biological systems and either extracting them or producing them synthetically. Food companies have core competencies in the formulation and marketing of food products. The strategic partnership affords both pharmaceutical companies and food companies the opportunity to develop new markets while building on their own internal strengths. Particularly in an age where few companies can start to develop new core competencies in a completely unrelated field, strategic partnerships are the most effective way to conduct business.

U.S. government laboratories, particularly USDA laboratories, frequently develop innovative technologies which can be applied by the food industry. Cooperative research is now possible between the scientists in USDA laboratories and industry. When these Cooperative Research and Development Agreement (CRADA) programs are established, research done and partially supported by industry is then available to the industrial supporter who has the first right of refusal on the technology. A good product example of USDA research reaching the market place is Oatrim oat derivative. Oatrim was developed at the USDA Northern Regional Research Center. ConAgra worked with scientists at the USDA and eventually Oatrim was commercialized by them as a low calorie ingredient and fiber source. The USDA gains some benefit from royalties and the corporate sponsor is able to effectively leverage research funds from the expertise and experience of the USDA researchers.

Several universities have developed centers surrounding core competency within the university. For example the Universities of Georgia and Wisconsin have formed research centers focused on food safety issues. The Center for Advanced Food Technology (CAFT) at Rutgers has built a center based on the physical chemistry of food where there is a strong competency. In both cases multiple industrial members provide support to the centers through yearly dues. These programs provide ready access to the information developed in the centers, a pipeline for new graduates to enter industry, and enhanced university-industry communication. Information from these centers can be brought to research teams either through corporate monitors learning at the university or by inviting university faculty members to actively participate in corporate development teams when their specific expertise is required.

Other external resources available are cooperative research with experts in universities or individual consultants. Generally these are established on a one-to-one basis between the researcher and the corporation.

SUMMARY

The organizational nature of the R&D environment is evolving away from the traditional hierarchical organization toward a team approach. Teams allow a reduction in the layers and roles of middle management. These middle managers can then be redeployed in the organization, utilizing their technical skills, experience, and expertise more fully. The key step in this research evolution is to develop high-performance teams for specific projects. These teams include members involved in the entire development and commercialization of the project. Research, marketing, engineering, production, regulatory, and packaging may be involved in the project from the inception. This allows the team to anticipate and solve many problems before they occur, rather than the traditional hand off from one group to another. The high-performance team succeeds because it is based on open communication. The communication includes careful listening to each other and supporting each other when there are either breakthroughs or breakdowns. The result is higher levels of commitment and trust throughout the project. Teamwork as described above accelerates project activities and provides a more pleasant working environment for all.

BIBLIOGRAPHY

Bridges, William. 1994. "The End of the Job," *Fortune,* September.

Cooper, R. G. 1993. *Winning at New Products, Accelerating the Process from Idea to Launch.* 2nd Ed. Reading, MA: Perseus Books.

Crawford, C. Merle. 1997. *New Products Management.* Boston, MA: Irwin/McGraw-Hill.

CHAPTER 10

New Food Products: Technical Development

STANLEY SEGALL

Translating concepts into prototype products that can be viewed, sniffed, tasted, and savored is a choreographed series of events. Beginning with the concept and its definition and proceeding to a product innovation charter and a product protocol which ultimately describes the physicochemical entity to be generated, food product development requires intimate blending of research findings, science, technology, imagination, experience, skill, and not a little bit of art. Development is just that, a process of initiation and advance, error, iteration, adaptation, and reiteration directed toward an elusive goal of a nearly perfect manifestation of the product concept.

THE NEW PRODUCT

WHAT exactly is a "new" food product? New products, can come in many forms depending on just who is classifying them. For example, although referring to only the first three types as "truly new," Hoban (1998) identifies what he refers to as the following six types of new products, calling them

- "Classically Innovative" (e.g., the first frozen juice concentrate)
- "Equity Transfer Products" (e.g., a restaurant deciding to market its distinctive salad dressing through supermarket outlets using its well-known restaurant name)
- "Line Extensions" (e.g., deciding to add a green color to an existing beer to exploit specialty marketing)
- "Clones" (e.g., producing and marketing your own, but nonunique,

217

version of a lemon and lime beverage to compete with a well-known competitor's product)

- "Temporary" (e.g., special bunny-shaped chocolates for Easter and sold only at that time)
- "Conversion" (e.g., replacing one size container in your line with another size of the same product)

There could be many other categories by which this list might be expanded. For example, a seventh category designated "Private Labels" (e.g., having another company package their item in your packaging but marketing it through your own distribution system could be added.

Regardless of how many categories, or how they might be classified, in almost every case a new food product means a product not previously marketed or produced by the organization for which it is developed or made available. Some products are really new in the most obvious meaning of the word, some are modifications of existing products, others are imitations or copies of competing or already existing products, and still others may be minor modifications in shape, size, color, or packaging. In some cases, this can even include products already available in the marketplace and obtained fully market-ready from sources outside one company and introduced, without modification or with only simple packaging or other minor modification, into another company's marketing and distribution network.

TYPES OF NEW PRODUCTS

Novel

The most obvious "new product" would be one which is novel, unique, and distinctly untried, unfamiliar, or even previously nonexistent. Very often products which meet this description fall into the category of invention, especially if they are innovations in technology, and are eligible for protection by patent. A patent more effectively locks out competitors and may be looked upon as a contract between the inventor and the government (representing the public) in which the inventor is guaranteed a limited term of exclusive use in exchange for providing a full written description and disclosure of the invention. In the United States this exclusivity period is normally 17 years from the time of the granting of the patent, although for certain items subject to lengthy delays by the United States Food and Drug Administration before obtaining permission to market, such as food additives, color additives, etc., additional time can be granted for the period of patent protection.

What can be patented? Exactly what can be patented is defined in the U.S. patent laws found in Title 35 of the United States Code.

The language of the patent law is subject to wide interpretation, and is perhaps best stated by Burton Amernick in describing the scope of patentability to "include anything under the sun that is made by man." This conclusion was engendered by a Supreme Court case (*Diamond* v. *Chakrabarty*, 1980) involving the patenting of a living organism, a genetically engineered bacterium developed for breaking down crude oil and thus useful in dealing with oil spills (Amernick, 1991). Despite this seeming universality, there are limits to patentability. Items such as natural laws, printed materials, naturally present materials, natural phenomena, and abstract ideas, etc., are not usually patentable.

On the other hand, in dealing with foods it does include things like useful compounds, material compositions, manufacturing methods, methods of altering or modifying natural product characteristics, and others, all of which could be considered potentially patentable. Protection of patentability is impacted by numerous items, for example, publication prior to application for patenting, different provisions of foreign governments, etc. Here at the outset of this chapter on new food product technical development, it is important to recognize that, strangely enough, it is not always obvious whether or not the result of a new product development project, or any step or part of the project, represents a protectable, unique, and novel product. Later review of the final and complete technical report on the project by the company's legal counsel is crucial to determining whether or not there are protectable elements to the product, process, or other aspects of intellectual property.

For this reason, as well as simply following one of the basic tenets of good laboratory practices, all work and discussions regarding any part of a development project must be faithfully and regularly (e.g., daily) documented, in writing, using a fixed page laboratory-style notebook. Scratch pads, scrap paper, backs of envelopes, laboratory paper towels, etc., make poor recording documents, but if they exist they should be saved. A loose-leaf notebook or a computer recording document might be useful as a working process practice but, unless they are backed up with a hard document or mechanism not easily subject to alteration, they might later prove problematic, since they are subject to after-the-fact alteration. The written document must be carefully dated and witnessed by a knowledgeable colleague.

For a more exhaustive treatment of patentability the reader should consult more specific literature on patent law, for example Amernick's *Patent Law for the Nonlawyer* (1986), or a patent attorney, or attorney dealing with protection of intellectual property.

Sometimes there are decisions within a company to forgo the protection afforded by patentability to avoid the concurrent necessity of public disclosure. In these cases it is still possible to retain the advantage of exclusivity, perhaps for even longer than that which might be gained from patenting. Companies can attempt to retain commercial advantage by use of trade secret protection. There is a great deal of variability in the legal protection of trade secrets and this is dependent upon prior case law and the state involved. Nearly anything, a formula, a process, a piece of equipment, a set of data, or anything else that allows one company to gain and maintain an advantage over competitors who do not have knowledge of the particular item or practice, can be viewed as a trade secret.

On the other hand, to be protected as a trade secret, the information must be so secret that it would not be easy to obtain by usual, proper, or legal means without consent of the owning party.

Information relating to trade secrets is protected against use of illegal means of disclosure. For example, bribing an employee in a competing company into revealing trade secrets, or deliberately hiring away an employee from a competitor with the intention of obtaining trade secrets from that employee, or, even without prior intent, utilizing information on a trade secret in the possession of a current employee who had obtained the information as a result of previous employment by a competing company, would be illegal in most jurisdictions.

To be fully protected, trade secret information should be really new, but it need not meet the extent of requirements for a patent. Very often, it is manufacturing methods and practices which are best protected by the trade secret method. Of course, utilizing the trade secret method does expose a company to the risk that their exclusivity could be ended by legitimate independent discovery of the secret by a diligent competitor.

One final item in this list of "really new" products may not actually be a product, though it often is associated with one, and that is the provision of a trademark or service mark. For a genuinely new product, even if it is patented, it is wise to protect the name or appearance of the product by use of a specific word, symbol, or combination of the two, to identify the product in such a way as to distinguish it from that of the competition, particularly if the products are similar. For example, aspartame (Nutrasweet) bears a distinctive trademark and thus distinguishes itself from similar non-nutritive sweeteners such as saccharin (Sweet 'N Low) or asulfame-K (Sweet-One). These marks can be registered in the U.S. Patent and Trademark Office, and once registered this status can be so indicated on the package.

Although not necessarily within the purview of our discussion of new products, it should be mentioned that a tradename is not the same as a trademark or service mark and is not ordinarily registerable unless it is used as a trade name while at the same time registered as a trademark. A trade name is usually used only to identify a company, unless also registered as a trademark or service mark.

COMPETITOR MATCHING

One of the most common types of "new" product development projects involves the need to have a product which directly meets a competitor's challenge. If your competitor is marketing a specialty pizza which has an unusual character, for example a spicy chicken topping or a strong basil flavoring, you may be called upon to provide your company with exactly the same product (even if the name and packaging are different). This would be a direct duplication. In many instances all that is required is a matching product.

In other cases this may really mean that what is required is essentially a duplicate product, though very often it may be something very much similar rather than an exact duplication. Slight but noticeable differences may exist simply due to the difficulty involved in exact duplication, or there may be deliberately built-in differences. In either event, this difference could be advantageous. Your pizza topping may indeed be a spicy chicken topping, but you could emphasize one spice more or less than the competitor to provide your product with a signature.

In many cases there is actually an advantage to having a slight, but distinct, difference. For example, if your company decides that a competitor's chocolate ice cream is pushing down your sales of chocolate ice cream, it may be wise to not just duplicate your competitor's chocolate flavor, but to go one better by providing a signature difference; perhaps a tangy note from the incorporation of just a touch of another flavor to the chocolate, perhaps a hint of orange, brandy, or coffee or other distinctive, or signature, flavor note.

In some cases a company may be marketing a limited line of items which logically requires expansion to a series of items. For example, there have been carbonated beverage manufacturers who traditionally marketed only one or two product items (e.g., ginger ale and tonic water, or just cola, etc.) for which they were well known, and that sold quite well, but in limited volume compared to the overall demand for carbonated beverages of other flavors. Such companies discovered that they were permitting competitors to gain a foothold with customers who were also looking for additional flavors.

Sometimes, just to protect your company's market position for its flagship products, and to compete successfully with competitors who might otherwise command a larger share of shelf space in the retailer's markets, it may be necessary for a food manufacturer to develop an entire line of products even if in the aggregate they may or may not be highly productive of profit in and of themselves. Of course, your company may discover that in producing and marketing the more extensive line of products required, not only does this technique make it more difficult for a competitor to command shelf space for his product, the added sales of your new beverage products may also improve the revenue stream of your company. Economies of scale begin to come into play as improvements in utilization efficiency of the marketing, manufacturing, and distribution systems add significantly to the bottom line.

Finally, we must not forget the "new and improved" product. It may only be a new container size or shape, or a new perfume addition to change the aroma of the product, or even the same product in a container that has improved characteristics or looks similar to a popular competing product, that is it has been adapted but it is still a "new" product. It has to be developed, tested, and added to the existing product line. By definition and in practice, if it has not already existed in the company product line, it is a new product.

In other cases there may be need for considerable technical study. For example, when it is decided that your food product is now going to incorporate wheat bran because someone states that there is evidence that this item reduces incidence of colon cancer in rats, or oat bran because another researcher states that this reduces circulating cholesterol levels, or ascorbic acid because it prevents product discoloration or supplies a needed nutrient, much more has to be done with product formulation than simply adding in the new ingredient. Stability, compatibility, texture, sensory acceptance, all these things and more must be extensively tested. Legal restrictions regarding health claims in a food product, as well as implications for label requirements, must be looked into in order to comply with various regulations.

Reinventing the Wheel

Besides the truly novel new product, the simply duplicative new product, or the adaptive new food product types mentioned thus far, there is a good possibility that the new product your company tells you is now required may in fact already exist. One of the less obvious but truly critical responsibilities of new food product development professionals is to know what is already available in the marketplace which might be readily adapted to fill the needs of your company.

Regardless of whether your company would reject or consider such an approach, you must be aware of what already exists in the market, and be sure it is taken into account during the discussion regarding the nature of a new product. There are times when it may be wise to take advantage of the prior development and existing manufacturing capability of another company to provide your company with a ready-made "new" product. It is always wise to look at the existing marketplace to see if there is a supplier who could quickly and economically provide your company with a product rather than incurring all of the risks, costs, and delays involved in beginning development from scratch. If you can locate an already existing source of a product, what is often called a "private label" manufacturer, the lead time from concept to availability can be reduced from a sometimes substantial delay of months or longer, to nearly zero. In addition, this can be a way to really test the market for the new item without the need for risking possible additional capital investment for development and manufacture. A major caveat is the danger of loss of product quality control and brand identity, but there are many ways for this to be overcome.

It is possible to request only the most minor alterations in color or size or shape or packaging to provide for integrating the ''new'' product into your existing line.

Variations on this approach have included purchasing and supplying the packaging independently, using a distinctive color in your company's product, developing an exclusive die or other piece of manufacturing equipment for use only with your product to provide exclusive brand identity, simply buying the potential supplier (sometimes called a merger), or working out a supply exclusivity agreement with the manufacturer. These practices are quite commonplace.

In the appliance industry it is not unusual for an equipment manufacturer to market an item under its own brand name while making the identical or near identical item under private label for a different even competing marketing company, or for another appliance marketer, or as a ''store brand'' item. This practice is obviously well known in the food product field where supermarkets both manufacture their own products and contract with major brand name suppliers to provide a ''store brand.'' Of course, under these circumstances it is still necessary for an entire set of technical control procedures to be imposed. Although that is a subject unto itself, it often falls to the food product development group to determine the parameters and specifications for the product and to develop, test, and implement the appropriate technical quality control procedures just as it would if the entire product were developed in-house in their own laboratory. In fact, this latter aspect is almost more important for a purchased new product than it would be if the entire product were manufactured and controlled in-house.

The lesson here requires that before launching elaborate development plans, don't ''reinvent the wheel'' until you are sure that company management has been made aware that a desired new product already exists, and only then consciously makes a decision and gives the order to proceed with a new in-house development.

In summary then, new food products take many forms. They can range from the truly unique and novel to the mundane and simply imitative. They can even be some other company's product adopted into your company in the thinnest of disguises. The important thing is that they represent an addition to your company's product line that is usually designed to improve or protect the competitive position of your company and/or to improve the overall bottom line.

THE CONCEPT

Where does the idea for a new food product originate? Companies themselves consider the most important sources of new ideas to be market research,

research and development, customers, and other consumer product companies. Sources such as university research, scientific journals, consultants, and trade events were rated as less important (Hoban, 1998).

On the other hand, even some of the more unlikely or lesser-rated idea areas can still be quite important. For example, the significance of nutrients as new product ingredients has been well documented in recent scientific literature (Pszczola, 1998). At an annual compound growth rate of 9.4% (1992 to 1996), flavored yogurts represent one of the fastest growing new product areas.

Where do the ideas come from? A simple answer would be to assume that new product ideas can spring from anywhere. While you might be tempted to think this a logical answer, experience shows that, with rare exception, a successful process is ordinarily much less fortuitous and much more deliberate. It might be said that suggestions for new products are formulated from 0.5% inspiration and 99.5% determination and perspiration.

There must be a corporate climate which encourages the mind-set needed to see the possibilities that might lead to a new product suggestion. The company has to provide clear incentives that encourage and reward employees for looking critically and carefully at current products, at changing market conditions, at competitor's products, at customer needs, at production techniques, at new ingredient materials, at new technologies, at changing demographics, at legal, safety, and environmental factors, in fact at anything going on around them, and relating all this to what might then result in an improved or new product.

Ideas can indeed come from just about anyone (see Chapter 6). They might originate with your customers, employees' professional colleagues, management, consultants, trade or professional meetings, technical literature, often the popular literature, and, surprisingly enough, even competitors. Perhaps we should be saying, especially competitors. The impact of competitive activity is often the principal driving force behind those "hurry-up, we need the item yesterday" cries familiar to all technical personnel. Remember, every one of your competitors is also charged with new food product development and therefore collectively represents a huge expansion of effort, ideas, and incentive.

Research and development groups must maintain constant surveillance on the activity and signals coming out of the market.

Competition is very often one of the strongest forces impacting on new product development activity. Remember, the competition is also your most severe and technically competent critic. They will find any flaws in your products and may point them out to you by their actions, even before you see them.

Monitor their actions and activity; it may be crucial to maintaining your own new or improved product activity. Every product marketed by a competitor has to be examined and evaluated to determine what, if any, response will be made by your company.

Although competition is an important source of pressure for new product activity, care must be exercised to avoid unthinking reaction to every competitive move. Ill thought-out or panic reaction to competitive pressure can detract from important and perhaps more significant development already in process in your company. Although response to competitive action must be maintained, care must be taken not to be so negatively reactive that your own projects suffer or die. All R&D programs operate with a limited set of resources and they must be used rationally, wisely, and systematically. Panic or knee-jerk reactions must be avoided. Decisions about new product development must be made with a clear idea of whether they are likely to fit with the company's objectives and its corporate culture.

Ideas by themselves, no matter what their source, are not always going to get you well thought-out possibilities for new products. There has to be a mechanism for gathering in suggestions and evaluating them on an organized basis. This can be as simple as the old-fashioned "suggestion box," or as institutionalized as a standing committee that seeks out ideas, reviews and evaluates them in a systematic fashion, and reports regularly to an appropriate decision maker, or just about anything in between.

It is the responsibility of the R&D team to translate good ideas, company approved ideas, into marketable products, but, R&D cannot operate as an independent agent in the development of new products. A well-run company has a series of decision points built into its organizational structure. Input must be sought from all the players in a company at every step along the way in the development of a new product. The role of the R&D team is to build what management decides is needed, and not necessarily, or always what may appear to them to be interesting in the short run. The R&D team must be careful to remember that they are a support organization and not the reason the company exists.

EVALUATION, INITIATION, AND REVIEW

Regardless of where new food product ideas are derived, a formal mechanism for their systematic evaluation must be present. In small organizations, particularly those headed by a strong founding or entrepreneurial figure, it is often simply the decision of one person. This may be very successful and is certainly very time efficient, but this method tends to lead to highly subjective decisions concerning which ideas to pursue or reject and is prone to a high risk of failure. A more rational approach involves review of needs and ideas by more than one person, preferably by a standing committee which meets on a regularly scheduled basis but also can be called upon immediately if an emergency, for example competitive activity, should arise.

Ideally, the composition of the committee should be relatively small and include representation from the key areas of the company, in particular marketing and sales, research and development, and production. In some cases, especially where interim commitment of supplementary development funds is necessary, representation from finance is included. The committee can meet as often as weekly, or as infrequently as monthly or quarterly, to go over suggestions from various sources and to review the company's perceived needs in the new food products area. The committee can also be called upon at any time to deal with sudden changes in the market, for example, the impact of a competitor's new or improved product, or the sudden action of a regulatory agency on a product or component of a product, or the unanticipated negative performance of a product.

Even unexpected changes in the availability or pricing of a product component might require a rapid new product response. The authority of the committee can vary from simply being advisory to an appropriate decision-making operating officer of the company, often the Marketing Head or the Chief Operating Officer, or if headed by a decision maker, can authorize the initiation of the new product process itself, usually assignment of responsibility, initial assessment of technical feasibility and preliminary estimation of a development budget. In some cases this same committee hears progress reports from R&D concerning on-going development projects.

Ideas for product improvements related to the manufacturing process often best come from the employees actually engaged in day-to-day manufacture. Many times these suggestions, no matter how important, are merely production improvements, but sometimes they can result in truly new products. An unusual example of this technique some years ago was the suggestion, by technicians originally working with the lyophilization of arterial sections for medical materials, that using the same equipment for lyophilization of lunch meats and fruit pieces produced a rather tasty and well-preserved and easily rehydratable dried food item. The name of the process was changed to freeze-drying, and of course, "the rest is history."

In a more conventional example, production line personnel in a breakfast cereal manufacturing plant producing large single serving–sized briquettes found during meal breaks that these made a nice snack item when broken into bite sized pieces. This eventually led to a suggestion resulting in simply manufacturing the original briquette as a party snack of small size to begin with, and this eventually opened an entire new market to what had previously been only a breakfast food producer. In both these instances, it took someone to observe and listen to the workers, someone who could see the value of their suggestion, and also "sell" the idea as one worth pursuing as a new food product.

THE POLITICS OF THE NEW FOOD PRODUCT DEVELOPMENT DECISION

New food product ideas can and often do originate with the technical personnel in a company, but this is not always the best source. This is because with this as a source, technical feasibility and production compatibility sometimes overshadow financial and marketing considerations. It is perfectly reasonable for ideas to come from the laboratory, or from production, but unless the marketing department has reviewed the suggestion, even if technical development is eventually successful, the result may be a wonderful product which will not or cannot be willingly or profitably marketed by the company.

In a well-organized company, it is generally the marketing department that plays the key role in characterizing and describing ideas which eventually translate into practical descriptions of desired new products, and this role is logical. Because they have to market and perhaps sell the product, it is the marketing department which often has the first and final words on whether to initiate or continue development of a new food product.

It is not at all uncommon for budgeting to be set up in such a way that the marketing department operates as a profit center and actually provides all or part of the development funds and recovers its expenditure when the new product becomes an income-producing part of the company line. In these cases, the performance of the marketing head is directly impacted by the fate of new products. Therefore these administrators have a strong voice in the critical decisions concerning initiation of a new product development project, and without their input, nothing happens.

In almost every situation, it is sales and marketing who are on the firing line where buyers (customers, potential customers, and consumers) show why they use certain products, describe modifications they would find desirable, point out undesirable features of company and competitive products, and express their opinions concerning what they would like to have as an ideal product.

It is an unwise and foolhardy R&D Director who today attempts to build a new product without first getting the marketing people on board. Production can often come up with great ideas for product and process improvement, but R&D by its very nature is constantly critical of its own products, and is and should always be looking for ways to "improve" products. While R&D can present really clever and innovative variations in products, it is marketing that has its fingers on the pulse of the marketplace, and it is marketing that has to market the product.

It is wise to bear in mind that the best and most unique products in the world are of no value if there is a lack of enthusiasm from marketing. Companies are in business to make sales and profits, not simply to make products. A wise and effective R&D leader, particularly one specifically responsible for building new products for the company, will keep this thought ever in mind and develop clear, open, strong lines of communication with marketing, and never commit to major budget expenditures for new or modified product development without consulting with the company's marketing arm.

Others do have a role in the decision making, so there are additional important company offices that must be considered. Production will eventually have to produce the item, and it is therefore critical to keep them informed as to progress and to solicit their input in order to smooth the way for an untroubled fit of the new product into the existing production facilities and schedules. Accounting will certainly have a lot to say about whether a product can be profitable, or whether one direction of development might be more financially advantageous than another. Purchasing has to be consulted since they will be responsible for arriving at the most cost-effective purchase system for component ingredients and materials. We have mentioned legal and regulatory considerations above. The individuals responsible for these areas must also be consulted and kept informed. In some cases, unless they are under the control of marketing, it may be necessary to call in the public relations and advertising departments to get their input.

Different companies may have still other areas which must be considered for input and information. Balancing all of the various constituencies and interests within the company, while keeping a careful eye on the competitive climate in the marketplace, calls for an R&D Director who can orchestrate all of the disparate elements necessary to assure success for a new product's development, and who is not merely a technical whiz in a white coat, but is a politically aware individual able to exercise all the skills of a trained diplomat while simultaneously working out the technical details involved in putting together all the components of a new product.

ELEMENTS OF PRODUCT DEVELOPMENT

Once a decision has been made to proceed with the development of a particular new product, a series of factors must be taken into consideration within the several elements of the project. These factors include the nature of the product, the types of expertise required, the composition of the development team, and the time-priority of this project in relation to other development projects already in progress. The elements themselves can be listed as follows.

THE NATURE OF THE NEW PRODUCT

Usually, one of the first limitations on any new product is selection of an upper limit for production cost because of the universal impact this has on every other aspect of a new product: cost of materials, cost of packaging, cost of new equipment, cost of distribution, and cost of marketing. The degree of difficulty likely to be encountered in the development of any particular new product, as well as the length of time the project is likely to take, will be heavily impacted by the extent to which the product is, or is not, a departure from items already in the company pipeline.

If the project involves modification of an existing product then the characteristics and behavior of the product are likely to be well known to the development team. Correspondingly less time needs to be allocated for the learning curve than if the project involved something totally novel to the company or especially if it is truly unique to the industry. The same might be true if the team will be working with familiar ingredients and raw materials rather than materials new to the company. If the latter, are the new materials readily available and compatible with available storage and handling facilities, and will new control and purchasing parameters need to be set? A decision has to be made as to whether there is a need to keep as close to existing materials and production and distribution methods as possible or to what extent there can be some or great departure from familiar or compatible ground.

If the new product is likely to be perishable but all the other company products are shelf stable, this will require new distribution methods compared to existing products, unless analogous items already exist in the product line. It is necessary to consider or specify parameters like item size, portion size, multiple or single serving, nutrient content, shape, moisture, lack of need for or extent of end preparation, microwavability, reconstitutability, the need to meet special restrictions (and whose) such as religious, environmental soundness, "organic" designation, "meatless," fat and/or sugar free, etc.

Must the product fit into currently available production, packaging, storage, transportation, marketing, and distribution systems, or can new systems be considered? Must the new product fit into the existing company regulatory compliance systems, or will new compliance requirements be tolerated? Will the new label present any special problems in terms of design and compatibility with the company's other items? Finally, there has to be consultation with the appropriate legal authorities in the company regarding consideration of the degree of intellectual property protection desired for, or inherent in, the new product. Or for that matter, the extent to which the nature of the new product is restricted or its development direction dictated by the need to avoid infringing upon the intellectual property rights of others.

THE DEVELOPMENT SYSTEM

To determine whether a given proposed new food product project should be carried out, the various elements which might be applied to any project must be considered. A systematic approach when deciding whether to carry out a specific new food product development project requires consideration of all or some of the following development steps:

(1) Evaluation:

a. As previously noted, ideas must be gathered, examined, evaluated and prescreened.

b. Those ideas deemed worth looking at further must be assessed for technical, safety, legal, and regulatory feasibility, the availability of appropriate technical facilities, and the likelihood of fitting in with the company mission (see Chapter 6).

(2) Parameter setting (sometimes called "Product Protocol"):

a. A definition or description of what the end product must be, or exactly what will constitute successful completion of the project or product, must be agreed to.

b. Upper and lower quality and shelf-life tolerances for all product characteristics must be set (see Chapter 13).

(3) Cross-functional technical operations: consideration of one or more aspects of a series of technical functions (see Chapter 9 and Chapter 11):

a. Ingredient selection
b. Formulating system
c. Manufacturing unit operation selection
d. Initial cost analysis
e. Physical and microbiological characteristics evaluation
f. Quality evaluation procedures (including testing protocols)
g. Safety evaluation (including initiation of Hazard Analysis Critical Control or HACCP development)
h. Functionality testing
i. Sensory testing
j. Stability and shelf life determination
k. Packaging and labeling requirements
l. Pilot production
m. Full legal and regulatory assessment
n. Semiproduction run and cost analysis
o. Consumer testing
p. Preliminary data evaluation
q. Full production run and cost analysis
r. Final selection of control and HACCP procedures
s. Test marketing
t. Full data evaluation

(4) Budgeting

Every potential project must proceed through evaluation (step 1) and should be permitted to go further only if the decision is positive. Not every project will require passing through all of the development steps noted above. The project can be rejected after consideration of step 1, or even step 2. There can even be a decision to reject based on consideration of any of the technical operations units (step 3), if it is determined that there is reason to expect that one of these operations cannot be successfully carried out or concluded satisfactorily. Based on a careful assessment of which these elements, particularly the technical operations (step 3), will be applied as part of the new product development, an estimated budget figure and a timeline can be attached to the project. By applying this information, the type and number of team personnel can be assigned and development costs further refined to complete the setting of the budget (step 4) for the project. At this point a rational decision can be made concerning final approval and initiation of a particular project.

THE BUDGET

The budget is one of the key factors required to decide whether or not to move from the evaluation step to further consideration of a development project. Besides its importance in determining the nature of the team required to carry out a development project, budget is also a critical factor influencing both the time required to develop a new product and the extent to which the new product is pre-market tested. For all these reasons, great care must be exercise in setting this figure. It may be true that a high budget can lead to more complete development of the new product and fewer glitches in the initial market introduction, but these advantages come at the cost of time delay in getting the product to market.

A delicate compromise must be reached between the desire of technical personnel to come up with a near perfect product and the anxiety of the marketing personnel to get the product to the consumer as quickly as possible. Obviously, a proper balance must be struck once the decision has been made that the initial idea should be pursued further. Clearly, "right sizing" in determining the budget for a new food product development project is the key to a successful management decision.

As important as this decision is, budgets developed before actual initiation should not be viewed as carved in stone. Prior to final review and formal approval of a technical development plan for the project, the budget should be considered as preliminary. A final budget requires that a detailed plan of the technical development operation has been drawn up by the R&D management and subjected to final review.

THE DEVELOPMENT TEAM

Development teams can vary in size from a single technical person, to a series of very elaborate specialty groups, or anything in between (see Chapter 9). Regardless of the number of functions to be performed, the development team must operate as a cohesive unit and not simply a series of disconnected technical groups. Once the decision has been made as to just what technical units need to be applied, the development team will consist of all of the members from each technical area. It is critical that each member of the team understand and participate in the gradual evolution of the project from idea to product.

The project must not be treated as a series of disconnected operations, with responsibility for each group ceasing when it perceives its particular portion of the task to be completed. What Takeuchi and Nonaka (1997) called "the old sequential, 'relay-like' approach in which the product is handed from one functional department to another will not meet the demands of today." They believe that product development is a "holistic" process, one in which, while the project is handled by each individual in the team, it is always treated as a team effort, a system they termed the "cross-functional team concept." Cross-functional teams favor cross-pollination of ideas during the project. Team composition is not necessarily fixed for the entire term of the project but can change as more or less expertise is needed for various tasks.

In some cases all the elements of the development team exist in-house, while for other cases the use of outside specialty groups may be quite appropriate and cost-effective. A flavor company might wish to develop a special flavor mixture for exclusive use by a potential customer that lacks its own flavor compounding expertise. If the form of the product is simply a mixture of essential oils and various aldehydes, ketones, esters, and alcohols, it may well assign the project to a single flavor compounder. In this case the expertise needed resides entirely in one person who, working alone, at least initially, compounds and formulates one or more products for evaluation by the prospect. This is a team of one. On the other hand, if the nature of the flavor product is more complex and also requires formulation in an emulsified water dispersible base, perhaps homogenization, perhaps further formulation into flavored and food colored syrups for soft drink use, then more than one type of expertise might be required. In this latter case there may be a need for specialists in emulsification technology and color stability to be added to the team. In either case, there may or may not be a need for sensory evaluation, for shelf-life study, for microbial safety, etc. If the expertise exists in-house then they would be added to the team. If they are not available in-house but are deemed necessary, then they can be added on a consulting basis or by turning to outside specialty or testing companies.

In recent years, depending on the nature of the substances used in compounding and formulating, it has been found there could also be a need to review legal, regulatory, and label status. Any or all of these could be done in-house or by partial use of consulting specialists, depending on how the company has organized management of governmental and legal areas.

If there is a need to determine consumer acceptance to a degree beyond the usual relatively small in-house sensory evaluation capability of many companies, then this rather extensive team would have to be part of the development protocol.

Together with concerns imposed by the nature of the product, any technical considerations that must be taken into account, and the availability of the particular expertise required, the size and make-up of the product development team is also impacted by costs. Part of the decision process in considering the set up of a particular new product project must necessarily involve the size of the budget assigned the project, a figure justified by the nature of the product, the number of variables requiring evaluation, the difficulty of its technical operation, the priority level given the project, and an assessment of the potential financial return likely to result from a successful project. Obviously, the greater the budget, the greater the number of elements that can be assigned to the development team.

In some cases, the decision may be made for various reasons (cost, availability of personnel or facilities, etc.) to utilize a smaller or more limited team. This can still lead to a successful result in developing the new product, it may just take more time or some elements of evaluation and testing may not be extensively carried out. For example, in smaller companies it is not unusual that a choice may be made to conduct sensory testing in-house rather than undertaking the more time consuming and more expensive full consumer testing, or there may be a decision to bypass some of the pilot plant runs or extensive market testing. Often, the nature of the innovation in an existing product to develop a new product is so minor that a large number of the technical operations steps can be bypassed or shortened considerably. Simply changing the shape or color of an existing product may obviate the necessity for an extensive development team or elaborate project protocol.

OUTSIDE HELP

Outside expertise can be obtained from a number of sources. A good pool of "no cost" help can always be found from ingredient and equipment suppliers. The limitations of this approach include a lack of confidentiality, a tendency for data favorable to the supplier's products to be emphasized, the presentation of only partial data or lack of access to data not necessarily favorable to the supplier's products but important for disclosing information on functional behavior, even if negative, a lack of experimental control, and an implied obligation to the supplier.

Despite these limitations, there are times when this approach can be used for preliminary or exploratory data, for a limited or very narrow item of information, and for exploration of noncritical or nonproprietary parameters.

Another source of outside technical assistance would be consideration of government agencies such as the National Center for Food Safety and Technology, a research facility of the U.S. Food and Drug Administration, which will partner with industry as part of a research consortium at the Illinois Institute of Technology's Center for Food Safety and Technology and can provide expertise and technical facilities in areas such as food safety, quality assurance, biotechnology, and food processing and packaging technologies. The Eastern Regional Research Center of U.S. Department of Agriculture in Philadelphia can enter into research and development agreements with companies to utilize its expertise and facilities for commercialization of technologies. The National Center for Agricultural Utilization Research in Peoria, Illinois specializes in developing new uses for agricultural commodities to make new added value products. Intellectual properties can be protected by patents and patent licensing, and cooperative research agreements provide for commercialization (Giese, 1997).

Another source of technical assistance can be found in universities. Many universities which house Departments of Food Science and Technology also have established technical centers or research institutes that form partnership arrangements with industry and offer a variety of technical facilities and technical expertise (Giese, 1997).

Finally, commercial research and development consulting firms are an excellent, if perhaps costly, source of high quality assistance at all levels, including technical, analytical, sensory, marketing, legal, regulatory, and others. While there may be some level of insecurity regarding confidentiality, in general legal and contractual considerations do assure a high degree of proprietary information protection. In many cases, particularly with the private consulting firms, you have a choice of selecting as much or as little technical assistance as you deem necessary. In some cases it may be useful to consider "farming out" the entire product development project. Going this far is justified only if you do not have the in-house expertise or facility to carry the project or your in-house product development capacity has been exceeded. The major drawback to this approach is usually that of expense, however such an approach could still be cost-effective if the alternative were no product, and the anticipated direct or indirect revenue from adding the new product to your company's line would quickly permit cost recovery on a timely basis and subsequently generate profits.

THE TECHNICAL DEVELOPMENT PLAN

After subjecting the new food product concept to careful scrutiny within the parameters laid down by the development system approach, a formal plan for the initiation and implementation of the technical development must be prepared. The plan should be a document which lays out the rationale justifying the development of the product, any history or significant factors bearing on the "go-ahead decision," any special aspects noted by various elements of the evaluation committee, a clear statement of the benefit the product will bring to the company, and document the commitment of the company management to the success of the project.

This plan will select the most appropriate technical steps likely to result in successful achievement of the development objective and include a detailed budget, account for resources needed, suggest tasks for the specific teams required, lay out significant review points or milestones in the project, and present a tentative timetable for intermediate and final goals. The Technical Development Plan should be reviewed by appropriate technical, production, marketing, and financial groups for final review and initiating decision. While it is expected that operation of the project will be guided by this plan, there must remain sufficient flexibility and openness of mind to alter any step or timetable as the project develops if the intermediate findings indicate that changes are required. This is not to say that the plan has little or no importance and can be easily deviated from, particularly when other areas of the company must be able to reliably plan for key steps like pilot runs for sample production and setting up any required consumer testing, etc.

The plan is, however, simply a document drawn up prior to actual experience in the development and operating world, and thus is simply a "best estimate" of the most likely path to success. If that real world experience dictates a need for accommodation and change, then, while the cost in terms of time and money of any proposed change must also be taken into account, it must be considered. After all, a plan is at best an intellectual exercise, and although management hopes that their technical judgment is up to the task of providing an accurate and correct predictor of success, it is still only a plan. It is the goal that is likely to remain fixed.

THE DEVELOPMENT PROCESS

INFORM

New food product development projects seldom come to the technical teams as a total surprise. R&D management often apprise their personnel of items that are in the process of consideration, in part as a technique for giving everyone a sense that they are part of the overall development team, and in part as a mechanism providing informal feedback to the committee for use in the evaluation process. This practice is often the best way to avoid speculation and rumor. Even where there are no regular information sessions, the "grapevine" often passes information along. For these or other reasons, there is usually no great surprise when management officially informs technical personnel about the initiation of a new food product development project. The technical team should be identified and then presented with a full briefing from the R&D management (preferably together with the marketing member of the evaluation or screening committee) as to the nature of the new project, its significance for the company as a whole and for them in particular, the reasons for its selection, the nature of any known or potential competition, the feasibility of the project, any particular technical challenges and particular attributes, and initial projections for major target dates (start date, initial "bench" prototype, laboratory or bench production, pilot production, pilot testing, shelf life studies, semiproduction run, safety studies, quality control procedures, consumer evaluations, initial production run, etc., and the desired final turn-over or completion date). The technical team should view the Technical Development Plan, be encouraged to offer their comments, and otherwise be encouraged to buy into the project.

INITIATION

The first phase of initiation usually involves a discussion session at which all team members are expected to discuss, criticize, suggest, and generally "brain storm" the project to identify where to begin, even if it is obvious, agree to or modify the technical steps needed, and review the target dates for their feasibility. Hopefully, they should come out of this initial session in substantial agreement so that they "buy-in" to the project and the development process.

The second phase of initiation can vary depending upon the nature of the project. Often it involves the assembling of ingredients, hardware, and equipment. Sometimes it may require information gathering before this can even begin, or this aspect can be carried out simultaneously. Information gathering can be as simple as reading the specification sheet developed by the evaluation committee or it can be as complicated as a full literature search, require identification of sources of information, perhaps call for a trip to the library or a nearby technical center or university, or wherever and to whomever the information can be located, consultation with supplier, review of competition products, etc.

Based on what can be determined from the various information sources and deliberations by the team members, additional information requirements should be identified, any additional equipment or ingredients shall be ordered or obtained, determination should be made concerning the possible need for any modifications in the original Technical Development Plan, need for any analytical or other preliminary work (modification of equipment or ingredients, etc.) should be identified, intermediate goals should be set, and, finally, initial specific action assignments made.

ITERATION

Trial runs are made starting with the initial formulation. Rapid evaluation is made on the resulting trial results using appropriate testing methods. Modifications are made in the formulation or the ingredients or the process, and the process continued through as many iterations as needed to reach a point where the resulting bench product is deemed ready for its first general assessment. Based on these preliminary evaluations, the iterative modification process continues until the result is judged to be a reasonable prototype. Methods of judgment at this stage might be simply "in-lab," but there should be progression to at least small in-house formal sensory testing using an organoleptic panel evaluation methodology (see Chapter 12).

It is not unusual to employ focus groups or small consumer panels to obtain information suggesting useful changes or confirming the direction of the development. Progress assessment must include marketing input so that agreement can be reached as to when the bench product has met the goal originally set for meeting the desired product definition. At this point preparations should proceed for moving to the pilot production stage and beginnings made toward defining the technical label declaration and the packaging need characteristics.

PILOT OPERATIONS

Although some products move directly from the bench to the production plant, this is a risky approach, one which the move to "short-cut" the development process and bring the product to market more quickly can prove costly in terms of lost material and technical time costs if something goes wrong. More importantly, although any technical mistakes can still be corrected, a major risk is loss of confidence in the new product by customers, marketing, and production. Nearly undetectable minor deviations at the bench level can become major defects at the production level. Many potential production problems can be avoided by using an intermediate stage before moving to a large scale-up. This can occur in a pilot plant, that is a small-scale facility equipped with full scale or smaller capacity commercial production equipment, where production runs of the product can be made on a near production run basis.

In some cases, particularly where no separate pilot production facility is available, product is manufactured in the actual production plant but on a reduced or limited volume basis. This is often referred to as a semiproduction run. The purpose of pilot or semi-production runs is to gain experience in larger than bench scale production of product with no or minimal interruption of the regular production of current products, and to do so on a modest scale. This practice permits R&D to minimize cost, translate scientific or laboratory terminology into production terms, determine any special equipment or production technique needs, fully test any new or modified equipment involved, provide for further modifications in formulation designed to improve the production process, more fully test quality control procedures, and accumulate larger quantities of product for further testing and evaluation. A more accurate assessment of costs can also be worked out.

At this stage plans can be formulated for the quantities of product needed for shelf life studies, microbiological studies, and, if desired, consumer sensory testing. The packaging requirements can be determined at this stage as well, although in many cases this could have also been carried out as part of the iteration stage. The pilot stage is a good place, if it hasn't been done earlier, to bring production personnel into the team. Besides familiarizing them with the product, their feedback in terms of assessing the likely behavior of the product in the full-scale production process, especially predicting where production "bugs" are likely to become a problem, and offering suggestions for any product changes to address this area, can be crucial to the eventual production of the final product on a full-scale basis. R&D must always remember that the production department will eventually take full responsibility for the product so the earlier they "buy in" to the product, the better.

PROJECT COMPLETION DETAILS

Having completed what R&D frequently considers the "important" phase of the project, the formulation of the new product, there is a temptation on the part of product development specialists to gloss over the remaining development processes, considering anything more merely "minor details." The usual anxiety on the part of marketing to scoop the competition frequently adds to these pressures. This is analogous to the busy chief of a surgical team, having performed what he/she considered the crucial part of an operation, dashing out of the operating theater while tossing a casual instruction back to the surgical assistants that the real work having been done, they can now finish up. Remember that old aphorism attributed to Yogi Berra, "it's not over 'till it's over!" A project is not "over" until everyone designated as having responsibility for the project has had a chance to review the results and formally sign off. Even after everyone is satisfied with the item finally produced, there are still important details critical to the commercial success of the product that have to be attended to before the book can be completely closed on the development. For example, even though not strictly part of the development phase, R&D personnel should expect to monitor initial or start-up production for a "shake-down" period. In addition, the following development considerations should be considered or satisfied, though not necessarily in the sequential order presented here. In point of fact, many of these and other prior development areas could be carried out concurrently. This is usually a matter of budget level, personnel available, time pressures, and corporate culture.

PACKAGING SELECTION

The exact nature of the packaging to be used for a product is dependent on a large number of factors. This topic is covered extensively elsewhere in the text in its own section (see Chapters 8, 15, and 16) so that no great detail will be given in this chapter, but packaging is a crucial element in any product and the brevity of its mention here should not be interpreted as a measure of its significance. It should be patently obvious that in the 20th and 21st century foods are not marketed and distributed in bulk at the consumer level. For the past 150 years or more, they have been contained and sold in some type of closed small unit package. Suffice to say that the packaging system selected must meet the technical needs of the product, the demands of the particular manufacturing process, the marketing objectives set for it, the intermediate and retail distribution system into which the product goes to market, and the handling, storage, and use conditions imposed by the final user, the consumer. Conditions such as compatibility or noninteraction with the product, appropriate gas transfer control, chemical and physical protective qualities, shelf life, microbiological and chemical safety, environmental compatibility, the expectations of the market (competitive or customary practices), filling and handling equipment compatibility, fitting into some over-all corporate practice, cost, visual design, etc., all these factors, and more, impact on the selection of an appropriate packaging system.

Packaging requirements must *not* be an afterthought left to a separate unit within the company. The need for packaging expertise must be considered as part of the team requirement process from the very beginning of planning for new product development.

SHELF LIFE AND CONSUMER TESTING (INCLUDING FAILURE LIMIT TESTING)

Samples of finished product are usually produced in the pilot or semiproduction facility, packaged in the container intended for commercial use, and stored for preset times under a selection of environmental conditions (also see Chapter 13). In any shelf life study, it is necessary to establish the limits of acceptability and failure for a product, otherwise known as tolerance limits. Once this has been identified, suitable testing methodology can be selected or developed. It is necessary to know how long and under what environmental conditions (e.g., temperature, humidity, light, etc.) a product can be held an still be acceptable in the marketplace. It is also necessary to know under what conditions the product is likely to fail, sometimes called failure limit testing. The definition of just what is acceptable, and what levels of acceptability are desired, should have been part of the initial agreed upon description (development system, step 2, parameter setting) of what constituted an acceptable product. This information is necessary to be able to set parameters of market life (when do you "pull" the product), and to determine how well your product performs compared to those of competitors. It is necessary to know whether problems of significant microbiological or chemical safety importance occur under expected marketing, distribution, and consumer storage conditions.

Ordinarily, sensory (color, taste, odor, texture) characteristics determine acceptability, however in some products additional or other factors may be the determining attributes.

For nutraceuticals or for foods with required or stated constituent content, for example the nutrient value of an infant formula or a product claiming specific quantitative content of n-3-fatty acids, acceptability may also be determined by ability to meet stated label contents or regulatory requirements.

If crispness is the required characteristic, as in potato chips, then judgment could be either or both analytical and sensory.

Where quantitative content of a specific ingredient or particular physical characteristics are the determining factors of acceptability or quality, judgment of acceptability can be made by analytical laboratory methods. In cases where acceptability or quality is determined by organoleptic or sensory qualities, human sensory perception methods must be employed (see Chapter 12). These can vary from small scale in-house panels of suitably trained and qualified taste experts, usually employing a limited number of judges (10 to 50), to large scale out-of-house consumer panels made up of large numbers (100 to 1,000 or more) of deliberately untrained consumers. Consumer panels are unwieldy, work best when only limited information is requested (yes or no, acceptable or unacceptable, etc.), and are time consuming and expensive. On the other hand, properly conducted, they yield very valuable information.

Very often, the consumer panel is used to determine initial acceptance or rejection of new products, whereas in-house panels mainly function for quality control after products reach the production stage. Of course, highly trained in-house panels are also used to assist in product development, but they are used primarily in an analytical manner, to get detailed information on specific or detailed aspects of the product. They do not function well in predicting over-all consumer acceptability. However, when trained to detect sensory characteristics previously determined to be associated with consumer acceptability, or technical limits selected to judge product quality, expert panels can be very effectively used to inexpensively evaluate product shelf life.

Sensory analysis and evaluation is a valuable and complex area with unique technical technique. Space limitations here preclude more than this abbreviated coverage which, for greater understanding, requires review of its own rather extensive literature.

REGULATORY COMPLIANCE, QUALITY CONTROL, AND LABEL REQUIREMENTS

All materials, ingredients, and procedures which become part of, or come into contact with, the product or its ingredients during manufacture must be continuously monitored to assure that all meet the test of being permitted for use in or on foods by the appropriate regulatory body, for example the U.S. Food and Drug Administration (FDA), or U.S. Department of Agriculture (also see Chapter 18). If the product is designed to meet FDA Standard of Identity for the particular product, the exact nature of permitted ingredients is very narrowly proscribed. The vast majority of processed food products require detailed label declaration of all ingredients. If the product is designed for marketing to particular consumer groups expecting that it meet special standards (e.g., kosher, vegetarian, organic, etc.), this becomes part of product requirements. This extends to all ingredients, to the contact surfaces of processing equipment, to all packaging materials, and includes all water. R&D must set up and monitor all production procedures to assure adherence to good manufacturing practice standards, including the training of production personnel and maintenance of accurate production records. In addition, appropriate procedures required for compliance with environmental regulations affected by the production of the new product must be developed and monitored. In an increasing number of cases, adherence to water use and limitation regulations must be monitored. This includes proper handling and disposal of solid and semisolid waste materials and processing water. It may even include fuel, power, and utility requirements.

It is incumbent on R&D to develop appropriate procedures for analytical, microbiological, safety, environmental contaminant, and sensory monitoring of product quality. Producing a physiologically safe food product is both good business and an absolute regulatory requirement. The means that analytical procedures must be developed concurrently with product formulation. Providing and utilizing such procedures in a systematic manner is part of a good manufacturing practices requirement. As such, quality control can be viewed as fulfilling a regulatory requirement, maintaining cost control, and assuring a continuing level of acceptance in the market. An R&D new product development project cannot be considered complete until these procedures have been appropriately developed, thoroughly tested, and are in place at the time production is authorized.

Quality control or quality assurance is usually considered part of the production monitoring process, but it must be considered by R&D during the development of new food products. Since no product can be any better than the ingredients from which it is prepared, all of the requirements and procedures noted for the finished food product must also apply to ingredients. Specifications for ingredients must be developed concurrent with formulation development. All ingredients must have an appropriate specification standard prepared by R&D so that the purchasing department can have parameters against which to judge and compare the cost and quality of materials supplied by vendors. This would include safety, physical, chemical, microbiological, and even sensory standards and the methodology to properly monitor these qualities. R&D must bear in mind that quality control is a dynamic process. The process does not end at the point of product turnover to production. The quality control process is one of ongoing quality improvement, i.e., process improvement, new equipment evaluation, cost monitoring, field monitoring, alternative and substitute ingredient evaluation, and even reformulation.

By or before the project has reached the pilot or semiproduction stage, R&D must assemble the information required for the label. For nearly all processed food products, this means a list of all ingredients arranged in descending order of quantitative predominance (by weight or percentage), starting with the quantitatively largest item, and called by the appropriate or permitted name. In addition, a nutrient declaration which meets the requirements of the 1990 Nutrition Labeling and Education Act (NLEA) and its amendments must also be prepared. If there is to be any health claim included as part of the label, special care must be exercised to assure compliance with regulatory guidelines (Porter, 1996) (see Chapter 16). If there is a desire to meet the specific requirements of special designations, such as kosher or other indicators of particular quality, these must be considered.

Even items like the size of print and placement of items on the label are subject to regulation. This can be carried out in-house if that particular expertise is available either in R&D or in the responsible legal and regulatory area of the company. Very often, this specialized requirement is best met by use of experienced professional outside experts who are well versed in the complex and somewhat arcane requirements of the act. This information must be supplied to the graphics personnel responsible for preparation of the printed label, however it is the responsibility of both the R&D personnel and the regulatory and legal personnel to review all proposed labels before final approval and use. Obviously, the marketing department has its requirements from a sales point of view and may require particular messages and items like the Universal Product Code (bar code) symbol for price scanning. Again, this subject is sufficiently specialized and complex to deserve its own separate treatment and is usually covered in chapters on packaging.

DOCUMENTATION

Early in this chapter we referred to the need to carefully document every step of the way in the development of new food products. When the dust has at least partially settled, and the new product looks like it is on its way, it is incumbent on R&D to look back on what transpired during the project, distill out the essence of what has been accomplished, and carefully write up a detailed report on all aspects of the project. At its very least, this is a history; at its best it documents the value of R&D to company progress. In part this is necessary to explain just what the company has received for its budgetary expenditure (or investment) and who has contributed, and how. It is also a technology transfer document. It is necessary to record the experience gained so that future projects can benefit from what was learned along the way; thus the report is a means of assuring technology transfer. Every new food product development project is different but every project is also the same. Lessons learned in one case are often transferable to another. This valuable asset should not be lost to the ephemeral memory of sometimes transient personnel. In addition, the project must be reviewed by the appropriate parties within or authorized by the company for any intellectual property ''gold'' that can mined from this rich source.

I place the documentation reminder both early and late in this chapter on new food product development in the hope that the reader will have learned what I term the central dogma of new product development: as important as they are, new product ideas are not products until a process of exploiting that idea has been developed and carried out, and that a new product development process is not truly complete until the product is on the market and the final report has been written.

APPENDIX: PLANNING AND SCHEDULING OF NEW PRODUCT DEVELOPMENT PROJECTS

New product projects are complex, time-consuming, and inherently risky. Projects involve many different activities, performed by different people across multiple functions in the organization. There is a necessary time sequencing to certain sets of activities, while other activities can be performed independently. Because it is vitally important to conserve both time and organizational resources, new product projects must be carefully planned and orchestrated. New product success in most cases is dependent on getting to market as quickly as possible, but not at the expense of skipping or short-circuiting vital steps in the development process.

Once a new product development project is approved, project planning must take place. The project team must determine the full range of tasks or activities to be performed, the order in which the activities must be performed, and the expected time to complete each activity. One of the most important steps to be completed in setting up the project is the determination of activities which, if completion of these is delayed, the entire project will be held up. These are termed critical activities. In addition, the personnel, financial, and capital resources required to complete each task must be determined, along with the administrative support needed to keep the project on time, within budget, and providing the deliverables required at each stage of the project. Responsibility for completing each task involved in the project must be determined and communicated. Finally, program budgets are drawn up and agreed upon.

Project planning network diagrams can be used to visually display the project. The essential elements of such networks are those discussed in the preceding paragraph. We can use network diagrams to both plan and control the project. Prior to commencing the project, the planning network can help project administrators plan an efficient schedule. Alternative planning networks can be evaluated to assess the impact on completion time and cost. Activities that can be performed in parallel can be identified, as can critical activities. By performing activities that are not time dependent at the same time, overall project completion time can be shortened. After commencing the project, administrators can compare actual time to complete activities versus projected time, and actual budgets to projected budgets, and identify variances. These variances indicate potential problem areas in both budgeting and project completion timetables. One of the uses of critical path analysis is the development of contingency plans to redeploy resources from "slack" activities (which are those activities that if we fail to complete on schedule will not hold up the entire project) to critical activities.

The development and launch of a new food product can be a very simple process, an extremely complex process, or can represent an intermediate level of complexity. The simplest processes are those involving simple line extensions such as new flavors for a line of isotonic beverages or chilled teas. The most complex processes are those for which new food and/or packaging technologies are being employed, and severe technical challenges must be met; plus, there is a need for significant development of and investment in new plant and equipment. Although both of these processes follow the same basic logic of a concept phase, a development phase, and an implementation phase, the number of activities at each phase is greater by orders of magnitude for the latter process. Figure 10.1 and Table 10.1 demonstrate the basic logic, sequence, and range of activities involved in a new food product development process.

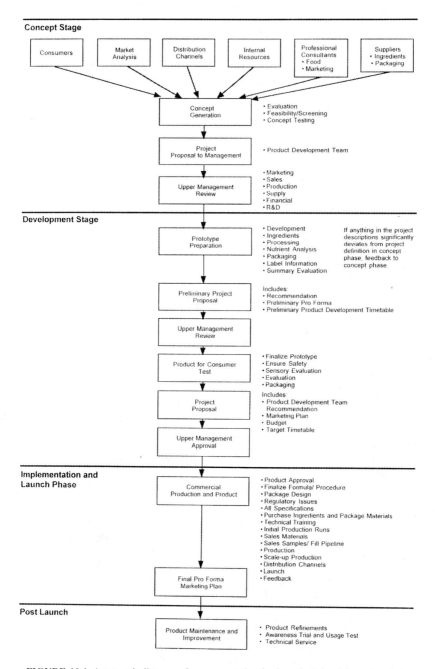

FIGURE 10.1 A network diagram of a representative food product development project.

245

TABLE 10.1. Representative New Food Product Development Project Planning Document.

Product Strategy

Objective: Develop and Implement New Food Product Strategy

Tasks	Resources Required	Responsibility	Budget	Time Required	Contingency
Marketing research inputs • Product requirements • Pricing • Characteristics of consumers, users, etc. • Size of markets • Growth rates	• Market, consumer, product, and industry experience information source(s)	• Marketing manager	$X	X months	• Develop new products without marketing information and subsequently determine markets
Develop basic marketing strategy • Identification of target markets	• Team of marketing and product technology	• Marketing manager • Product development management	$X		• Develop new products without marketing information and subsequently determine markets
New product strategy: • Focus groups/research from consumers	• Marketing research • Focus group facilities • Willing consumers	• Marketing research manager	$X	X months	• Develop new products without marketing information and subsequently determine markets
• Selection of food product mix based on marketing research inputs	• Team of marketing and product technology	• Marketing manager • Product development management	$X	X months	

TABLE 10.1. (continued).

Tasks	Resources Required	Responsibility	Budget	Time Required	Contingency
• Determination of new product requirements from users					
• Selection of new products based on both marketing and product requirements					
• Prices/costs					
• Properties					
• New product development strategy:	• Technical • Product development	• Product management • Marketing manager • Technical manager	$X	X months	• Develop new food products without strategy or plan
• Imitate existing product	• Marketing • Product development				• Modify existing products
• Improve on relevant existing product properties					
• Duplicate commercial product to price on more economic basis					
• Innovate totally new products					
• Extend current product line					
• Internal development					
• External development					
• Use existing company production equipment					
• Product development planning	• Technical staff • Project planners	• Project planner	$X	X weeks	• Proceed with minimal or no planning
• Objectives and tasks					
• Times and schedules					
• Establish mileposts (stages/gates)					
• Pro forma costs and budgets					

247

TABLE 10.1. (continued).

Tasks	Resources Required	Responsibility	Budget	Time Required	Contingency

• Resources required
   • Human
   • Physical
   • Consumables
   • Physical facilities
	• Technical management				

• Financial					
• Presentations to and approvals from management					
• New product development		• New product management • Product development • Technical manager			• Performed by external agency

• Implementation of plan
   • Laboratory
   • Pilot plant
			$X	X years	

• Establishment of test protocols	• Technical staff	• Technical manager	$X	X months	• Use FDA protocols

• Evaluation of samples versus
   • Existing products
   • Alternative iterations
	• Internal laboratory	• Technical manager	$X	Ongoing over X years	• Independent external laboratory

• Evaluation by consumers	• Users/finished products	• New product management • Marketing manager • Sales manager	$X	X months	• No alternative

TABLE 10.1. (continued).

Tasks	Resources Required	Responsibility	Budget	Time Required	Contingency
Production procedures • Equipment • Materials • Operating parameters • Measurements and controls	• Technical management	• Production manager	$X	X months	• Add-on to company's existing line
Production specifications • For all new products	• Technical quality management	• Quality assurance	$X	X month	• Use technical staff
Quality assurance protocols	• Technical management	• Quality assurance	$X	X month	• Use technical staff
Cost development	• Financial analysts • Cost inputs • Capital requirements	• New product management	$X	X weeks	
Establish procedures to monitor product use in field and to feedback information	• Field technical staff	• Quality assurance	$X	X weeks	• Use technical staff

TABLE 10.1. (continued).

Pricing

Objective: Establish Pricing for New Food Product

Tasks	Resources Required	Responsibility	Budget	Time Required	Contingency
Establish product costs for new product • Materials • Inventory • Labor • Utilities • Projected variables • Projected indirects	• Production • Purchasing • Technical • Accounting/ financial prior cost development data and information	• New product management	$X	X months	• Set price based on competitive product pricing • Set arbitrary low introductory costs
Capital investment required	• Engineering • Production • Financial	• New product management	$X	X months	• No ROI computation required
Operating capital	• Accounting	• New product management	$X	X weeks	
Develop total costing	• Accounting	• New product management	$X	X weeks	• Set price based on competitive product pricing
Determine prices	• Marketing research	• New product management	$X	X months	• Point-in-time pricing
Determine lower and upper limits of pricing	• New product management	• New product management	$X	X weeks	• Price based on competitive product pricing only
Determine margins to generate appropriate returns for business venture	• New product management	• Accounting • New product management	$X		• Let market pricing dictate margins

TABLE 10.1. (continued).

Tasks	Resources Required	Responsibility	Budget	Time Required	Contingency
Ensure that pricing generates cash flow to drive the new product	• Accounting	• New product management	$X		
Establish pricing strategy • Launch • Steady state	• New product management	• New product management	$X	X weeks	Price based on competitive product pricing only

Communications

Objective: Communications Strategy for New Food Product

Tasks	Resources Required	Responsibility	Budget	Time Required	Contingency
Identify target markets for the new product category	• From marketing research and strategy • Marketing research	• New product management • Marketing manager	$X	X weeks	• No communications program
Characterize target markets by needs and perceived needs	• Marketing research	• New product management	$X	X months	
Determine marketing strategy and objectives	• Marketing data	• New product management	$X		

Develop communications strategy based on marketing strategy
• Objective
  • Launch
  • Steady state
• By product
• By market
	• Marketing strategy
	• Inputs on targets from research
	• Communications manager
		• New product management
		• Marketing management
			$X	X months	• No communications strategy
					• Tactical only

TABLE 10.1. (continued).

Tasks	Resources Required	Responsibility	Budget	Time Required	Contingency
Message to each target market	• Writer(s)	• New product management	$X	1 month	• Single message to all
Communications category	• Communications manager	• Marketing manager	$X	X months	• General—no targeting
• Media conference	• Advertising agency	• Communications manager			• Internal resources
	• PR agency				
• Hard copy pieces					
• Media releases					
• Trade promotion					
• Trade show stand	• Trade show management				
• Trade show presentations	• Trade show management				
• Advertising	• Advertising agency				
• Schedules	• Communications manager				
• Art	• Art				
• Budget	• Financial budget person				
• Launch					
• Steady state					
• Measurement of results	• Marketing research				
• Modifications of plan					
Implementation	• Communications manager	• Communications manager within group	$X	• Launch—X months	
	• PR agency			• Reminder—on-going	
	• Advertising agency				
	• Trade show management agency				

TABLE 10.1. (continued).

Distribution

Objective: Develop Distribution Channels to Move Product from Production to Retailer/Consumer

Tasks	Resources Required	Responsibility	Budget	Time Required	Contingency
Select location(s) of production	• Production	• New product management • Production management	$X	X months elapsed	• Contract with co-packer
Determine products to be made at each site, if more than one production site is selected	• Production		$X	X months elapsed	• Produce to order; still requires distribution
Identify users to be served • By type of product • By volume • Frequency of delivery • By location	• Marketing • Marketing research	• New product management	$X	X months	• Respond to user orders

Develop logistical plan to balance
• Production
• Inventory
  • Production
  • Distribution
• Delivery
	• Distribution manager	• New product management	$X	X months	• No plan, not acceptable

Determine distribution policy • Direct delivery • Brokers with control of distribution • Costs for each • Delivery to users for each • Geographic dispersion • Construct alternative strategy for each • Select one or more	• Distribution manager	• Distribution manager • New product management • Marketing management	$X	X months	• No policy, not acceptable • Go directly to distributor • Use company's current distribution

253

TABLE 10.1. (continued).

Tasks	Resources Required	Responsibility	Budget	Time Required	Contingency
Identify distributors, if distributor system is selected; brokers; distribution warehouses; etc., depending on strategy selected	• Distribution manager • Marketing research	• New product management • Distribution manager	$X	X months	• Use predetermined list • Use company's distribution system
Communicate with distribution channels	• Purchasing • Distribution manager	• Distribution manager	$X	X months	
Negotiate with distribution channels	• Purchasing	• Distribution manager	$X	X months	
Establish interactive electronic data interchange between distribution channel and production	• EDI manager	• New product management • Distribution manager	$X	X months	• Paper transactions • Telephone transactions
Select method for product movement, e.g., truck • Use external distribution • LTL • Full truckload • Independent • Contracted • Own fleet	• Distribution manager	• Distribution manager	$X	X weeks	
Negotiate and contract with carrier	• Distribution manager • Purchasing manager	• Distribution manager • Purchasing manager	$X	X months	
Implementation	• Distribution manager	• New product management • Distribution manager	$X	Ongoing	

TABLE 10.1. (continued).

Sales

Objective: Develop Sales Strategy for Product to Buyer

Tasks	Resources Required	Responsibility	Budget	Time Required	Contingency
Marketing strategy • Target market products • Pull strategy with users • Distribution intermediaries	• Marketing plan • Marketing manager	• New product management • Marketing manager	$X	X months	• No marketing strategy—not acceptable
Sales strategy	• Sales manager • Marketing manager	• New product management • Marketing manager	$X	X weeks	• Go directly to one predetermined sales strategy
• Employ company's sales staff • Employ separate sales force • Independent sales • Broker sales force • Distributor sales • Evaluate alternatives					
• Determine costs for each		• New product management			
• Project probable sales outcome for each		• Sales manager			
• Select best alternative(s)	• Sales manager • Research • Financial analyst	• Sales manager	$X	X months	
Identify target markets to be contacted by sales • Purchasing agents • Retail	• Research • From communications plan	• Sales manager	$X	X months	• Respond to general inquiries

TABLE 10.1. (continued).

Tasks	Resources Required	Responsibility	Budget	Time Required	Contingency
• HRI • Distributors • Others • By name, location, telephone, fax, e-mail, etc.					
Determine message to be employed for each target market category	• Marketing manager	• Marketing manager	$X	X weeks	
Determine type of sales person desired for each target category	• Human resources • Sales manager	• New product management	$X	X weeks	
Determine training required for sales persons for each category • Product • Industry • Company's policies/culture • Message	• Marketing manager • Sales training • Sales manager		$X		• No formal training program
Develop sales training program • Product benefits • Marketing support • User needs • Competitor counters	• Sales training • Sales manager	• Sales manager • Marketing manager	$X	X months	• Direct field training by sales and/or marketing manager
Determine number and type sales person for each target category • Centralized • Geographic dispersion • Inside • Field	• Sales manager	• Sales manager	$X	X weeks	• Inside sales only

TABLE 10.1. (continued).

Tasks	Resources Required	Responsibility	Budget	Time Required	Contingency
Collateral materials • Brochures • Videos • Internet home page • Etc.	• Communications manager • Advertising agency • PR agency	• Sales manager	$X	X months	• Use existing collateral materials
Develop budget for sales	• Sales manager • Financial analyst	• Sales manager • New product management	$X		• No budget—not acceptable
Hire required sales staff	• Human resources • Sales manager	• Sales manager	$X	X months	• Transfer from inside company • Transfer from technical • Use brokers • Inside sales only
Sales training	• Sales training • Physical facility • Sales staff • Technical management • Marketing management • Sales manager	• Sales trainer • Sales manager	$X	X weeks	• Field training only
Sales staff dispersion • Target market • Geography	• Sales manager	• Sales manager	$X	X weeks	• Broker sales force • Distributor sales force • Inside sales
Inside sales and customer service • Procedures • Responses • Staffing	• Sales manager	• New product management • Sales manager	$X	X months	• Directly to production scheduling
Implement		• Product management	$X	Ongoing	

BIBLIOGRAPHY

Amernick, B. A. 1991. "Protection of Intellectual Property," in *Food Product Development from Concept to Marketplace*, Graf, E. and I. S. Saguy, eds., New York: Van Nostrand Reinhold, pp. 365–378.

Giese, J. 1997. "Technical Centers Facilitate Food Product Development," *Food Technology*, June, pp. 50–54.

Hoban, Thomas J. 1998. "Improving the Success of New Product Development," *Food Technology*, January, pp. 46–49.

Pessemier, E. 1982. *Product Management Strategy and Organization*, 2nd Ed. New York: John Wiley & Sons, pp. 20–23.

Porter, D. V. 1996. "Health Claims on Food Products: NLEA," *Nutrition Today*, 31:35–38.

Pszczola, D. E. 1998. "The ABCs of Nutraceutical Ingredients," *Food Technology*, March, pp. 30–37.

Takeuchi, H. and I. Nonaka. 1997. "The New Product Development Game" (*Harvard Business Review*, January-February 1986). In *Managing Teams in the Food Industry*, P. Hollingsworth. *Food Technology*, November, pp. 75–79.

CHAPTER 11

Innovative New Food Products: Technical Development in the Laboratory

ALVAN W. PYNE

Long considered the dual fountains of new food products, kitchens and laboratories are essential elements but not the totality. Here is where the blending of superb comprehension of scientific principles marries with creativity and daring—the synergy of chefmanship and food science. The prototype is only one piece, which, if acceptable, must subsequently be translated into a product that can be commercially produced, packaged, and distributed to consumers who would receive a true reflection of the original promise of the concept. Implementation in a factory with specifications, quality assurance tools, and packaging represents a challenge for those team members who have been involved from the beginning but are now responsible for producing the product.

INTRODUCTION

NUMEROUS books and articles have been written on the subject of new product development and many seminars and workshops offered to help marketing and technical managers understand this nebulous area of business and how to succeed at it. Food product development is a demanding, fast-paced, high-risk/high-benefit part of a business that requires dedicated inputs from multiple sources, e.g., marketing, food scientist/technologists, engineers, plant operations, outside suppliers, and financial and business management, all working closely together to achieve tightly defined objectives. Such a food product development team must work under difficult time constraints and pressures, following disciplined schedules all designed to achieve a timely competitive advantage in the marketplace. This is certainly the case for innovative new product development. It is acknowledged, however, that most product development activities are carried out to deliver line extensions and brand expansion of existing product lines, a process requiring far less risk and development effort than that expended for development of innovative new products. Line extensions generally are based on development of new flavors, new packaging regimes, package sizing, i.e., basically a variation on a theme.

In this chapter, we focus on truly innovative product development as differentiated from the more common line extension and brand expansion activities, although the basic steps required to take products into the marketplace are similar.

The three fundamental truths to understand about this game of new product development are

- The probability of success is low.
- You never really know whether you have a success until you have taken the plunge to find out if it is going to sell.
- The rewards, if successful, can be very great.

New product development is tough work. If you are going to play the game of new product development, you have to have the constitution for it, and a willingness to roll the dice, back your hand, analyze the game, and know when to quit. Participation in the game requires the willingness of management to risk considerable funds to accomplish committed and agreed upon objectives.

The process of new product development is at best a delicate and tortuous path (see Figure 11.1). New product development in any organization must have a bonafide champion, must be nurtured and encouraged and must be supported by top management. In general, marketing leads and the technical

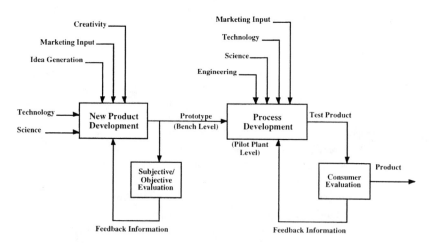

FIGURE 11.1 Interaction of key elements in new product development.

function follows working as a team to develop and commercialize new products.

In addition to requiring the assembling of a tightly knit multidisciplinary team to develop and deliver the product, there must be the proper consumer input, which requires essentially bringing the consumer into the product development picture at the very early stages of the project. This is achieved by performing appropriate studies and panel testing protocepts with a variety of target consumers to identify what the consumer needs or wants or to assemble a reasonable story to convince the consumer that he/she really need the proposed new product.

Generally, the product concept has been defined based on the inputs of both marketing and technical personnel. The technical development department is challenged with the task of translating the desired consumer product attributes into technical terms and identifying and working on approaches to ensure that the types of attributes are delivered within the product according to the defined cost parameters, shelf life considerations, and desired flavor and texture attributes. It is imperative to carry out such activities, keeping in mind that the main objective is to deliver a product that will be acceptable to the consumer and possess a fully defined point of difference from the competition. To ensure that this is the case the team has to be well aware of the competitive products in the market or the potentially competitive products that could be introduced to the marketplace.

SETTING THE STAGE

PROJECT/PRODUCT PLANNING

New product development activity should be integrally tied into the business planning process whether it is at the division level or the corporate level of a company. There should be a solid commitment at the highest level of management to support the activity with the required resources. Provision should be made for timely review of progress reports and identified milestones to keep the product planning process on schedule.

MARKETING RESEARCH BRIEF

In general, marketing will provide a brief to the technical development group based on the following information provided in Table 11.1. For a new product, the anticipated points of difference from competition should be clearly defined. The projected method of preservation and distribution, e.g., shelf stable, dehydrated, heat processing, refrigeration or freezing, or a combination

TABLE 11.1. New Product Development. Phase 1: Technical Feasibility Study—Information Requested from Marketing.

A. General product description
   1. Concept
   2. Anticipated consumer need and type of use for product
   3. Closest competition on market today
   4. Anticipated points of difference from existing competition
B. Marketing position of product
C. Anticipated route of distribution—refrigerated, frozen, shelf-stable
D. Targeted selling price for given unit size
E. Targeted plant cost per unit[a]
   1. Anticipated number of units to be produced for
     (a) test market
     (b) commercial production
F. Anticipated packaging
G. Anticipated shelf life requirements
H. Anticipated production facilities
I. Anticipated date of production introduction
   1. Test market
   2. Roll-out
J. Extent of anticipated line extension possibilities

* Key piece of information.

should be established. These will impact product safety, shelf life, and packaging considerations.

THE TECHNICAL FEASIBILITY STUDY

In response to the marketing research brief, the technical development group carries out a technical feasibility study providing the information delineated in Table 11.2. The probability of technical success and any technical red flag areas should be stated. The recommended processing and types of equipment should be identified. In general, the written technical feasibility study should provide enough information to reach a decision to move on to the next step of in-depth product development.

SELECTING AND ORGANIZING THE TEAM

SELECTION OF TEAM MEMBERS

The degree of success of the new food product development activity is highly dependent on the appropriate selection of team members. Not only should the members be competent and highly skilled in their particular discipline, be it technical or nontechnical, but they should also work harmoniously

TABLE 11.2. **New Product Development. Phase 1: Technical Feasibility Study—Expected Results from Technical Feasibility Study.**

A. Review of prior art/patent literature
B. Overall assessment of chances of technical success within specified framework
    1. Potential technical red flag areas
    2. Estimated cost/time requirement to circumvent red flag areas
C. Provide a range of prototype products based on ingredient cost/quality relationship
D. Breakout of cost formula
E. Ingredient specifications
F. Estimated shelf life
G. Technical recommendations on types of packaging alternatives
H. Preliminary estimate of production requirements
    1. Processing considerations
    2. Types of equipment required
I. Comparison of alternative production facility candidates based on expected production requirements
J. Estimated cost, time, and extent of further development work required based on preliminary assumptions

together supporting one another in pursuing defined objectives. Generally, the team leader is selected from the marketing department, but as the progress of the new food product development unfolds, this leadership may be shared to a greater degree with the technical management as the product development activity moves into process development, process engineering, pilot plant, and eventual commercialization.

DEFINING THE BUDGET

The new product plan is generally prepared outlining the objectives or rationale for the new product development, the expected impact on the business, and the probability of marketing and technical success. This provides a basis for identifying the required budget and the resources for carrying the new product development activity forward. The product plan will also identify agreed upon schedules, time sequence charts, Gantt charts, and the mutually agreed-upon milestones (gates) and essentially provide the road map to bring the product forward to full commercialization.

It is imperative that realistic budgets be formulated and adhered to as the product development moves from concept to commercialization. In small entrepreneurial companies, new food product development schedules may be compressed. The levels of bureaucracy are not as cumbersome, and a sense of urgency can be more firmly established. The tendency to use outside sources to supplement in-house resources may be easier to establish within smaller companies. In many cases, the innovative spirit required to bring the products forward at a fast pace may in fact be more resident in small, more flexible and nimble companies than the larger companies. In fact larger companies are increasingly trying to emulate the smaller companies in developing the desired flexibility and lower bureaucracy level in introducing new food products.

PHASE I: PROTOTYPE DEVELOPMENT

Following a brief given by marketing, a R&D or equivalent technical group generally prepares prototypes using the resources of research chefs, test kitchens, and food scientist/technologists working at bench level to develop preliminary initial formulae. The objective is to come up with prototype products that can be presented to marketing to initially determine if they are in line with marketing's expectations. The final jury, however, is not comprised of marketing or business or technical people, rather of consumers. It is imperative that the input of the consumer be factored into the initial preparation of these prototypes, and that repeated testing of formulae be performed using consumer panels to get the feedback from the consumer as to the acceptability or desirability of these prototypes (see Chapter 12). Based on the feedback from this testing, adjustments then must be made to optimize the sensory properties of the food product, that is, the flavor levels, sweetness level, the amount of seasonings, texture, thickness, color, etc., of the product.

SYNERGY OF CHEFMANSHIP AND FOOD SCIENCE

An approach used by some companies is to have chefs in the laboratory prepare the desired prototype product utilizing ingredients and recipes to deliver a product prototype that is deemed acceptable by marketing and confirmed by preliminary consumer panel testing. These chefs, working in conjunction with food scientists/technologists, modify the recipe and eventually together come up with a food formulation that is based on the numerous chef preparations and the practical considerations of the food technologist in developing a product at a reasonable cost that will perform under the desired shelf life constraints. Regardless of the approach, one thing must be kept in mind constantly, and that is the product must taste good, and so the flavor systems and the flavor delivery systems are of paramount importance in development of the prototypes, and eventually the finished product, to ensure the primary desired product attributes.

INITIAL FORMULATION

Perhaps this activity can best be envisioned by some examples. Let us assume that the defined task is to deliver a superior low-fat salad dressing product that will possess superior mouthfeel, texture, flavor definition, and balanced flavor compared to the known competitive products on the market today. Initially, a comparative evaluation of the existing competitive products in the marketplace will be carried out identifying the strengths and weaknesses of such products. The technical group will then conduct a search to identify emerging technologies that could provide the superior product attributes that are desired. Technologies could come from internal development capitalizing an ongoing long term research program designed to identify and develop low-fat texturizing systems and/or utilization of external technologies identified as potential candidates to deliver the desired product attributes. Increasingly, R&D organizations are probing the potential of bringing in the external technologies to augment their internal development programs or to capitalize on an identified technology to provide a proprietary positioning. Regardless of where the technology comes from, the important challenge is harness it as quickly as possible, fit it into the matrix, and use it to deliver the proposed and desired product attributes.

Such technologies can come from a variety of food and/or nonfood industries, applying the concept of lateral thinking to utilize a particular technology in a new and innovative approach. For example, certain drug delivery systems can be utilized and defined to provide flavor delivery systems. Microencapsulation technologies can be utilized to provide the desired point of difference in releasing a flavor on demand at the point of preparation or the point of consumption.

If at all possible, the technology employed should be proprietary technology, either on the basis of acquiring the technology, licensing of the technology, or achieving a patentable position.

The food product development team will, in many cases, require development people to work closely with food ingredient suppliers, in some cases bringing in the suppliers as members of the team, or testing a number of ingredients that were developed specifically to achieve required product attributes.

SHELF LIFE CONSIDERATIONS

By this time the decision will be made as to how the product is to be marketed, that is, shelf-stable, frozen, or refrigerated. It is advisable to present the product to the consumer in this format taking into consideration the method of preparation. For example, if one is designing a dehydrated soup mix it is very important to establish at the outset the time and convenience of the preparation of the resulting quality of the final prepared soup. Based on the refinements of the prototypes, intensive sample preparation at the bench level is undertaken to develop a product as close as possible to the objectives and the product attributes. At this point, decisions have been made with regard to the specific ingredients to be used, the cost constraints, and the parameters for the product preparation at the consumption venue. Once the modes of preservation and marketing are identified, shelf life testing should be initiated to determine what organoleptic and biological problems might be encountered under such conditions. Accelerated shelf life testing can be carried out where the temperature of the product is raised to determine, for example, if there could be any substantial problems with regard to chemical interactions, autoxidation of oils, and any potential microbiological problems. Once the formula has been agreed upon, then an extensive shelf life study should be carried out prior to test marketing of the product (see Chapter 13).

PRODUCT SAFETY

The format for delivery of the product to the consumer must be carefully defined. Should the product be shelf-stable? Is it to be frozen, refrigerated, dehydrated, or heat processed to deliver the desired product quality aligned with the desired product safety? Above all, the product must be safe for consumption under the defined processing parameters, product preparation, and product consumption parameters.

TESTING OF THE PROTOTYPES

Once initial prototypes have been prepared and tested in the laboratory, it is necessary to return to the consumer with these prototypes to see if, in fact, these prototypes match the concepts that were originally discussed with the consumer groups. An approximate costing of the prototype formula is prepared prior to this testing to ensure that the delivered product will be within the guidelines of the cost parameters identified for the product. Panel testing is used to obtain feedback concerning confirmation of the concept, the overall product acceptance, the level of enthusiasm expressed for the product, some indication of intent to purchase and repurchase, and the flavor intensity levels and mouthfeel and texture parameters (Chapter 12). This information is used to refine and fine-tune the formula. Generally several sensory panel tests are required to get it right.

PHASE II: BENCH LEVEL FORMULATION

SELECTION OF INGREDIENTS

A delicate balance must be maintained between the cost of selected ingredients and the desired quality of the product. The projected cost of the product is defined in conjunction with the anticipated margin and selling price of the product. The formula must be fine-tuned to ensure that the anticipated ingredient from the suppliers will perform and that sufficient quantity of the ingredient is available within the constraints of the cost structure of the product. For example, it is extremely risky to formulate a product with an ingredient from a flavor house or food supplier that is currently only available at bench level or pilot plant quantities. The danger is that the ingredient supplier stream may not be adequate to keep up with the test market and eventual nationwide or worldwide roll-out.

PANEL TESTING

Once in-depth product development is initiated it is necessary to go back to the consumer with these samples to see if, in fact, they match the concepts and prototypes that were originally discussed and presented to the consumer groups. Panel testing is used to obtain feedback concerning confirmation of the concept (Chapter 12). Read out is obtained concerning the overall product acceptance, the level of enthusiasm expressed for the product, some indication of intent to purchase and repurchase, the flavor intensity levels and mouthfeel, and texture parameters. This information is used to refine the formulation. Generally several sensory panel tests are required to optimize the formula.

PHASE III: PROCESS DEVELOPMENT

Once bench level samples and formulae have been agreed upon by the product development team, and proper consumer input has been factored into such development, the focus shifts from the bench level to the pilot plant to scale up the formulation and carry out test runs based on the selection of the equipment and unit operations required to produce the product. Generally a decision is made as to whether the method of production will be batch, semibatch, or continuous, and the selection of the equipment will be in line with that decision. The objective is to obtain the best quality product with the most economical throughput through the pilot plant operations. At this point the food engineer has to work very closely with the food scientist/technologist to ensure that the desired quality is preserved and that the resulting product off the pilot plant line is as close as possible to the standard samples. Inevitably, with the scale-up problems involved, the product off the pilot plant line will have to be fine-tuned to obtain as close a match as possible to the bench samples. Many times, it is impossible to come up with an exact match and compromises must be made in order to come up with a reasonable product that is as close as possible to the originally defined criteria.

Many food companies will use the pilot plant produced product for test marketing prior to making a decision to invest capital to modify or build a plant to produce the new product. An alternative is to use toll processors (i.e., contractors) to produce test market quantity to get enough data to justify further capital investment. Generally, further panel testing is required to fine-tune the initial finished product off the processing line.

The product development team will put together an estimated FOB plant cost based on the selected ingredient formulation throughput and fixed and variable cost required to produce the product. Specifications will be firmed up. A quality assurance (QA) manual will be drafted. If the product is to be produced in an existing plant, the product development team will take up residence at the plant working with the plant personnel to ensure a smooth transition to pass the responsibility of producing the product to the plant personnel. The product development team leader will work closely with the plant personnel to ensure that he/she is fully satisfied that information concerning the product and the knowledge required to produce the product has been adequately transferred to the plant level.

PROCESS OPTIONS

Once the bench prototypes have been prepared and approved, translation from bench preparation to pilot plant production involves identification and selection of required processing equipment to produce the product. If the new product is patentable, a preferred approach would be to apply for and try to obtain a patent involving both the product and process if indeed novelty is involved in the processing of the new product. Primary equipment required, such as various mixing vessels and pumps, are generally available at the pilot plant or at the commercial production site. Novel processing equipment can be custom designed by having R&D and process engineers work closely with the equipment suppliers. In some cases, the required processing equipment may actually be built from scratch by the process/engineering department. Throughput designs, flow rates, ease of cleaning the equipment, computer operation of the equipment, etc., are all key considerations in designing and laying out the required processing line.

PRODUCT SPECIFICATIONS

Once the product formula has been agreed upon and the method of product has been agreed to, the accepted practice is to prepare a product "write-out" that will detail all of the specifications for the product, the formula of preparation, the processing conditions, and the QA, and/or quality control (QC) specifications for ensuring the expected quality of product. The R&D department, working in close conjunction with marketing and the process engineers and the plant management, generally prepares it. It constitutes essentially the "Bible" for the products with no deviation from the product write-out permitted without express approval and sign-off of marketing, R&D, process engineers, and plant management. R&D people will travel to the plant to assist the plant personnel in producing the product and will be in charge for the first test market production runs after which time the responsibility will be turned over to the plant management with R&D assisting in a consulting capacity. The QA department, generally at the division level or at corporate level, is responsible for preparing all quality assurance delineation for the new product introduction, and the QC department of the plant is normally responsible for the day-to-day monitoring and testing of the daily production of the product at the plant.

PHASE IV: TEST MARKET PRODUCTION

Test market production of a new product may be carried out using toll processors or using pilot plant facilities at the company's process development facilities or utilizing existing processing lines in the company's plants.

TOLL PROCESSING

If this approach is selected it is necessary to diligently screen the potential contract processors to determine the suitability of the processing facility to ensure that good manufacturing practices are in place, that the facility is acceptable as a food processing site, and that adequate controls are in place to ensure the confidentiality of the test market production. A contract must be negotiated that will provide the desired insurance of consistent product quality and safety, which requires microbiological monitoring and testing of the products as specified by the R&D department. Moreover, the R&D department will be on the site working closely with the contract processor to ensure that the proper product quality is obtained. In negotiating the contract with the toll processor it is imperative to ensure that the cost parameters are clearly understood, responsibilities are delineated, liabilities are properly established, and that both management of the food company and the toll processing firm are committed in writing to provide the required input to ensure a proper test marketing success.

PILOT PLANT PRODUCTION

If this mode is selected generally the R&D people will work on line with the pilot plant personnel in making sure that the established processing line is operating correctly, that all controls are in place, and that consistent product can be produced from this site. The monitoring of the test market production is provided by the R&D department utilizing in-house analytical and microbiological facilities supplemented by outsourcing, if necessary, to carry out some of the required test monitoring.

TEST MARKET PRODUCTION IN AN EXISTING PLANT

If this approach is selected, R&D management must work closely with process engineering and plant personnel to either produce the product on an existing or modified processing line or a newly designed and constructed processing line. In either case the product write out is followed very precisely. Any changes that are made must be agreed upon by all parties involved, and the personnel involved are committed to be on site as long as necessary to ensure a smooth roll-out of the initial test market production.

COMMERCIALIZATION/LIMITED ROLL-OUT

Once the test market production has been successfully completed and the test market has resulted in positive response, the product development team transfers the responsibility for commercialization of the product to plant operations. The hand off is facilitated by the product development team ensuring an orderly transfer of all the product information in the form of a product write out. The product write out is essentially the ''Bible'' for the product, and contains specific details with regard to the ingredient specifications, the product formula, the processing conditions, the quality-control and quality-assurance procedures required, in short, everything to ensure that the product is correct and will have the desired product quality. Sometimes the commercialization is initially limited to a regional launch, followed by a national roll-out, and in some cases the company will move ahead with an initial national roll-out. The plant manager assumes full responsibility for continual production of the new product.

TECHNOLOGY SKILLS AND PRODUCT TRANSFER

Given that the product is a success, the company may see fit to launch the new product in other markets and even set up the new product in various countries to exploit the technology more fully to obtain increased sales. Once the required information is furnished to the new location, the product development team may be required to work with personnel and management at these locations. The product usually has to be fine-tuned to the palates of the particular market involved. Ingredients may have to be substituted, but, in general, the intent is to ensure the adequate transfer of the product knowledge, the skills, and the technology to capitalize on the investment made by the company.

PRODUCT AUDIT/FOLLOW UP

Senior management is extremely interested in the potential success of the launch of the new product. Comparisons will be made between the projected return on investment, the sales volume, and the repeat sales. All of this is rigorously tracked to determine if the company has a successful food product on their hands, or not. Lessons learned from such an audit can be valuable providing a basis for honing skills in pursuing the next new product and to carrying it forward to commercialization.

In some cases, the product development team will follow the product and actually provide the resources for a new business unit or a newly formed division based on how successful and broad based the product commercialization is. In other cases, the product development team will regroup or be partially reorganized to tackle the daunting challenge of doing it all over again and bringing out another new product that hopefully will be a success for the company.

Costing of the prototype formulation is prepared prior to this testing to ensure that the delivered product will be within the guideline of the cost parameter identified for the product.

One of the most difficult tasks in setting up a new product development effort is to achieve a satisfactory interaction between people of very diverse backgrounds, training and personalities. Experience shows that many times the technical individual starts his/her career in industrial new product development with the perception that new product development really centers around the technical activities involved in the process. Perhaps this is because the technical individual has not had that much exposure in general to business training and does not have an understanding of the total process involved in developing and marketing new products. At any rate, it is not too long before one realizes that in order to be effective in corporate structure, one must provide meaningful inputs to the management responsible for directing the overall effort.

It is obvious by now that new product development involves a multi-disciplinary approach where such teams must work together and communicate frequently to accomplish agreed upon objectives. There must be a continuous exchange of information and activities and there must be mutual trust and respect established between the individual members of the team.

The team effort referred to at the beginning of this chapter is much more important in carrying out product development to develop truly innovative products that will have a defined point of difference from the competition in the marketplace.

TIME CONSTRAINTS

The product development process is indeed an intense, complicated activity and operation to manage. Selection of the product of the product development team is crucial not only in terms of identifying the multifaceted competency of each member, but also their compatibility and their ability to interact and work together under very tightly controlled time constraints. The product development team works in an environment that is really a microcosm of a small business and functions best with a certain degree of entrepreneurship infused into the group.

SHELF LIFE TESTING

Early on, as indicated above, the decision must be made as to how the product is going to be distributed. It is important to translate the prototypes into the desired mode of preservation and distribution as soon as possible to determine what effect the mode of processing and preservation will have on the finished product quality.

TEST MARKET PRODUCTION IN A NEWLY DESIGNED AND BUILT FACILITY

In rare cases this approach is used to establish a greenfields plant to produce the new product. In this case all of the foregoing must be carefully integrated to ensure that the production facility is acceptable and that consistent quality product will be rolled off of the processing line.

PACKAGING

An integral part of the product is in fact its package. The package design for the product is of paramount importance to ensure the compatibility of the package with the product characteristics providing for attractive marketing, safety, and quality assurance of the packaged product. The R&D packaging department is closely involved in product development and their input is very important to the whole procedure of product development and eventual test market roll-out. Details of this interaction and input are covered in Chapter 8 of this book.

PROJECT PLANNING

It is imperative to establish at the outset the ground rules for how the product process will proceed. This requires determining that the project leader of product development will be rotated to the marketing, R&D and process engineering, and production groups as the project moves from concept to commercialization, or whether one project leader will be in charge of all phases of this development. Whatever the decision it is important to establish who the key players are going to be and what their roles will be and the established ground rules as to how the project product development will be carried out. In most cases, the senior leader for the product development will be selected from the division of corporate marketing staff and the leadership role will be shared at least initially with the key management in the technical department. The product development team must be clearly identified, must be adequately communicated to corporate management, and be firmly supported by the management at all levels. The members of the product development team must feel that it is their baby in fact, so that the team operates in the spirit of entrepreneurship ideally under a corporate umbrella and functions as if they were bringing on a new business for the company, particularly if the product is considered to be truly a innovative. A very rigorous approach generally is not required to develop line extensions or brand expansions but in the case where a major innovative product is desired, and the commitment is there to commercialize such a product, the endeavor must be fully recognized and supported by management as a major initiative during the planning cycle of the company. Generally, some kind of a time sequence chart is prepared using the required inputs of the team, e.g., a Gantt chart that will identify what particular elements of the product development schedule will be executed and by whom, and completion dates firmly identified and agreed up by the members of the team. The project coordinator may be appointed to ensure that all inputs are dovetailed and project is kept on schedule. Progress will be reviewed at agreed upon milestone dates, interim reports will be prepared and circulated to members of the team and to the supporting managements. Senior management will be involved from the standpoint of granting approval and lending support to the team effort and will be apprised on the basis of timely reports and presentations as the development process unfolds.

BUDGETING

However, if the new product involves the input of substantial emerging technologies and innovative thinking, the resources for such must be harnessed very early in the process and can be available to management as needed. Product development is a very demanding process. Extreme and tight planning are required to establish what the expected lead-time would be in terms of introducing the new product. In fact how is the new product going to be protected, by filed patents or trade secrets? Such a decision is really a function of the company's business objectives and the timing involved. In any case it should be expected that once the new product is launched the competition will try to come out with me too products either by trying to get around the patent if the patents have in fact been issued, or to come out with their rapidly developed versions of a similar product. Part of the new product may involve the licensing of technologies. From other sources it must clearly be identified to what the lead-time is and the commitment and the expiration date of such an underlying patent. If the new product involves the development of a new ingredient based on certain "mother" patents, it is desirable to carry out the work to ensure that the patent portfolio can be enlarged and extended by coming up with new applications and new innovations building on the so-called mother patent.

ONGOING DEVELOPMENT

Once the product is launched, depending on the degree of initial success, it may be necessary to reformulate the product to provide for a fine-tuning of consumer preferences, pricing strategies, and or line extensions of the new product.

DISPOSITION OF THE PRODUCT DEVELOPMENT TEAM

The product development team which has been assembled on an ad hoc basis in order to deliver the product to the market place may follow the product into commercialization depending on the degree of innovation of the product, whether it is an ingredient or new food product, and in some cases a new division or new business unit may be organized to support the commercialized product. In that case members of the product development team may follow the course of success of the product and be employed as an integral part of that emerging business unit. Otherwise they may revert back to their respective organizations and begin the arduous task of developing another new product, which will require similar disciplines that were adhered to in bringing out the previous product.

GOVERNMENT REGULATIONS

If the product in fact is a new ingredient, it may be necessary to travel along the tortuous route to getting government approval, that is FDA approval, for utilizing the ingredient or to assume the risk of utilizing the new ingredient in the new product. At any rate, substantial documentation of the potential risk involved in using the ingredient, required toxicology studies, etc., are all required. All of this must be carefully prepared and be available to the regulatory agency for the company seeking approval for the ingredient or for defending its position in launching the new product. Primary concern, obviously, is that of product safety, whether it is from a perceived microbiological threat, chemical interaction, preservation, etc. The company launching the new product assumes liability for the product. However, all facets of the product development team are involved in sharing such liability, including the toll processor, ingredient suppliers, etc. It can be seen that new product development can indeed be a risky game, but the rewards can be very high. For those who play it well, it is indeed a satisfying experience to reach the marketplace and achieve a truly competitive advantage as compared to line extensions and brand expansions.

CHAPTER 12

R&D-Driven Product Evaluation in the Early Stage of Development

HOWARD R. MOSKOWITZ

Among the many steps in food product development is the monumental task of measuring the target consumers' response to the concepts, prototypes, pilot plant outputs, initial production line efforts, and ongoing packaged products. Each phase requires different information to be provided to the formulators, refiners, optimizers, marketers, and their associated team members. The variety of data to be generated from consumers means that a range of test protocols should be employed in several venues: laboratory, office, central location, consumer homes, etc. Tests may be descriptive, ranking, scaling, qualitative, semiquantitative, or quantitative depending on the stage of development. As with all testing intended to be predictive of consumer behavior, sensory evaluation is not perfect, but with the proper direction it can be a powerful tool to guide new food product development.

INTRODUCTION

THIS chapter deals with consumer-based tests of products at the early stage of development. By early stage we mean R&D development, where the developers has many available options and can use research to guide product changes. The ultimate objective is to achieve a desired sensory profile and/ or level of consumer acceptance.

WHAT HAPPENS IN CORPORATIONS TODAY IN THE PRODUCT DEVELOPMENT PROCESS?

Over the past decade developers, and indeed general managers, have come to realize the critical importance of consumer feedback at the early stage of

277

development. In years gone by, product development was often conducted in an information vacuum. The prevailing attitude was experts or top management knew pretty much what was best for consumers, and that the development process would be just fine if the developer simply combined management dictates with creative intuition. When products were scarce, such arrogant indifference to consumers did not matter. Product shortages guaranteed that consumers would purchase the items that were offered. Today things are quite different. The market confronts the consumer with a plethora of products. Retailers have limited shelf space. To insure commercial success it is vital that throughout the development process consumer feedback be used to keep the product acceptable and thus to afford it the best opportunity to succeed.

Given this growing importance of consumer feedback as a key to product success it should come as no surprise that the field of early stage product testing has become the focus of attention by management, marketers, and developers alike. Increasing numbers of articles deal with the consumer evaluation of products, and with consumer guided product development. Whereas two decades ago such early stage testing was subsumed under the rubric of "flavor" or "texture" evaluation, today the testing has earned a place for itself in the armory of manufacturer's tools for ongoing quality under the rubrics of "total quality," "house of quality," or "the voice of the consumer" (Amerine et al., 1965; Moskowitz, 1994).

THE HISTORY AND GROWTH OF THE EARLY STAGE TESTING INDUSTRY

As intuitively reasonable as early stage testing for development may seem today, it was not always the case. Only in a few industries did developers realize the importance of guidance. A major step of historical importance was the establishment of product evaluation panels by the Quartermaster Corps for the U.S. Army. The U.S. Army funded ongoing consumer testing of its products to ensure adequate military acceptance. To this end the military founded a food acceptance laboratory, located first in Chicago and later at the U.S. Army Natick Laboratories in Massachusetts. Product testing was institutionalized as a standard step both in the development of military rations, and for the assessment of commercial food items procured for army use. The acceptance testing procedures slowly diffused into industry as well. Evaluation, here, was often used for finished products rather than for early stage development, although the test protocols developed at the Quartermaster Corps were later adopted for commercial product developmental.

As might be expected the story is less clear in the commercial sector, due to the different corporate cultures, the different needs of developers, and the influence of companies such as the Arthur D. Little Inc. Food and Flavor group on corporate test procedures. In the 1950s the early stage testing had been confined to expert panels (Cairncross and Sjöstrom, 1950). Occasionally a (then small) market research company might be commissioned to conduct a consumer test of a product, but for the most part the testing business was informal and undeveloped. Consumer tests were often done by cigarette companies, where a single share point might mean several millions of sales dollars. However, for most products consumer evaluation would be done less formally, less extensively, less expensively. The testing business simply was not as developed, because demand in most categories exceeded supply.

By the late 1960s, however, sensory analysts in the R&D laboratories and marketing researchers reporting to the marketing department had established the need for early stage testing to guide development. Recognition of the need for testing was not at that time, however, due to economics. The economy was still growing and consumer demand was still rising. Rather the recognition was a sociological phenomenon. Practitioners were transforming the testing business from an informal, clerical task to a profession, loudly proclaiming this developing professionalization in the scientific literature and in trade meetings.

In the early 1960s the American Society for Testing and Materials (ASTM) in Philadelphia recognized the importance of such guidance testing by forming Committee E-18 for sensory evaluation. Most members of this committee had not been formally trained in product testing, but rather picked up skills and knowledge through practice (on the job training), and through the few manuals on testing that were available. Later that decade sensory analysts belonging to the Institute of Food Technologists (IFT) formed the Sensory Evaluation Division to recognize and to promote product testing. A similar pattern of professionalization would emerge but delayed by 10 to 15 years in the health and beauty aids industries.

By the 1970s the testing business was in full swing, as company after company introduced formal procedures for early stage product evaluation. Most companies used expert panels, investing considerable sums of money to develop and maintain these panels. Some other companies recognized the need for consumer research, and commissioned outside research companies to provide rapid, low cost tests of prototypes with consumers. (These are called research guidance tests.) The tests began to supplant the standard expert panel data, as developer after developer recognized the need for consumer input, rather than experts.

By the 1980s the early stage testing business for development guidance began to boom. Developers began to rely more and more on consumer data, whether of a simple or a more sophisticated nature. Many developers could not easily apply the data provided to them by practitioners who used expert panel techniques. The data from experts did not link to consumer testing and could not easily predict consumer acceptance. All too often the reports were long on statistics and short on application. Consumer research data, however, provided an easier vehicle for development, perhaps because it was less sophisticated, less technical, less statistically oriented. In the main this consumer research provided a report card to management.

By the 1990s early stage testing experienced its greatest spurt. Management in company after company recognized the importance of consumer acceptance, the need to maintain sensory quality, and the need to cost reduce products so as to maintain or increase margins. By this time the tools were in place for guidance research to play a key role in maintaining product quality, and helping ensure success. Testing was institutionalized in many companies. Expert ''profile'' panels were used less often, and in some cases disbanded entirely. Developers began to look more and more at consumer guidance for early stage development.

ON THE DEVELOPING IMPORTANCE OF CONSUMER INPUT AT THE EARLY STAGE

A casual perusal of the scattered literature of sensory analysis and product testing will reveal that over the past thirty years studies of consumer response to products have achieved more and more interest, whereas studies involving the objective classification of a product's sensory attributes, and the profiling by experts, has diminished in importance. This is not to dismiss the relevance of expert panel data and profiling but rather to emphasize that interest shifted from description to marketplace acceptance, and the attempt to predict that acceptance as early as possible in the development cycle.

There are three reasons for this trend:

(1) Bottom line orientation: manufacturers have begun to realize that it is consumer acceptance, first and foremost, which determines product success. All other measures in research are surrogates for consumer acceptance or attempts to look at factors that could influence acceptance. Thus when research budgets are cut or else maintained at a constant level, it is consumer acceptance tests that must be funded above all other forms of sensory analysis research.

(2) Speed to market: shortened product life cycles make it ever more vital to get into the market quickly. Training expert panels takes time. More and more these panels are being used for sensory-based quality assurance and control, rather than for product development. Developers are going right to the source of product acceptance—the consumers themselves—and eliminating as many nonessential steps as possible in the introduction of a product to the market. Even rigorous acceptance tests are occasionally eliminated in favor of a few focus groups to confirm that the product is on target.

(3) Overhead cost of evaluation: expert panels are expensive to maintain. In an era of downsizing and rightsizing the expert panel becomes a target for cost

cutters because often it cannot be shown to affect the bottom line for the company.

SOME KEY GENERAL ISSUES

No matter what product is being developed, no matter how fast the time track of development may be, certain considerations repeatedly arise regarding the rationale for the test, the methods used to conduct the test, and the analyses. This section deals with these issues from a broad perspective.

THE PERENNIAL CONTROVERSY OF PANELISTS: EXPERTS VERSUS CONSUMERS

There are two major schools of thought for early stage testing. One school comprises those who believe that development should be guided by experts. These experts should have a sense of consumer tastes, and equally important be able to communicate these consumer desires to product developers. This school of thought believes that the expert possesses a superior ability to guide development. There is no solid support evidence for this claim of superiority. Rather, the claim comes from the tradition of the expert wine taster, the brewmaster, the flavorist. Although many practitioners today would not go so far as to espouse this position in the 1990s, the position that the "expert is king" has held sway for a long time in product development, and continues to do so in a number of companies.

Among those who prefer to use experts at the early stage are the flavor researchers (Caul, 1957; Clapperton et al., 1975) and sensory researchers who do training (Meilgard et al., 1987; Stone et al., 1974). These practitioners feel that the early stage development process is best served by individuals who possess a common vocabulary, anchored to physical standards. The training can take months, weeks, or even days. The flavor profile (Caul, 1957) relies upon the consensus opinion of a small group of well-trained experts. With the growing interest in quantitative methods, researchers at Arthur D. Little, Inc. (developers and chief proponents of the flavor profile) have modified the method to accommodate statistics. The modification is called PAA (Profile Attribute Analysis). The QDA method (Quantitative Descriptive Analysis) (Stone et al., 1974) is fundamentally more quantitative, and relies upon statistical analyses of the results, such as means and analysis of variance.

At the opposite extreme are researchers who use consumers for almost all tasks. The author (Moskowitz, 1983; Moskowitz, 1985; Moskowitz, 1994) has been among the most vocal of these consumer-oriented researchers. Moskowitz and a host of other market researchers and experimental psychologists have argued that a well instructed consumer can perform as well as an expert. Indeed recently Moskowitz (Moskowitz, 1996a) developed a reverse engineering technique which allows the product developer to relate profiles of ratings assigned by experts to profiles of ratings assigned by consumers, and vice versa. A consequence of this development is that it does not matter whether the early stage data are obtained from consumers or from experts. One data set can be transformed into another data set by straightforward equations. Figures 12.1(a) through 12.1(c) show how well correlated are data from consumers with data from experts for a gravy sauce product. High correlations between expert and consumer ratings for a single attribute across products suggests that consumers and experts are interchangeable, with the caveat that as a result of training experts have a better descriptive vocabulary and sharper tuned ability to describe products. For many common sensory

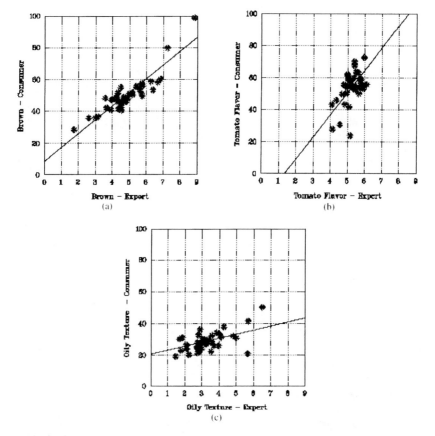

FIGURE 12.1 Scatterplots of consumer versus expert ratings for 37 gravy products, for three attributes: (a) brownness of color (appearance), (b) strength of tomato flavor (taste/flavor), and (c) oiliness of texture (texture/mouthfeel). The experts used an anchored 1 to 9 scale, the consumers used an anchored 0 to 100 scale. All correlations exceed 0.90.

attributes experts and consumers appear to do equally well (Moskowitz, 1996a; Moskowitz et al., 1979).

TEST VENUE: HOME USE VERSUS CENTRAL LOCATION TESTS

Much of the traditional early stage flavor and texture work was done in an unstructured, informal fashion in the development laboratory, using one's colleagues as the panelists. Before the testing field grew into a profession with standards it was common to "test" a new formulation by tasting it oneself, giving some to one's laboratory associates, or to one's clerical staff, soliciting opinions about direction (viz., was the product and thus the developer on target), and then moving on to make the requisite changes. Developers didn't concern themselves with formalized instructions, did not worry about the validity of what they were doing, and generally did not worry that their informal tests "at the bench" were anything but useful to guide them.

This informal "benchtop" testing has evolved into three distinct and formalized ways to test products. The first method is the "central location intercept test." The interviewer has a booth in a well trafficked area (usually an indoor shopping mall). The interviewer intercepts potential panelists, screens individuals for appropriateness for the test, and then invites the person to participate. The intercept test must be very short, permitting the researcher to have a panelist evaluate two or at most three products. For more products the interviewer must pay the panelist or use more panelists, with each panelist testing a partial set of products. Intercept tests are very popular because they are fairly cheap. They do not, however, allow for the best quality control of the interview because the quality of the interviewing staff varies from market to market, the turnover in interviewers is high, and the interviewer is rushed for time in each interview in order to make quota. Panelists don't want to participate because they are not paid, and because the interview often interferes with other tasks that they must perform when they go to the shopping mall. The interview becomes more of a nuisance than anything else.

The second method is the "central location pre-recruit" or more simply (in the English style) "hall test." For hall tests the researcher recruits a consumer to participate, gives the consumer several products to test in some type of randomized order, and then obtains ratings of the product(s). The hall test can be conducted with a fairly high degree of control, the interview can be observed, and the results come in quite quickly. The panelist can test many products, not just one or two as is the typical case in, and the limitation of, the central location intercept test. In the most advanced hall tests the consumers participate using computers linked together by a local area network, so that the results are instantaneously computed after the last panelist has finished.

The third tradition of testing has the consumer evaluate the product at home under the more ordinary venue where consumers typically consume the product. The consumer is either called afterward by phone to report their rating, or invited back to a central location for a personalized interview. The home use test can accommodate several products, as long as the researcher takes care to ensure that adequate controls are maintained. Controls include separating the products by a day or more, dictating the order of products to be tested at home and creating a questionnaire that the consumer understands. Home use tests are often used for products that the consumer has to prepare, where consumption is relevant rather than just taste alone, and when there is the possibility that with repeated consumption the consumer may become bored with the product. Repeated consumption of the same product at home can ferret out the boredom factor.

Which test venue is better? In early stage testing there is no right nor wrong location for the test. Whichever method does the job is best, but always with an eye to efficiency and the cost/benefit ratio. Under the most normal of circumstances home use tests provide more valid data, but the data is most expensive, comes in more slowly, and may not enable each panelist to test every product. From the statistical viewpoint it is ideal to have each consumer test every product. The power of testing products in their natural consumption state (viz., at home) is offset by the weakened statistical power of the home use test since for large scale tests the panelist cannot evaluate all of the products.

There are other differences among the test venues. These differences can become quite serious because they affect the scores received by the products, and thus affect the decisions made when directing product development. Home use tests may inflate the scores of products, whereas hall tests may deflate these scores. Often products which score well at home score poorly in central location. A decision to "go with a product" because it scores well in home use tests may mislead the developer, who always looks for positive results. Part of this difference between home use tests and hall tests can be traced to the number of products tested, and the frames of reference brought to bear by each test venue. In home use tests the researcher tests only one or a few product. Consumers do not have a frame of reference provided by other products, and tend to uproot products, especially the product tried first. If there is one product in the home use test then 100% of the products are tried first. If there are two products, then 50% of the products are tried first. In hall tests, however, where consumers have many products they may up-rate the first product, but accurately rate the second and other products. Consequently, in hall tests only a portion of the products tested are subject to the up-rating bias. If the hall test comprises eight products, then only 12.5% of the products will be tested in the first position.

A brief comparison of the pros and the cons for hall testing versus home use testing appears in Table 12.1. It is left up to the reader, faced with the actual real world problem, to decide. For the most part hall use tests and central location tests will yield the same or similar answers—winning products

TABLE 12.1. Central Location versus Home Use Tests: Advantages versus Disadvantages.

Aspect	Central Location	Home Use Test
Control over ingestion and evaluation	High	Low
Test site versus typical consumption	Unnatural	Natural
Amount of product to be evaluated	Limited	Unlimited
Number of different products tested	Many	Few
Measure satiation and wearout	No	Yes
Mix many concepts and many products	Yes, easy	No, hard
Number of panelists/product for stability	20–50	50–100

will remain winners, losing products will remain losers. There are, however, certain products which may appear to be winners in a short test but may become less acceptable when the panelist is forced to live with them on a daily basis. In these cases it is vital for the test to be home use because a central location hall test may provide incorrect answers. If time and resources permit, an alternative strategy uses hall tests to screen down to a few products, and then follows with a home use test of the winning (and perhaps modified) products in a home use test.

RESEARCH OBJECTIVES: WHAT DOES THE TEST MEASURE?

Product sensory and acceptance tests are instruments, only as good as the researcher who uses them, the validity of the test design, and the appropriateness of the questionnaire and the panelists. The test must be designed to accord with the goals. If the objective is to describe the characteristics of the product, then the test should comprise a descriptive analysis of product attributes, whether these be appearance, aroma, taste/flavor, texture, or any combination thereof. The panelists describe their perceptions, and may rate the characteristics on a scale to show the degree to which the characteristics are present in the product. If the objective is to determine whether or not two products differ from each other, then the test should assess difference, either directly or indirectly. The test should allows the researcher to state with confidence that two products either differ or do not (all-or-none statement), or measure the degree to which products differ from each other (scaling). If the objective is to determine the amount of a characteristic present, then the test is a profiling/scaling one, in which consumers assign numbers to reflect the degree to which a characteristic is present in the product. (This may at first appear to be similar to descriptive analysis or classification. There are differences. Descriptive analysis concentrates on the attributes which describe the product, and is primarily a task of classifying. Rating the attribute is secondary. Profiling/scaling requires panelists to use a predefined scale to rate the products, and is primarily a quantitative task.)

SCALING—THE PANELIST AS A MEASURING INSTRUMENT

Closely allied with the previous topic of what the test measures is the issue of how does the panelist measure his impression, and by so doing, quantify the intensity of the impression.

Description

The first task requires the panelist to delve into his/her conscious mind and mental lexicon in order to discover the correct descriptors of a product (descriptive analysis). The panelist simply reports what he perceives (viz., introspection). The issue of intensity of perception need not be included in description, and indeed descriptive analysis as yet need not have any numbers attached to the final report. However, researchers often combine descriptions with numbers in order to show how much of an attribute is present in a product. This combination provides a richer data. Table 12.2 shows an example of descriptive analysis of a gravy from both experts and consumers, respectively.

Classification

The panelist must sort a product or a set of products into two or more classes. Studies involving discrimination and threshold work are instances of classification. Threshold tests require the panelist to classify a single stimulus into one of two classes—perceivable or not perceivable. Discrimination tests require the panelist to classify two or more products as belonging to one of two classes—same versus different. Paired comparison tests requires the panelist to classify two stimuli into one of three classes: product X has more of a characteristic than does product Y, they have equal amounts of the characteristic, or product Y has more of a characteristic than does product X. Table 12.3 shows the results from a paired comparison test of two sausages on three attributes: overall liking, liking of appearance, and liking of flavor, respectively. Table 12.3 shows some of the raw data (which product was preferred).

Ranking

The panelist must place a set of products into a rank order, depending upon the magnitude of a characteristic. For instance, panelists may rank six candies in terms of degree of liking, from the candy liked most to the candy liked least. Ranking can be done for any attribute along which the panelist discriminates. Paired comparison can be viewed as a special case of ranking. When confronted with two products in a paired comparison test the panelist ranks the two to show which product is liked more. Conversely, ranking can be

TABLE 12.2. **Example of Descriptive Analysis Results for Gravy: Consumer Profiles versus Expert Profiles for Samples A–C.**

	Consumer			Expert		
Panel Product	A	B	C	A	B	C
Appearance						
Brown	57	43	62	5.9	4.4	6.6
Flecks	59	30	73	4.8	2.9	5.9
Tomato pieces/amount	22	42	64	2.3	3.6	5.7
Tomato pieces/size	18	35	40	2.3	3.6	4.2
Vegetable pieces/size	11	15	38	1.1	1.4	2.7
Aroma/Flavor						
Aroma	45	44	57	5.2	5.3	5.4
Flavor strength	61	57	76	6.0	5.9	6.3
Tomato flavor	55	59	49	5.2	5.6	5.0
Meat flavor	13	5	6	2.2	1.1	1.1
Mushroom flavor	8	4	6	1.0	1.0	1.0
Onion flavor	25	20	36	3.1	2.9	3.3
Green pepper flavor	11	9	30	1.0	1.0	1.0
Vegetable flavor	18	18	36	1.0	1.1	1.8
Herb flavor	52	36	70	NA	NA	NA
Spice flavor	NA	NA	NA	4.8	3.3	5.5
Oregano flavor	NA	NA	NA	3.1	1.8	3.3
Basil flavor	NA	NA	NA	3.1	2.6	4.5
Pepper burn	NA	NA	NA	1.2	1.1	2.6
Black pepper flavor	19	14	36	2.1	2.0	3.2
Garlic flavor	30	20	34	2.9	2.4	2.7
Cheese flavor	11	32	5	1.1	4.8	1.0
Salt taste	25	24	23	3.9	4.1	3.4
Sweet taste	37	28	19	4.8	3.5	3.1
Aftertaste	48	44	70	2.4	1.8	2.3
Sour taste	26	36	54	3.9	3.8	3.5
Oily flavor	26	23	24	1.1	1.6	1.7
Texture/Mouthfeel						
Crisp vegetable texture	35	34	40	1.1	1.2	2.7
Oily mouthfeel	29	24	25	3.8	2.9	4.0
Thickness	53	41	47	5.9	4.6	3.8

Consumers used an anchored 0 to 100 scale, experts used an anchored 1 to 9 scale. Numbers in the body of the table show the intensity of the descriptive attribute.

viewed as a set of paired comparisons, in which the panelist compares each pair of samples to determine which sample in the pair deserves the higher rank.

Scaling

Scaling requires the panelist to act as a measuring instrument. The panelist assigns numbers to the products to represent the amount of a characteristic

TABLE 12.3. Data from Paired Comparison of Two Sausages (1 = New Product Wins, 0 = Current Product Wins).

Panelist	Total	Appearance	Flavor
1	1	0	0
2	0	1	1
3	1	1	0
—			
29	1	1	0
30	1	0	1
Average	0.37	0.53	0.30
Difference	−0.13	0.03	−0.20

present. The numbers can fall on an interval scale (e.g., similar to the centigrade scale), where equal scale intervals are presumed to be psychologically equal. (For temperature all intervals of 10 degrees are equal, no matter where on the scale the interval is measured.) However, ratios of ratings are not meaningful. The numbers can fall on a ratio scale, such as the scale erected by means of magnitude estimation (Stevens, 1975). Magnitude estimation allows the panelist to assign ratings to stimuli so that the ratios of their ratings reflect subjectively perceived ratios.

Stevens (1946) presented a classification scheme for scaling that clarifies the distinction. This scheme appears in Table 12.4. The scheme shows the types of scales, the permissible transformations of scale data, and the meaning of the scale.

TEST CHOREOGRAPHY—ACQUIRING DATA FROM PANELISTS

Early stage testing can be done simply or elaborately. At the very simplest

TABLE 12.4. Stevens' Classification System for Measurement.

Scale	Necessary Operation to Create the Scale	Permissible Transformations	Permissible Statistics
Nominal	Determination of equality	$X' = f(X)$ where $f(x)$ is any one to one transformation	Number of cases, mode, contingency correlation
Ordinal	Determination of greater than or less than	$X' = f(X)$ where $f(x)$ is any increasing monotone function	Median, percentiles, rank order
Interval	Determination of equality of intervals or differences	$X' = aX + b(a < > 0)$	Man, standard deviation, product-moment correlation
Ratio	Determination of equality of ratios	$X' = aX (a < > 0)$	Geometric mean, coefficient of variation

Source: Stevens, 1946.

level the panelist can simply choose which one of two products he/she prefers. The panelist tests the two products, and chooses the more preferred. At a more complex level the panelist can evaluate a set of products, rating each on attributes. Depending upon the research objective the specific activities will change. Choreography of the test ensures the property quality of field execution by requiring the panelist to follow a prescribed set of activities in a prescribed time frame.

Table 12.5 shows the types of activities that a panelist might go through in a complex, choreographed session. During the session the panelist will rate a number of different products on attributes. The choreographed session, while appearing quite detailed, actually enables the field work to proceed smoothly. Evaluation sessions that are more free form often degenerate into chaotic messes, especially when the panelist must test many products, when the

TABLE 12.5. **Choreography of Activities for Central Location Product Test.**

Step	Activity
1	One week prior to the study pre-recruit panelists to participate for a 4 hour test (or for as many hours as the test will require). The panelist must fit the screening requirements to participate.
2	Panelists show up for the test session in groups of 10–25, depending upon the logistics of the study. Panelists do not know each other. The session is set up with this size group to facilitate the field work, and to acquire data efficiently, yet in a cost effective fashion.
3	Moderator introduces panelist to the task by explaining the purpose of the test, explaining complex actions that may be required, etc. This orientation makes the panelist feel comfortable.
4	Panelist completes a short practice exercise, designed to introduce the panelist to the study. This practice exercise may consist of evaluating a different product on another set of attributes.
5	Moderator begins with the first product, which has been randomized to be different for the various panelists. The moderator guides the panelist through the first product, explaining the attributes.
6	After the panelist has finished the product, the panelist shows the rating sheet to an attending interviewer, who asks the panelist 1 to 2 questions about the product just rated. Neither the panelist nor the interviewer knows the right answer. The questioning procedure maintains panelist interest.
7	The panelist waits a prescribed period, and then continues to the next product, completing step 6 above, for each product, and waiting the prescribed period of time.
8	Other longer breaks are interspersed in order to maintain sensitivity and break up the pattern.
9	Once the panelist finishes the evaluations, the panelist then completes an attitude and usage questionnaire.
10	The panelist rates relative importance of attributes.
11	The panelist is paid for participating and dismissed.

samples have to be prepared freshly, and when the panelists have no feedback whether or not they are doing the right thing.

PRODUCTS AND FATIGUE—HOW MANY PRODUCTS CAN/ SHOULD A PANELIST TEST?

There is some controversy in the sensory analysis literature, and even more heated controversy among practitioners, regarding the number of products that a consumer can test without becoming fatigued. Some practitioners aver (often vehemently) that the consumer cannot test more than one or two products before becoming fatigued. These researchers feel that after the second product has been tested the consumers are confused, as they try to sort out in their minds the differences between products, and as they try to recall the sensory impressions of the products in order to make their decisions. These conservative researchers feel that the panelist should remember all the sensory information in order to do the evaluation correctly. There are other researchers who believe that a well instructed panelist can rate many products, especially if the panelist is well motivated and rests between samples to ensure ongoing sensory sensitivity.

These two camps of researchers trace their heritage to different sources. The former, conservative researchers often come from traditional marketing research and sensory analysis backgrounds. The latter researchers come from psychophysics and experimental psychology, where it is not unusual to have panelists test a dozen or two dozen stimuli in order to discover the functional relation between stimulus intensity and perceived sensory magnitude. Why this difference between groups of professionals with extensive experience in product evaluation?

Reason 1—Incomplete Knowledge about Sensory Fatigue

One reason for the disagreement is that there is no compelling data in the technical literature to show how repeated evaluations affect the sensory system, at least in a form acceptable to basic scientists and accessible, and at the same time acceptable, to applied market (viz., consumer) researchers. We know that there are residual tastes such as the lingering aftertaste of artificial sweeteners. These cannot be easily washed away. It takes a few minutes for the bitter aftertaste to diminish. Similarly if a person is exposed to a fragrance such as lavender for a period of several minutes, the fragrance disappears, as the sensory system adapts. Let a few minutes pass by in fresh air and the panelist regains sensitivity. Our sensory system is quite robust and fairly immune to loss in sensitivity, provided that a few minutes (e.g., 5 to 10) elapse between the testing of one product and the testing of another. Those who aver that the panelist loses his sense of taste (or ability to perceive flavor) assume that this loss lasts a long time, rather than dissipating in a few seconds.

Reason 2—Stated Loss of Sensitivity/Acceptance with Repeated Evaluations

Sometimes panelists themselves report that they can no longer taste product differences. Quite often this is due to boredom and disinterest in the evaluation, especially if the panelist is not being paid to participate. Yet give the panelist money and the panelist will stop complaining, get back to the task, and provide good data. Comparisons by the author of paid versus unpaid consumers in the early 1970s showed that both groups of panelists gave similar data, but that the unpaid consumers complained more about the task, and about loss of sensitivity.

Another factor to consider is the reduction of product acceptance over repeated evaluations, due to satiety. As a person continues to ingest a product the blood chemistry changes, the glucose level rises, and the fat in the product is absorbed in the digestive system. This leads to ongoing changes in product acceptance, or the so-called satiety and sensory specific satiety effects (Vickers, 1988; Rolls and McDermott, 1991). Consumers stop liking the product. Satiety can affect the product acceptance, and lead to erroneous answers, if the products are consumed too rapidly, and if satiety is allowed to build up.

Table 12.6 shows the suggested maximum number of products that a consumer can evaluate in a central-location pre-recruit session, provided that there is sufficient time to recover one's sensitivity. The column "wait" shows the suggested minimum interval between products in order to maintain physiological sensitivity and to prevent panelist confusion and complaint. Table 12.6 comes from the author's experience. The numbers are only approximate, but give a sense about the order of magnitude.

CLASSIFICATION SYSTEMS FOR FLAVOR AND TEXTURE

Our first substantive area deals with the language of food and drink. In the previous sections we alluded to the first stage of guidance testing as classifying one's sensory perception of product characteristics. The importance of classification was first popularized by researchers at Arthur D. Little, Inc. in Cambridge, Massachusetts, who recognized that to properly guide development the researcher had to understand what the consumer was experiencing, the nature of the sensory characteristics, their magnitude, and their order of appearance. The Flavor Profile is a process by means of which a panel is trained to recognize the different sensory aspects of a product, and then chooses reference stimuli to act as examples of these attributes. The pioneering work of the

TABLE 12.6. **Maximum Number of Products that Should Be Tested (Max), Minimum Waiting Time between Samples (Wait), Pre-Recruit, Four Hour, Central Location Test.**

Product	Max	Wait	Comments
Carbonated soft drink (regular)	15	7	Little adaptation or residual exists
Pickles	14	10	Garlic and pepper aftertaste build up
Juice	12	10	Watch out for sugar overload
Yogurt	12	10	
Coffee	12	10	
Bread	12	4	
Cheese	10	15	Fatty residue on tongue
Carbonated soft drink (diet)	10	10	Aftertaste can linger
French fries	10	10	Fatty residue on tongue
Cereal (cold)	10	10	
Milk-based beverage	10	10	
Soup	10	10	Can become filling
Sausages	8	15	Fat leads to satiety
Hamburgers	8	15	Amount ingested has to be watched
Candy—chocolate	8	15	
Croissants	8	10	Fat leads to satiety
Salsa	8	10	Longer waits necessary for hot salsas
Cereal (hot)	8	10	Amount ingested has to be watched
Ice cream	8	10	Fat leads to satiety
Mousse	6	15	Combination of fat + sugar yields satiation
Lasagna	6	10	Amount ingested has to be watched

Arthur D. Little, Inc. group was later followed by similar, albeit modified, descriptive methods, such as the QDA (Quantitative Descriptive Analysis) (Stone et al., 1974) and the Sensory Spectrum (Meilgaard et al., 1987). All three approaches are process oriented. They guide and enable the developer or sensory researcher to create the appropriate descriptor system for the product being studied. They do not provide a fixed list of terms per se, primarily because in the description of flavors a myriad of attributes often emerge, many of which are unique to the particular product.

Communicating descriptive analyses can be difficult because the reader may be faced with a wall of numbers in the table. Creative ways have been developed to plot the descriptive analysis results, and by so doing communicate these results in an easy to understand way. Figure 12.2 shows an example of the ''spider-plot'' for a product, and a modified formulation. The plot provides an easy way to represent attribute data for two or more products. The attributes correspond to vectors originating from the center, the length of the vector corresponds to the intensity of the attribute. The plot visually highlights the differences between products on attributes.

Dravnieks and his colleagues (Dravnieks, 1974) attempted to create a standardized thesaurus for describing flavor, which appears in Table 12.7.

There are many other schemes to classify flavor, with these schemes deriving from specific industries. Table 12.8 shows a system for beer (Clapperton et al., 1975). Table 12.9 shows a system for fragrance provided by Haarman & Reimer, Inc., an ingredient and flavor/fragrance supplier serving the food and personal products industries.

Texture, in contrast to flavor, has had an easier time, perhaps because there simply are fewer attributes to describe texture. Szczesniak and her colleagues at the General Foods Corporation in White Plains, New York, proposed a list of terms to describe texture (Szczesniak et al., 1963). These appear in Table 12.10. These terms deal with the mechanical and the geometrical aspects of food. There are fewer nuances in texture, so the challenge for Szczesniak was to create the basic list first and then provide reference samples to show gradations of each texture attribute.

When all is said and done the most important thing to keep in mind about any descriptor system is that it is an attempt to boil down the rich experience of a product into a usable set of terms and scale values. Description by itself does not tell the developer what to do. It simply shows the developer what the panelist is experiencing for the particular product. It is the developer's job to translate descriptions of the characteristics of products into operationally meaningful activities which alter the product in the proper, consumer driven, direction.

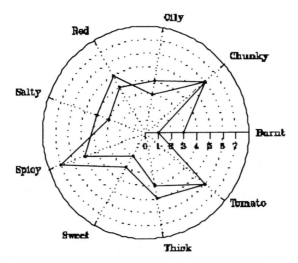

FIGURE 12.2 Polar coordinate diagram of two products on attributes (nine point scale). The products are plotted so that the length of the vector is proportional to the intensity of the attribute. The attributes are arrayed around a circle, separated by vectors with equal angles between them. The angular separation has no meaning other than for purposes of making the display more readable.

TABLE 12.7. **Dravnieks' Classification Scheme for Odor.**

• Eucalyptus	• Beer-like	• Wet paper-like
• Buttery	• Cedarwood-like	• Coffee-like
• Like burnt paper	• Coconut-like	• Peach (fruit)
• Cologne	• Rope-like	• Laurel leaves
• Caraway	• Sperm-like	• Scorched milk
• Orange (fruit)	• Like cleaning fluid	• Sewer odor
• Household gas	• Cardboard-like	• Sooty
• Peanut butter	• Lemon (fruit)	• Crushed leaves
• Violets	• Dirty linen-like	• Rubbery (new)
• Tea leaf-like	• Kippery (smoked fish)	• Fresh bread
• Wet wool, wet dog	• Strawberry-like	• Oak-wood cognac
• Chalky	• Stale	• Grapefruit
• Leather-like	• Cork-like	• Grape juice
• Pear (fruit) like	• Lavender	• Eggy (fresh eggs)
• Raw cucumber-like	• Cat urine-like	• Bitter
• Raw potato	• Bark-like	• Dead animal
• Mousse-like	• Rose-like	• Maple syrup
• Pepper-like	• Celery	• Seasoning—meat
• Bean-like	• Burnt candle	• Apple (fruit)
• Banana-like	• Mushroom-like	• Soup
• Burnt rubber-like	• Fish, cigarette smoke	• Fried fat
• Geranium leaves	• Nutty (walnut)	• Urine

MEASURING THRESHOLDS

One of the traditional methods in flavor measurement (and a mainstay of sensory analysis) is the threshold study. The threshold is the lowest level of a stimulus at which the stimulus can be detected as being present (detection threshold), or that the quality of the stimulus can be recognized (recognition threshold).

A century ago, before experimental psychologists developed scales to measure sensory intensity they sought valid measures of sensory intensity. The often unspoken but operative assumption was that the human being could not act as a valid measuring instrument. Thus they needed a surrogate measure of subjective intensity. It seemed reasonable to use the lowest physical level of a stimulus as a measure of its intensity. They assumed (albeit incorrectly) that two stimuli with different thresholds, X and Y, would differ in their sensory intensity by the ratio Y/X. The higher the threshold the lower the sensory intensity. (Thus the common but incorrect notion that saccharin is hundreds of times sweeter than sugar, when the reality is that the threshold for saccharin is hundreds of times lower.) In the applied world researchers often paid lip service to the threshold value, but except for those studies dealing with off odors/tastes the concept of threshold did not prove particularly useful in applied product development. Table 12.11 presents

TABLE 12.8. **Classification System for the Sensory Characteristics of Beer.**

	First Tier Term	Second Tier Term
1	Spicy	
2	Alcoholic	Warming, vinous
3	Solvent-like	Plastic-like, can liner, acetone
4	Estery	Isoamyl acetate, ethyl hexanoate, ethyl acetate
5	Fruity	Citrus, apple, banana, black currant, melon, pear, raspberry, strawberry
6	Floral	Phenylethanol, geraniol, perfumy, vanilla
7	Acetaldehyde	
8	Nutty	Walnut, coconut, beany, almond
9	Resinous	Woody
10	Hoppy	
11	Grassy	
12	Straw-like	
13	Grainy	Corn grits, mealy
14	Malty	
15	Worty	
16	Caramel	Primings, syrupy, molasses
17	Burnt	Licorice, bread crust, roast barley, smoky
18	Medicinal	Carbolic, chlorophenol, iodoform, tarry, bakelite
19	Diacetyl	Buttery
20	Fatty acid	Soapy/fatty, caprylic, cheesy, isovaleric, buytric
21	Oily	Vegetable oil, mineral oil
22	Rancid	Rancid oil
23	Fishy	Amine, shellfish
24	Sulfitic	
25	Sulfidic	H_2S, mercaptan, garlic, lightstruck, autolyzed, burnt rubber
26	Cooked vegetable	Parsnip/celery, dimethyl sulfide, cooked cabbage, cooked sweet corn, tomato ketchup, cooked onion
27	Yeasty	Meaty
28	Ribes	Black currant leaves, catty
29	Papery	
30	Leathery	
31	Moldy	Earthy, musty
32	Sweet	Honey, jammy, oversweet
33	Salty	
34	Acidic	Acetic, sour
35	Bitter	
36	Metallic	
37	Astringent	
38	Powdery	
39	Carbonation	Flat, gassy
40	Body	Watery, characterless, satiating

TABLE 12.9. Fragrance Classification According to Perfumers at Haarman & Reimer, Inc.

	General Class	Sub-Classes
1	Citrus	Classic
		Modern
2	Green	Fresh
		Balsam
		Floral
3	Floral	Fresh
		Floral
		Heady
		Sweet
4	Aldehydic	Fresh
		Floral
		Sweet
		Dry
5	Oriental	Spicy
		Sweet
6	Chypre	Fresh
		Floral-Animal
		Sweet
7	Fougere	Classic
		Modern
8	Woody	Dry
		Warm
9	Tobacco/leather	
10	Musk	

TABLE 12.10. Descriptive Attributes for Textural Properties.

Textural Properties	Geometrical—Related to Particle Size and Shape	Geometrical—Related to Orientation
• Hardness	• Powdery	• Flaky
• Fracturability	• Chalky	• Fibrous
• Chewiness	• Grainy	• Pulpy
• Adhesiveness	• Gritty	• Cellular
• Viscosity	• Coarse	• Aerated
	• Lumpy	• Puffy
	• Beady	• Crystalline

Source: Szczesniak et al., 1963.

TABLE 12.11. **Taste Thresholds for Chemicals.**

	Chemical	Level
Sweet	Sucrose—D	0.0100
	Sucrose—R	0.0170
	Glucose	0.0800
	Na saccharin	0.00023
Acid	Hydrochloric	0.0009
	Nitric	0.0011
	Sulfuric	0.0010
	Formic	0.0018
	Buytric	0.0020
	Oxalic	0.0032
	Lactic	0.0016
	Malic	0.0016
	Tartaric	0.0012
	Citric	0.0023
Salt	Lithium chloride	0.025
	Ammonium chloride	0.004
	Sodium chloride	0.010
	Potassium chloride	0.015
	Magnesium chloride	0.015
	Calcium chloride	0.010
Bitter	Quinine sulfate	0.000008
	Quinine hydrochloride	0.00003
	Nicotine	0.000019
	Caffeine	0.0007
	Urea	0.12
	Magnesium sulfate	0.046

D = Detection threshold; R = Recognition Threshold.
Source: Pfaffmann, 1959.

a list of thresholds, of interest more for historical reasons than anything else (Pfaffman, 1959).

HOW DO TWO OR MORE SAMPLES DIFFER FROM EACH OTHER?

Of more practical interest is the difference threshold, and of more generality the measurement of differences. Differences are relevant because quite often manufacturers want to maintain the status quo in their product (viz., produce a new product that is perceived to be identical with a currently marketed product, even though some key ingredients have been changed for cost or availability reasons).

TABLE 12.12. Difference Thresholds—Just Noticeable Difference (% Change Needed to Notice that Two Stimuli Differ).

Sense Modality	% Change
Deep pressure from skin and subcutaneous tissue	1.29
Visual brightness	1.66
Lifted weights	1.92
Loudness (1,000 Hz tone)	8.33
Smell of rubber	9.01
Cutaneous pressure	14.0
Taste of a salt solution	20.0

Source: Boring et al., 1935.

DIFFERENCE THRESHOLDS

At the same time that researchers were investigating the perception of the lowest level that one could perceive (the absolute threshold), they were also measuring the smallest change that one would have to make to a physical stimulus in order for the panelist to state that the two samples differed from each other. This difference they called the differential threshold, or just noticeable difference. Table 12.12 shows these difference values, expressed in relative amounts.

DISCRIMINATION TESTS IN REAL WORLD APPLICATIONS

In the world of application difference tests and the just noticeable difference find their greatest use for ingredient substitutions, and for stability tests, where the objective is to determine whether or not two samples differ perceptually. The requirement to measure and thus discover small differences has resulted in the array of procedures known as discrimination tests.

The discrimination test is basically very simple—present the panelist with two or more stimuli, and instruct the panelist to:

(1) State whether two stimuli appear to be different or are the same.

(2) Select the odd or different sample from three or more stimuli, all of which are the same except for the one odd sample.

(3) Sort a set of stimuli into two or more sets, with the stimuli in the same set being identical, and stimuli in different sets being different from each other.

The stimulus to be varied is simple (viz., ingredient A versus ingredient B, or ingredient A at current versus new levels). Thus many flavorists and food processors are interested in the differential threshold values, because these values have economic meaning (viz., in terms of the latitude allowed in changing a product's ingredients, or method of production).

Discrimination testing and differential thresholds find their greatest use early in the development cycle, with in-house panels. If, for instance, the manufacturer wants to substitute high fructose corn syrup for sucrose in a blend for a beverage sweetener, then the developer will run a discrimination test. The developer will create a set of samples (e.g., two samples made with the current sweetener and one sample made with the new sweetener). The researcher will give these samples to a panelist, whether this be consumers or in-house experts, and instruct to panelist to identify the odd sample (viz., the sample that differs from the other two). If the panelist fails in this task, and if this failure to discriminate occurs with a majority of the panelists then the developer feels that the change in formulation can be made with little risk that the consumers will notice the difference.

FROM DIFFERENCE THRESHOLD TO MAGNITUDE OF PERCEIVED DIFFERENCE

Differences among products become far more relevant when we expand the notion of differential threshold to the idea of the magnitude of overall perceived difference. In the real world the odds are that any two products will differ from each other on a host of characteristics. Therefore, for the most part the highly precise measure of difference threshold becomes irrelevant, even though the idea of different/same is quite relevant. It is not that a consumer can discriminate (or fail to discriminate) between two products, but rather to what degree do these two products differ from each other? Furthermore, to what extent can two or more products differ from each other noticeably before the consumer classifies these products as belonging to different classes? We are here expanding the notion of difference threshold to so that difference means "belongs to different classes." Two samples can differ noticeably on many dimensions yet be classified as being the same product. An example of this is pizza. No two pizzas are exactly alike. Any consumer can perceive differences between two samples of pizza, simply because of the nature of the product. Yet the consumer will classify two pizza samples as being the "same" versus "different." On what basis?

The problem of perceived difference is important for developmental work for the following two reasons:

(1) At the very simplest level perceived difference reduces to the differential threshold—can the consumer distinguish between the two samples?

(2) If the developer is trying to create a new product, or match a product to an existing one, the final outcome may be that the old and new products can easily be distinguished, but are "close enough" so that the consumer would not report any difference.

How does the researcher measure perceived difference? Happily the procedure is far less tortured and complicated, and far more direct than discrimination testing. The most direct way is to have the panelist rate the degree to which two samples are similar or different, using a scale. The panelist acts as an integration device, considers all aspects of the product, and then assigns the rating. Table 12.13 shows an example of this type of datum. The study comprises seven different sausages. A to G, which were compared in pairs and scaled on overall dissimilarity on a 0 to 100 scale. The higher the number in Table 12.13 the more different the two sausages were perceived to be. The researcher does not ask the panelist to describe the reasons for the dissimilarity ratings, but rather simply uses the ratings.

Once the panelist becomes a measuring instrument new rules of the game emerge. It is no longer a question of whether or not two products are the same, versus are different, but rather at what rating value of perceived difference does one classify two products as belonging to the same class, versus belonging to different classes. The problem becomes one of identifying a location along a continuum (of perceived difference) which separates two or more samples.

Direct (or even indirect) ratings of difference between products produce another by-product known as a "product map." Product maps are visual representations of the products in a geometrical space, with the property that products close to each other on the map are perceptually similar and products far away from each other are perceptually different (Boring et al., 1935). Product mapping is the direct outgrowth of assessing the overall difference among products that are perceptually distinct. Figure 12.3 shows a product map for the 7 sausages based upon ratings of overall difference between pairs of products (see Table 12.13). A variety of statistical procedures have been published which enable the researcher to process the matrix of difference values among pairs of products, and by so doing fit the products into a geometrical space of relatively low dimensionality (e.g., locate the product on 1 to 4 axes).

TABLE 12.13. Direct Ratings of Overall Dissimilarity (viz., Differences) between Pairs of Sausages on a 0–100 Point Scale.

	A	B	C	D	E	F	G
A	0						
B	17	0					
C	81	67	0				
D	23	24	64	0			
E	32	20	54	11	0		
F	42	59	114	56	66	0	
G	73	79	76	63	70	83	0

Scale: 0 = identical; >100 = extremely different.

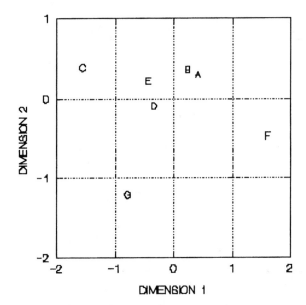

FIGURE 12.3 Multidimensional scaling map of 7 sausages in two dimensions, based upon direct ratings of the overall dissimilarity of appearance between each pair of products. Each pair of products was directly rated on a 0 to 100 point scale (see Table 12.14). Products far away from each other (high on dissimilarity) lie far away on the map.

Another, more popular method instructs the panelist to profile the products on multiple attributes using a scale, and then cumulating the differences between the two products across the different attributes. Table 12.14a shows a database for six sausages profiled on 15 attributes. Table 12.14b shows the average distance (city block) between each pair of sausages. The distance (or perceived difference) consists of the average of absolute differences across the 15 attributes. From this average the researcher can classify the products which fall into one group versus the products which fall into another group. For instance, product 710 is "close" to (or not too different from) product 714, and far away from product 726.

As the interest of developers broadens beyond the narrow focus of ingredients, and into the realm of expected consumer reactions to product formulations, the role of difference analysis becomes increasingly important. Two competing forces always exist, which underscore the importance of measuring differences for practical business applications. The first force is the ever-present need to reduce the cost of goods by substituting ingredients or changing processes. Does this change bring in its wake perceived differences? Does the perception that the revised product now differs from the original product diminish consumer acceptance and thus potentially reduce sales? Here discrim-

TABLE 12.14a. **Profile of Six Sausages on Sensory Attributes.**

Product	710	714	716	717	724	725
Dark	40	50	46	61	24	42
Even	68	65	64	73	69	69
Red	40	52	45	64	23	41
Wet	82	75	74	75	31	60
Thick	40	45	48	47	41	44
Flavor	67	61	62	61	48	54
Aroma	51	52	55	53	35	50
Spicy	62	55	54	51	42	46
Meaty	53	46	59	59	55	59
Salty	61	57	56	46	45	48
Aftertaste	68	55	52	51	39	49
Firm	41	31	58	57	57	66
Gritty	28	23	30	29	17	21
Juicy	79	78	67	69	45	57
Greasy	50	47	42	43	26	32

ination testing and difference measurement act as brakes on the development process. They reveal whether or not a change can even be noticed. The second force is the ever-present need for new products, and the requirement that a product be different, but not too different, in order to enjoy market success. Here the emphasis is on progress and change. Noticeable differences between products mean that the product developer has succeeded. The measurement of such differences indicates how new the product appears to be versus current products in the market.

PERCEIVED INTENSITY—THE PANELIST AS A MEASURING INSTRUMENT

The previous section introduced the idea of measurement. This section

TABLE 12.14b. **Cumulative (City Block) Distances between Pairs of Sausages.**

	P710	P714	P716	P717	P724	P725	P726	P727
P710	0.0							
P714	6.4	0.0						
P716	7.6	5.8	0.0					
P717	10.3	7.5	4.6	0.0				
P724	18.2	18.2	15.1	16.3	0.0			
P725	11.3	10.5	6.7	8.3	9.5	0.0		
P726	18.4	17.7	14.8	16.6	4.6	9.9	0.0	
P727	13.8	12.1	8.6	10.4	8.3	5.1	6.5	0

continues that topic, dealing in depth with the panelist as a measuring instrument. When researchers began in earnest to use subjective measurement tools, the first attempt was to create a scale. The initial scales comprised a limited set of categories. There are many different types of scales that one can use, and many variations in the way that scale data is treated. We will look at the range of measuring procedures, from the most primitive method, ranking, to the most sophisticated, ratio scaling.

RANKING PRODUCTS IN ORDER

We begin first with ranking, wherein the panelist simply ranks the products in terms of degree of an attribute. Table 12.15 shows results from a candy study. The rank order is obtained by averaging the results from a group of panelists. Panelists find ranking quite easy to do. If the panelists disagree with each other then the average ranks will all lie close together because some people will have ranked the products one way, whereas other panelists will have ranked the products another way. The ranks will cancel each other out. In contrast, if the panelists agree with each other then the ranks will be more spread. Table 12.16 shows a worked example for six products by a group of consumers who agree with each other, versus another group of consumers who disagree with each other. The left side shows the average from five panelists who agree, and the right side shows the average from five panelists who disagree. The right shows the average from five panelists who disagree with each other. Table 12.16 reveals that the products can end up having the same final ranking, although for some panel data the final ranking is clear, whereas for other panel data the final ranking does not do justice to the differences among panelists.

TABLE 12.15. **Rank Order of Six Candies (A–F) on Liking by Nine Panelists, and the Final Ranking.**

Panelist	C	A	B	F	D	E
1	2	1	3	4	6	5
2	3	4	6	5	4	1
3	3	2	1	4	5	6
4	3	4	2	1	6	5
5	1	4	5	2	3	6
6	2	1	3	5	5	4
7	3	4	2	6	1	5
8	1	3	2	5	4	6
9	5	1	4	2	4	6
Mean rank	2.6	2.7	3.1	3.8	4.2	4.9
Final rank	1	2	3	4	5	6

1 = Worst; 9 = Best.

TABLE 12.16. Mean and Individual Ranks of Five Panelists Who Agree (101–105) versus Five Panelists Who Disagree (201–205).

Panelist	A	B	C	D	E
101	1	2	3	5	4
102	2	1	3	4	5
103	1	3	2	5	4
104	1	3	2	4	5
105	1	2	3	4	5
Average	1.2	2.2	2.6	4.4	4.6
Final rank	1	2	3	4	5
Panelist	A	B	C	D	E
201	3	1	4	2	5
202	4	3	2	1	5
203	2	4	1	5	3
204	1	2	3	5	4
205	1	4	5	3	2
Average	2.2	2.8	3	3.2	3.8
Final rank	1	2	3	4	5

Ranking leaves much to be desired. First, in terms of test administration and panelist behavior the consumer has to return again and again to the products, to assure himself that all of the products line up in the proper order. There is the implicit demand that all products be compared to each other in order to ensure that the ranks assigned are correct, and that no products are out of order. Second, on the statistical side the ranking procedure is quite weak because the rank order data do not permit analysis by standard and powerful statistics to which the researcher is accustomed. Third, ranking just shows order or merit but does not show the degree to which a product possesses a specific characteristic. We know only that one product has more of a characteristic than does another when it is ranked higher on that characteristic. Although ranking produces weak data the method is seductive because at first glance it looks so easy to do in the field, it is intuitively obvious, and appears to pose few problems. At the analytic end, however, there is precious little that can be obtained from the results without invoking extensive analysis.

RATING PRODUCTS ON A CATEGORY SCALE

The next higher level of measurement is the rating scale or interval scale. Panelists use the scale to show the degree of feeling about a product. It is fair to say that the rating scale is today the most popular measuring tool, primarily because it is easy to use, the data can be rapidly and easily analyzed by personal computers, and the statistics allow for powerful analyses which lead to great insights and powerful conclusions.

The typical category scale comprises a limited number of categories. A glance through the sensory analysis and marketing research literature reveals all sorts of scales. Some researchers aver that the scales must be long (e.g., comprise 100 categories), whereas other researchers with just as much vehemence announce that the only valid scale is one whose individual categories can be named, and therefore this latter group insists on shorter scales (e.g., 5 to 7 points).

Despite these "researcher-driven" disputes, for the most part category scales of any type perform reasonably well when put to the test, provided that they comprise a sufficient number of categories. The issue about which scale is best is moot, and probably will never really be answered. Most reasonably long scales (e.g., 5 to 7 points or more) discriminate among products, and thus are useful measuring instruments. Most scales of different lengths will correlate highly with each other when used to measure the same products, although when plotted against each other they may not describe perfect lines. The upshot of all of the arguments is that the researcher should use a scale with which he is comfortable, but make sure that the scale has a sufficiently large number of points that it can capture product differences.

There are paradoxes, however, when using a category scale of finite length. These paradoxes show up most dramatically for short scales (e.g., 5 points or fewer). For instance, it is easy to show differences in the color of seven candies with a 10 point color scale ranging from dark to light. It's a little more difficult to show such differences in the same seven candies if the candies are all demonstrably different from each other, but the panelist can only use a 3 point scale ranging from dark to light. For development work it is advisable to use an easy to administer, easy to analyze scale, anchored at both ends, such as a 0 to 100 point scale.

RATIO SCALING

Although for the most part category scaling does a fine job in research, there are instances when the researcher would like a more powerful, less scale-dependent measure. (The results from category scaling critically depend upon the number of points used in the scale.) Furthermore, basic research scientists always search for more powerful types of measurement to quantify subjective reactions. These researchers are not satisfied with category scales for fundamental scientific work because of the biases and the dependence of results on the length of the scale.

In 1953, S. S. Stevens at Harvard University reported that he was able to create scales of perceived intensity by instructing panelists to adjust the intensity of one stimulus continuum (e.g., sounds) to the intensity of stimuli from another continuum (e.g., lights) (Stevens, 1953). Although at first glance this might seem a bit academic, the objective of the research was important—to erect valid ratio scales of perception. Later Stevens would generalize this approach (Stevens, 1966) so that panelists would adjust the perceived intensity of one continuum to match the perceived intensity of another continuum (e.g., adjust the loudness of sound to match the brightness of light). When the panelist adjusted the magnitude of numbers Stevens called this special case the method of "magnitude estimation." Magnitude estimation would subsequently lead to a burst of activity in sensory analysis and psychophysics, which would become important to product developers.

The initial objective of magnitude estimation was to erect a ratio scale of perceived intensities. The outcome, however, was more profound. Stevens instructed panelists to match numbers to the perceived taste intensities of different solutions, the perceived roughness of grit, the perceived darkness of colors, etc. The scaling exercises using magnitude estimation generated repeatable data. The curves relating the intensity ratings to the physical stimuli described power functions of the form:

$$\text{Magnitude estimate} = k(\text{physical intensity})^n$$

The exponent n was repeatable from study to study and showed how the sensory system transformed physical stimulus intensities to sensory magnitudes. Table 12.17a shows a list of power function exponents relevant to food developers.

The power function exponent tells the researcher the relation between physical stimulus level and sensory perception. Furthermore, given the power function exponent and a change in the physical stimulus, the developer can estimate the likely change in sensory response, and vice versa. This information helps to select appropriate ranges for product development. For example, when the exponent n, is lower than 1 (as it is for perceived intensity of odors versus concentration, for viscosity versus centipoises, etc.), then a wide range of physical magnitudes is shrunk to a narrower sensory range. When n is higher than 1 (as it is for hardness versus physical indentation, for sweetness versus sugar molarity, for saltiness versus molarity), then a wide range of physical magnitudes is expanded to an even wider sensory range (see Table 12.17b). Because the sensory system processes stimulus information differently for tactile thickness versus sugar sweetness, the same sensory range is created by different ranges of a thickener versus a sugar solution.

MEASURING LIKING—THE PANELIST AS A JUDGE OF QUALITY

If a product doesn't taste good then the odds are that it will not succeed in the marketplace. The goal of development is to create a product that truly does well in the market. To this end the key evaluative criterion that developers

TABLE 12.17a. **Exponents for Power Functions—Taste, Odor, Texture.**

Sensory Continuum	Exponent
Taste	
Sweetness—sugar	1.3
Sweetness—saccharin	0.6
Sweetness—aspartame	1.0
Saltiness—sodium chloride	1.3
Sourness—citric acid	0.7
Bitterness—quinine sulfate	0.6
Smell	
Amyl acetate—in liquid	0.13
Anethole—in liquid	0.16
Butanol—in liquid	0.31
Citral—in liquid	0.17
Methyl salicylate—in liquid	0.20
Texture	
Force/indentation ratio—rubber	0.6
Dynes/cm^2 (handgrip)	1.6
Thickness versus viscosity	0.4
Chunkiness—hamburger	0.55
Roughness—versus sandpaper grit	− 1.5

Source: Moskowitz, 1983.

TABLE 12.17b. **Relation between Power Function Exponent and Stimulus/ Perception Changes of Given Ratios.**

Exponent	If Stimulus Change =			If Perception Change =		
	1.5×	2.0×	2.5×	1.5×	2.0×	2.5×
	Then Perception Change =			Then Stimulus Change =		
0.2	1.1	1.1	1.2	7.6	32.0	97.7
0.4	1.2	1.3	1.4	2.8	5.7	9.9
0.6	1.3	1.5	1.7	2.0	3.2	4.6
0.8	1.4	1.7	2.1	1.7	2.4	3.1
1.0	1.5	2.0	2.5	1.5	2.0	2.5
1.2	1.6	2.3	3.0	1.4	1.8	2.1
1.4	1.8	2.6	3.6	1.3	1.6	1.9
1.6	1.9	3.0	4.3	1.3	1.5	1.8
1.8	2.1	3.5	5.2	1.3	1.5	1.7
2.0	2.3	4.0	6.3	1.2	1.4	1.6
2.2	2.4	4.6	7.5	1.2	1.4	1.5
2.4	2.6	5.3	9.0	1.2	1.3	1.5
2.6	2.9	6.1	10.8	1.2	1.3	1.4
2.8	3.1	7.0	13.0	1.2	1.3	1.4

use is liking. The higher the liking rating the more likely it will be that the product will succeed.

Liking is quite easy to judge, as the section on scaling demonstrated. Indeed there are a plethora of scales. The important thing about scaling liking is not the scale used, but rather whether the scale has norms associated with it, and whether the product is tested on a blind basis or a branded basis, respectively.

When it comes to measuring the degree of liking pronouncements and dogma become even more severe—with some researchers insisting that the scales be even with the same number of points to denote "good" as "bad" in order to avoid bias. Other researchers feel that in a commercial setting most of the products will be acceptable because they are designed for consumption. Furthermore, these researchers feel that it is more important to differentiate among the different levels of good products by means of more scale points than it is to balance the scale. Finally, some researchers insist that the scale must have a point for "neutral" (viz., a middle category), whereas other researchers insist that the scale should have no neutral point at all, thus forcing the panelist to decide whether he likes or dislikes the product. Table 12.18 shows a list of scales that have been proposed over the years.

WHAT DO LIKING RATINGS MEAN?

A scale is only a measuring instrument. Developers want to know what the scale point means. On a 1 to 9 point scale of liking, for instance, what does it mean to say that a product scores an average of 5.5, whereas another product scores an average of 7.0? Is this a statistically meaningful difference? What behaviors are to be associated with products scores of 5.5 versus 7.0, respectively? Does the consumer select products scoring 7.9 more often than products scoring 5.5? If not, then the scale differences are an artifact with no real meaning, and in fact with a great deal of potential to confuse.

Let us begin this detailed analysis of liking scales by the statistical analysis of liking ratings assigned to seven yogurts. The panelists each tested every yogurt in a randomized order, rating the degree of liking using the nine point hedonic scale (Peryam and Pilgrim, 1957). Table 12.19a shows raw data from the first six panelists. Table 12.19b shows the mean and standard deviation from the full panel of 30 consumers, for each yogurt, respectively.

The first question that a researcher would ask is whether or not the seven yogurts differ in the degree to which the consumers like them. It is not sufficient to look at the differences between the yogurts and stop there, because the difference may be due to chance alone, or may actually be due to repeatable differences. A standard analysis looks at the variability due to the means versus the random variability. The ratio of the two variabilities is an index number called the F ratio, which has a defined statistical distribution. By computing the F ratio, and knowing the number of observations for the means and the number of observations for the error term, the researcher can determine

TABLE 12.18. **Category Scales—Liking.**

Two point scale	Like
	Dislike
Three point scale	Acceptable
	Dislike
	Never tried
	Like a lot
	Dislike
	Do not know
	Well liked
	Indifferent
	Disliked
Five point scale	Very good
	Good
	Moderate
	Tolerate
	Never Tried
	Very good
	Good
	Moderate
	Tolerate
	Dislike
Fact scale	Eat every opportunity
	Eat very often
	Eat frequently
	Eat now and then
	Eat if available
	Don't like—eat on occasion
	Hardly ever eat
	Eat if no other choice
	Eat if forced
Nine point hedonic scale	Like extremely
	Like very much
	Like moderately
	Like slightly
	Neither like nor dislike
	Dislike slightly
	Dislike moderately
	Dislike very much
	Dislike extremely
Purchase intent scale	Definitely would buy
	Probably would buy
	Might/might not buy
	Probably not buy
	Definitely not buy

TABLE 12.19a. **Ratings of Consumers on Seven Yogurts (Partial Data Set).**

Panelist	A	B	C	D	E	F	G
101	6	5	5	6	5	8	2
102	5	7	4	5	2	3	1
103	8	6	6	5	3	5	3
104	5	6	3	5	4	3	3
105	3	7	7	6	6	6	3
106	6	4	2	5	4	5	3

the probability that the observed F ratio for the yogurts comes from chance alone, or represents a true effect (viz., that the difference is not due to chance). This method is the analysis of variance, and is part of the arsenal of research tools available for statistically analyzing the data. Table 12.19b shows the analysis of variance table for these seven yogurts, and reveals that indeed there are differences among the yogurts in degree of liking that were captured by the hedonic scale. That is, by using the scale to represent overall liking panelists told us that they liked some yogurts more, and other yogurts less. (We deduce that there is a significant difference among the yogurts by considering the F ratio, which is significant, meaning that it would be very rare to observe this F ratio if the products were really identical to each other, and differed only by random chance.) Finally, Table 12.19d shows a test of least significant differences using the Tukey test. The difference in liking ratings for pair of yogurts can be analyzed to determine whether or not this specific pair of yogurts differs or is the same. There may be an overall significant effect (viz., yogurts do differ), but on a pairwise basis only some yogurts will differ, whereas others will be the same.

From this simple, but instructive exercise the reader can begin to see the power brought to early stage development by scaling. Rather than force the panelist to rank products, a category scaling method gives the panelist a yardstick, and has the panelist "have a go" at the products. The panelist can scale one product, five products, 100 products or more. The products can be rotated, and indeed each panelist need not scale all of the products, but rather

TABLE 12.19b. **Summary Statistics and Analysis of Variance.**

Yogurt	Mean	Standard Deviation
A	5.0	1.21
B	5.1	1.13
C	4.9	1.58
D	5.1	1.03
E	4.3	1.48
F	4.1	2.04
G	3.8	1.41

TABLE 12.19c. **Analysis of Variance.**

Source	Sum-of-Squares	DF	Mean-Square	F Ratio	Prob/Chance
Yogurt	116.781	6	19.463	9.281	0.000
Error	425.700	203	2.097		

need only rate a limited number. No memory is involved, since the panelist reacts to each product as it is presented, trying his best to locate the product on the liking scale.

STATISTICAL TESTS VERSUS NORMATIVE INTERPRETATIONS

Researchers use two different approaches by which to assess the meaning of liking ratings. One approach uses statistical difference testing and inferential statistics, which we saw above for the yogurts. By performing the standard tests of difference the researcher can determine whether the numerical difference between products observed in the study would repeat itself in subsequent tests or whether the difference observed is simply due to chance. The statistician does not talk about the external meaning of the difference but rather concentrates on the repeatability of the results.

Researchers often talk about the statistical significance of differences, but these statements about statistical significance do not necessarily have anything to do with the performance of products in the marketplace. Small differences in ratings of liking can become highly significant (viz., not be due to chance alone) when the researcher vastly increases the number of ratings for each product. The greater the number of ratings per product the more significant will be the same numerical difference.

Consider, for instance, the data shown in Table 12.20, pertaining to the liking ratings for cereal prototypes. By themselves these numbers simply show that cereal prototype A is liked more than prototypes C, D, E, etc. The statistician can tell us whether or not the differences is statistically significant—

TABLE 12.19d. **Probability That the Difference between Products Is Due to Chance.**

	A	B	C	D	E	F	G
A	1.00						
B	1.00	1.00					
C	0.20	0.20	1.00				
D	0.50	0.50	1.00	1.00			
E	0.00	0.00	0.62	0.28	1.00		
F	0.00	0.00	0.38	0.13	1.00	1.00	
G	0.00	0.00	0.04	0.01	0.83	0.96	1.00

TABLE 12.20. **Scores for Cereals, Significant Differences, and Normative Values for Products on the Moskowitz/Jacobs, Inc. 0–100 Point Scale.**

Product	Liking and Interpretation
A	66 = Very good
B	62 = Very good
C	60 = Good, may need a little work
D	56 = Good, probably needs work
E	54 = Good, probably needs work
F	43 = Poor
G	35 = Very poor
H	31 = Very poor
I	27 = Exceptionally poor
Norms	**Least Significant Difference = 3.2**
70+	Excellent, needs no work
61–70	Very good, can use slight improvement
51–60	Good, needs improvement
41–50	Fair, needs a lot of improvement
40–	Poor
Norms	**Number of In-Market Product in this Range**
70+	1
61–70	4
51–60	6
41–50	2
40–	2

viz., if we were to repeat this experiment 100 times would we see prototype A beating prototypes C, D, E, etc.?

The analysis of liking through norms is often more germane to product development. Normative data enables the researcher to consider the numbers in light of existing products whose performance is already known. The norms provide a means by which to interpret the data. For example, what does a rating of 56 mean? On an absolute basis what do the different points on the scale mean? Does a rating of 60 on the 100 point liking scale insure success, whereas a lower rating (even slightly lower) of 58 forecast that the product will fail. Table 12.20 (middle) shows norms for the 0 to 100 anchored liking scale used by the author since 1981, in more than 1,000 studies, along with interpretive comments about locations on the scale. Descriptions such as these also help the developer to understand what the scores really mean, whether the product is a winner, or a real dud.

DIRECTIONAL RATINGS

Over the past four decades product developers have sought more information from product tests. From the need for guidance, developers have come up with a hybrid scale, neither sensory descriptive nor liking, but a mixture of the two. This scale is the so-called "directional scale." The panelist tastes a product and answers a question similar to the following: "How would you describe this product? 0 = not sweet enough, 50 = just right, 100 = too sweet?" The particular attribute and scale can change, but the key thing to keep in mind is that the scale has an anchor in the middle called "just right," with one end representing too much of an attribute and the other end representing too little of the attribute. The directional scale is neither a sensory scale nor a liking scale, but rather a mixture of the two. The consumer first has to register the sensory amount of an attribute, and second assess the degree to which the product departs from one's own self described ideal.

Directional scales are quite popular among developers because they appear to be easy to use. For a complex product it is not unusual to ask the panelist to rate the product on 10 to 30 directional scales. From the scales the developer can get an idea of the degree to which a product is off-target on specific characteristics. Table 12.21 shows a profile of a complex product, American Noodles + Beef, on 14 directional scales. Panelists find the directional scales easy to use, and at a superficial level product developers can easily interpret the scale results.

Despite the superficial simplicity of directional scales there are some problems which, when they emerge, seriously misdirect the developer who relies upon the scales to guide development. For certain attributes, especially appearance and texture, panelists can tap into their ideal level, and have no problem using the scale to identify problem areas and the direction needed to fix the

TABLE 12.21. **Example of Directional Scales for American Noodles + Beef (+ = Too Much; 0 = Just Right, − = Too Little).**

Directional Attribute	Score	Comment
Amount of mushrooms	−33	Far too little
Onion flavor	−27	
Aroma strength	−16	
Meat flavor	−13	
Amount of meat pieces	−8	Too little, but not too far off target
Size of meat pieces	−7	
Flavor strength	−6	
Amount of pasta	−5	
Thickness of gravy	−3	On target
Saltiness	−3	
Darkness of gravy	−1	
Firmness of pasta	8	Too much, but not too far off target
Spiciness	11	Too strong
Tomato flavor	12	

problem. For other attributes, especially taste and flavor, there may be some situations in which panelists say one thing but really don't mean it. Here are two examples:

(1) Covert social demands: some attributes are to be minimized for "political correctness." For instance a beverage should not be overly sweet (even if the consumer likes overly sweet products). Sweetness produced by sugar is assumed to be caloric, and public wisdom has it that excessive calories are no good. The same goes for saltiness and greasiness. It is simply not appropriate to want greasy foods, especially when grease is associated with fat and poor health. Therefore, products can always be made "less greasy," "less fatty."

(2) Confounding and misdirection: some attributes may be confounding and misleading. In a fruit flavored beverage, one can lower the sweetener level considerably. Panelists will report that the "flavor" is too weak. Yet adding flavoring (the superficial response to this directional result) will simply produce a tarter, more bitter taste, which is still too low on "flavor." Consumer reports flavor, but the key ingredient is sugar. Thus if the panelist says that the flavor is too low the developer may add flavoring. The panelist will continue to say that the flavor is too low, and the developer will take this direction seriously, and add more and more flavoring until the product becomes unpalatable. The real answer is to add sugar. Similarly with chocolate flavor. Take away the fat and the chocolate flavor drops. Consumers will demand more chocolate flavor, but the real key is to add fat.

BEYOND MEASUREMENT TO PATTERNS

In applied research the objective is to use sensory analysis and early stage testing in order to guide development. Often this guidance requires the developer to identify and capitalize on a pattern relating liking to ingredients or sensory attributes, respectively. The nature of this relation provides a guide that can be used to direct product modifications.

UNIVARIATE PATTERNS AND MODELS—GENERAL PRINCIPLES AND AN EXAMPLE FROM MEAT

As a formula variable or as a sensory attribute increases from low to high, overall liking will change. However, the relation will not be a straight line but rather a curve, similar to an inverted U [see Figure 12.4(a)]. As the independent variable increases, liking will first increase, then peak at an intermediate maximum, and then decrease. The particular pattern will vary

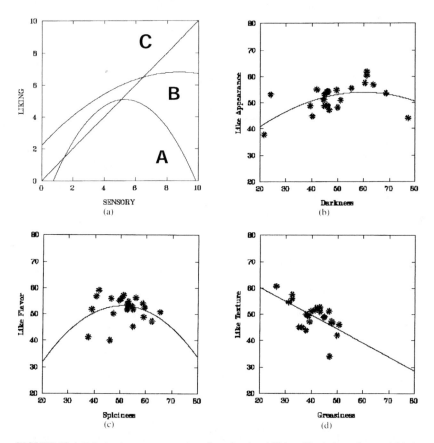

FIGURE 12.4 Relation between sensory attribute level and liking. The independent variable is the sensory attribute level, the dependent variable is liking. (a) Schematic for three types of relations: inverted U, increasing, asymptotic, respectively. (b) Liking of appearance versus perceived darkness for processed meats. Each point corresponds to an in-market product. (c) Liking of flavor versus perceived spiciness for processed meats. (d) Liking of texture versus perceived greasiness for processed meat.

by the product, the sensory attribute, the range of the independent variable tested in the study, and the panelist himself (Moskowitz, 1981). In contrast to the liking curve, perceived sensory intensity will increase, either on a linear basis, or begin to asymptote, but rarely if ever will decrease. Figure 12.4(b) through 12.4(d) show three of these patterns for a set of processed meats.

Data of the sort shown in Figures 12.4(a) through 12.4(d) provide a great deal of insight to the developer. Table 12.22 shows answers to six most often asked questions, using the curves developed for the attributes. These questions range from the nature of the relation between sensory and liking, to the direction of change for a product, to the expected increase in acceptability.

TABLE 12.22. Directions for Product Improvement, Using the Relation between Sensory Attribute and Attribute Liking.

	Like Appearance versus Darkness	Like Flavor versus Spiciness	Like Texture versus Greasiness
Nature of the relation?	Inverted U	Inverted U	Linear down
Where is my product on the X axis?	20	60	33
Where is the optimum?	60	50	25
How should I change the product?	Increase		No
Is the change important?	Yes	Moderately	No
How many points in attribute liking will I gain?	~15	~5	~2

MULTIVARIATE PATTERNS—THE CATEGORY MODEL FOR A MEAT PRODUCT

Overall liking is determined by the interaction of many sensory attributes. We saw above that as a sensory attribute increases, liking first increases, peaks, and then drops down. The pattern becomes more complicated with two or more attributes. The situation becomes almost impossible to deal with when we realize that we have to marry two domains—sensory attributes and liking—and that in the sensory domain alone there may be dozens of attributes. Which attributes should serve as the independent variables from which we will predict liking?

Happily there is a method by which the researcher can integrate all of the sensory attribute data in a product category and relate that data to overall liking. The procedure begins with a data set comprising products as rows, sensory attributes as columns, and overall liking as yet an additional column. The procedure uses factor analysis (principal components) to reduce the sensory attributes to a set of independent factors, and then builds a model relating those factors, their squares, and significant pairwise interactions to overall liking, and to each sensory attribute, respectively. Each attribute generates its own equation. From this model it becomes possible to

(1) Optimize the product, by finding out which combination of factors maximize liking, and from that optimization identify landmark products to serve as development examples. This approach provides concrete guidance to the developer because it reveals which particular sensory attribute levels correspond to the optimal product.

(2) Optimize liking, subject to sensory or image constraints

(3) Discover what combination of factors, and therefore what sensory profile, corresponds to a target profile. This target profile may be a profile of ratings of image characteristics assigned to the product.

Table 12.23a lists the steps to create the category model. Table 12.23b shows the equation for overall liking versus factor scores. Table 12.23c shows results from six optimization exercises, where the goal was to maximize an attribute, subject to constraints on the factor scores, and constraints on the range of sensory attributes that the optimum product is expected to possess. Table 12.23d shows the results from "reverse engineering." Reverse engineering involves the discovery of the combination of independent factor scores (viz., independent variables) which, in concert, generate a profile as close as possible to the goal profile specified by the developer. The consumer rated attribute profile is the goal in reverse engineering. Table 12.23e shows landmarks for product development, using the expected sensory levels as landmark levels, and then products in the test set as exemplars for product development.

The key point to take away from this topic is that, using sensory analysis or consumer profiling of products on both sensory and liking/image attributes, the product developer can create a database and a category model that together provide early stage guidance. The products need not be systematically varied. What the developer must do, however, is to discover the pattern relating sensory attributes to liking. Once a convenient and mathematically sound system is developed the researcher and the product developer can navigate through the world of existing, in-market products in order to discover what combination of sensory attributes corresponds to business opportunities. The system uses in-market products, takes into account interactions and nonlinearities, and creates a usable model for ongoing development.

THE DESIGNLAB PARADIGM FOR EARLY STAGE DEVELOPMENT

Thus far this chapter has dealt with the basics and advances in early stage product evaluation. Moskowitz (Moskowitz and Gofman, 1996; Moskowitz, 1996b) developed an integrated system called the "DesignLab Paradigm" to accelerate early stage development. DesignLab combines standard tools of product and concept development, with computers to present concepts and acquire data. The DesignLab environment allows data to be gathered from many people rapidly, and the results analyzed almost instantly.

Table 12.24 presents DesignLab's development paradigm in outline form. The approach begins with concept testing, identifies the "hot buttons" for a concept, optimizes the concept, and inserts the optimized concept(s) into a product evaluation questionnaire for the (immediately) subsequent product test. The consumer evaluates many products in the category, rating each on sensory, image, liking, and fit to the concepts just developed. At the end of the evaluation the data is analyzed, allowing the developer to create a category model, similar to that shown in Table 12.23. This model then becomes the

TABLE 12.23a. **Sequence of Steps to Create the Category Model.**

Step	Activity	Rationale
1	Create a product x attribute matrix.	Basic input for the model
2	Isolate the sensory attributes.	Prepare these for factor analysis.
3	Factor analyze the sensory attributes, extracting a limited number of factors.	Factors will be used as independent variables in regression models.
	Use principal components method.	Extract a limited number of factors, because of degrees of freedom in regression modeling.
4	Rotate the factor solution to a simple one (e.g., quartimax).	Optional, if you want to interpret the factors
5	Save each product in terms of its factor scores.	Locate each product in a smaller space, where the dimensions (factors) are statistically independent and parsimonious (few factors).
6	Adjoin the factor score matrix to the original data.	Augment the matrix in preparation for modeling.
7	Create equations using linear, square, and statistically significant cross terms.	Independent variables = factors Dependent variables = attributes One equation for each attribute Each subgroup of consumers = attribute
8	Retain the model, calculate statistics.	Shows how well the equations fit the data
9	Use the model to optimize.	
10	Optimize—discover the combination of factors that maximize a criterion variable (e.g., liking), within constraints provided by other variables.	Stay within the range of factor scores tested, and (to maintain reality of the solution) stay within the envelope of sensory levels achieved in the study.
11	Reverse engineer—discover the combination of factor scores that maximize closeness of fit of a goal profile of attributes (established by researcher) to the estimated goal.	Research provides a goal. Computer adjusts factor scores to find a set which yield a profile close to the goal.
12	For either step 10 or 11, estimate the full profile of attributes corresponding to the optimum set of factor scores.	Use step 7 (equations) to estimate the profile, given the factor scores.
13	Identify examples of products with the requisite sensory levels.	Use the data base created at the start of the project.

TABLE 12.23b. Parameters of the Regression Model for Overall Liking versus Factor Scores.

Dependent Variable	Overall Liking
Squared multiple R	0.83
Adjusted squared multiple R	0.76
Regression equation:	
Constant	55.6
$F1$	-2.75
$F2$	-0.48
$F3$	2.36
$F1 \cdot \text{"}F1\text{"}$	0.34
$F2 \cdot \text{"}F2\text{"}$	-1.44
$F3 \cdot \text{"}F3\text{"}$	-2.25
$F2 \cdot \text{"}F3\text{"}$	-1.09

guide to development. The developer and the marketer know what to communicate about the product, what the product should have, and what product landmarks in the competitive frame should guide development. What is important about the DesignLab paradigm is that it creates the database quickly for both concepts and products (viz., in a matter of 1 to 4 hours), thus quickly speeding up the development process. Furthermore, the DesignLab paradigm is independent of the developer's state of knowledge. As long as the researcher can create concept elements, select pictures, select in market products, and developer a battery of product and liking attributes, the DesignLab paradigm will return with concrete direction for development.

A CASE HISTORY—CREATING A NEW FLAVORED COFFEE

The best way to illustrate the DesignLab paradigm is by means of a case history. This case history deals with the creation of a new flavored coffee, by discovering "hot buttons" or persuasive elements for a product concept, optimizing the concept, and then fitting a product to that concept.

Concept Development

Concepts comprise statements about what the product should be (product concept), or why the consumer should buy the product (positioning concept). Developers use product concepts to guide their product creation. In the DesignLab paradigm the researcher creates a reservoir of concept elements. These are phrases and pictures, which can be considered as elements of the concept. There may be dozens or hundreds of elements, dealing with aroma, flavoring, reasons for drinking the coffee, pictures of the product, etc.

TABLE 12.23c. **Factor Scores and Expected Profile for Six Optimization Tasks (Subject to Constraints).**

Maximize	Max Liking	Fit Con. 1	Fit Con. 2	For a Meal	For a Meal	For a Meal
Constraint	None	None	LD > 55	None	Kid < 40	Kid < 40
F1	−1.85	1.52	−1.29	−0.07	−0.01	−0.06
F2	−1.15	−2.27	−2.34	2.05	2.60	−1.17
F3	0.89	−0.54	0.55	1.14	−1.45	0.27
Like total	62	43	55	46	41	55
Like appearance	54	31	38	53	51	47
Like aroma	56	47	52	50	56	50
Like flavor	59	46	56	47	45	56
Like texture	67	25	54	42	41	48
Dark	39	28	23	75	70	36
Even	63	54	55	74	74	56
Red	37	27	21	75	71	37
Wet	44	57	35	67	72	59
Thick	43	42	41	53	42	44
Flavor	48	63	51	66	63	58
Aroma	39	48	39	58	58	49
Spicy	42	63	47	55	60	51
Meaty	60	49	55	59	50	55
Salty	45	60	48	46	54	51
Firm	69	31	56	69	67	53
Gritty	17	22	17	39	43	24
Juicy	45	71	46	57	54	63
Greasy	28	45	28	38	41	39
Aftertaste	39	67	46	58	61	51
Similar to chicken	50	47	55	40	43	51
Similar to turkey	54	42	50	41	41	50
Similar to red meat	40	49	47	57	55	50
For lunch	52	46	45	75	77	42
For a meal	29	72	46	81	67	52
For kids	33	49	25	79	40	39
Expensive	39	67	51	71	68	51
Traditional	38	55	40	63	49	43
Fit concept 1	37	79	56	69	61	51
Fit concept 2	35	67	44	78	50	45

Concept Creation

A computer program creates combinations of these elements, using experimental design. The computer program uses a file of "restrictions" as well, with that file listing incompatible pairs of elements. This list ensures that no concept is illogical because it has paired elements that do not go together. Each consumer tests a unique set of concept elements (60 in total), combined

TABLE 12.23d. **Factor Scores and Expected Profile for Four Reverse Engineering Tasks (Subject to Constraints).**

Fit	Directionals = 0		Profile A		Profile A		Profile B	
Constraint	None		None		LD > 50		LD > 55	
$F1$	−0.4		−1.1		−0.1		−0.8	
$F2$	−0.9		2.9		−1.2		1.2	
$F3$	1.8		−1.8		1.5		0.7	
	Est	Goal	Est	Goal	Est	Goal	Est	Goal
Like total	55		40		55		55	
Like appearance	49		51		47		58	
Like aroma	52		51		51		53	
Like flavor	55		44		56		54	
Like texture	57		38		55		50	
Dark	49		67		43		63	
Even	58		69		57		67	
Red	49		66		43		63	
Wet	55		58		54		65	
Thick	48		40		46		49	
Flavor	58		55		59		57	
Aroma	52		52		51		52	
Spicy	51		55		53		48	
Meaty	62		49		60		59	
Salty	53		50		55		44	
Firm	63		75		60		70	
Gritty	22		46		22		32	
Juicy	54		42		56		56	
Greasy	36		34		37		36	
Aftertaste	50		55		52		48	
Direct—color	1	0	13		−4		12	
Direct—red	−1	0	14		−5		11	
Direct—wet	4	0	5		3		9	
Direct—thick	−2	0	−3		−4		−5	
Direct—spicy	−2	0	3		0		−4	
Direct—meat	−5	0	−11		−5		−2	
Direct—salty	9	0	10		11		1	
Direct—firm	−1	0	8		−4		8	
Direct—gritty	0	0	14		0		6	
Direct—juicy	−1	0	−9		1		−2	
Direct—greasy	6	0	6		6		6	
Similar—chicken	45		41		47		43	70
Similar—turkey	48		42		48		48	
Similar—red meat	49		51	70	50	70	50	
For lunch	53		70	70	47	70	64	30
For a meal	41		59	70	43	70	58	70
For kids	32		30	30	30	30	60	70
Expensive	42		54	30	42	30	57	70
Traditional	41		46	30	40	30	50	70
Fit Concept 1	44		56		48		49	
Fit Concept 2	38		43		39		56	

TABLE 12.23e. Landmarks for Product Development from Reverse Engineering.

Constraint	Profile A LD > 50	Landmark Product
Dark	43	#725
Even	57	#726
Red	43	#727
Wet	54	#727
Thick	46	#735
Flavor	59	#752
Aroma	51	#735
Spicy	53	#790
Meaty	60	#735
Salty	55	#716
Firm	60	#798
Gritty	22	#725
Juicy	56	#729
Greasy	37	#729
Aftertaste	52	#716

into 100 concepts. The design for each consumer is such that the researcher can use that consumer's own data to create a regression model showing how each element tested by the consumer drives acceptance.

Concept Evaluation

Each consumer rates concepts on interest and (optionally) other attributes. Here the panelist rated the concepts on interest, and two other attributes: when they would most likely drink the coffee (morning versus evening), and why they would drink the coffee (more for a snack versus more for at the end of a meal). Panelists find this type of evaluation easy to do. Each panelist rates a completely unique set of combinations. Thus, the DesignLab paradigm explores a large number of possible concept elements in an efficient manner.

Concept Modeling—A Full Utility Matrix for Each Panelist

The experimental design and the ratings provided by each panelist enable the researcher to create a model for each panelist showing how every element tested drives the rating of each of the three attributes, respectively. Furthermore, a separate program enables the researcher to estimate the utilities for elements not directly tested by the panelist (Moskowitz, 1996c). Table 12.25 shows an example of the utility matrix averaged across panelists.

TABLE 12.24. Schematic of the DesignLab Paradigm for Rapid, Early Stage, Consumer-Driven Development.

Step	When	What Happens	Elapsed Time (min)
	Prior to the Fieldwork		
1	1 week ahead	Select concept elements, create restrictions, dimensionalize	
2	1 week to 1 month ahead	Select competitor products, create prototypes	
3	1 week to 1 month ahead	Select package features, scan into computer	
4	1 week ahead	Recruit 25–50 consumer panelists	
	Field Work at the Test Site		
5	5 minutes	Orient the consumers in the task	5
6	20 minutes	Obtain ratings of test concepts	25
7	10 minutes	Obtain results, models, optimize concepts, insert optimized concepts into the product questionnaire, while at the same time consumers take a rest break	35
8	120 minutes	Obtain ratings of test and competitor products	155
9	off line	Analyze product ratings, create category models, optimize	
10	10 minutes	Consumers take a rest break	165
11	20 minutes	Obtain ratings of packages	185
12	60 minutes	Conduct focus group with consumers (optional)	245

Creating New Concepts in Real Time

Based upon the models from 25 to 50 panelists who participate in the same session, and whose data are aggregated within a minute or so, the developer creates new concepts combining current elements. Figures 12.5(a) and 12.5(b) show two concepts developed in this way, emphasizing different aspects of the coffee. The concept model allows the developer to create many different and novel combinations of features to communicate different aspects of the coffee and thus set the stage for different expectations about the product.

Product Testing

After a short rest, the panelists return to the test area, and evaluate existing flavored coffees from the United States, and around the world. They rate each

TABLE 12.25. **Examples of Utilities for Concept Elements.**

Type	Negative Values (−) = Positive Values (+) =	Hate Love	Snack Meal	Evening Morning
T	Bitter	−9	1	3
T	Coffee in a tea bag	−8	4	2
T	Individual coffee bags	−6	4	6
G	European in cafe	0	−5	−5
T	Robust flavor	0	5	4
G	Coffee in spoon	0	2	2
T	Time to reflect, time for oneself	0	−6	5
T	Fresh and natural taste	0	3	0
T	Available in chocolate macadamia	11	2	−4
T	Available in irish cream	11	−1	−5
T	Served at the finest coffee shops	11	−5	−5
T	Available in orange brandy	12	4	2

T = text element; G = graphics element.

coffee on a variety of attributes, including sensory, liking, image, and fit to the three concepts just created in step 5.

Category Modeling

This modeling follows the category modeling approach described above in Table 12.23. This time the category is flavored coffee. The products comprise in-market products, the concepts are those just developed a few minutes before, possibly changed by the developer after the concept optimization stage.

Step 8—Optimization

The optimization follows the approach described in Table 12.23. The optimization algorithm searches for the location in the map corresponding to a user defined objective, and then estimates the full sensory profile of the product corresponding to the optimum. Finally, since the model uses actual products as inputs, the researcher can call up the original matrix of sensory attribute x product, and for optimal sensory attribute identify a landmark coffee product having the requisite sensory level. Figures 12.6(a) and 12.6(b) present a schematic view of this approach.

Optimization may look for the following:

(1) *Holes in the category:* a hole is a location in the factor map where there are no current flavored coffees. Where is the hole? How acceptable is the hole? What is the sensory profile of the product corresponding to the hole? What is the image profile of the product corresponding to the hole?

Specialty flavors

The enticing smell of freshly brewed coffee wafting up to your bedroom

Fresh and exotic flavor

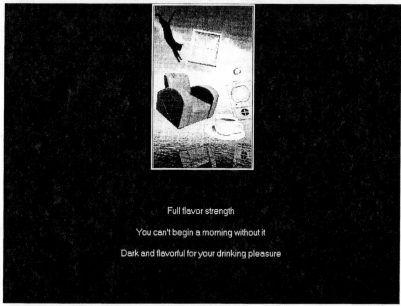

Full flavor strength

You can't begin a morning without it

Dark and flavorful for your drinking pleasure

FIGURE 12.5 Two concepts for flavored coffee, created by maximizing purchase intent and image aspects of the product. These concepts are then inserted in the product evaluation and treated as attributes.

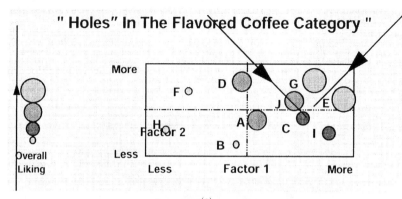

" Holes" In The Flavored Coffee Category "

More

F ○ D G E

J

Factor 2 A C

H I

B ○

Overall
Liking Less

Less Factor 1 More

(a)

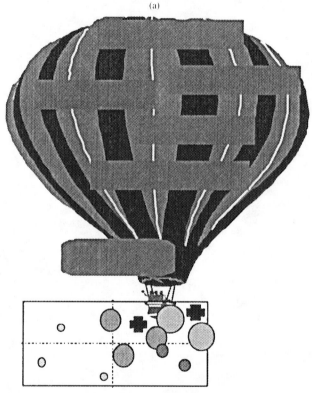

(b)

FIGURE 12.6 (a) Schematic of the DesignLab output. The result of the product analysis is a factor map in which the coordinates are the factors for the sensory attributes. The map may have one to five dimensions. Here we show a map with only two dimensions. Each snack food in the category comprises a location in that map. The objective of the DesignLab paradigm is to find holes in the category, corresponding to product opportunities. An opportunity may be a location where liking is highest, where liking is high and the product fits a concept, etc. (b) Schematic showing how the location of the a product in the factor space generates a full profile of the product, estimated liking, and the concept.

326

What concept generates an image profile matching the image profile of the hole?

(2) *Acceptable products:* locations in the map which are highly liked, obtained by maximizing liking.

(3) *Products which fit a concept:* products which maximize perceived fit to the concept, subject to constraints on liking.

(4) *Products which fit an image:* through reverse engineering and goal fitting, the location in the map corresponding to a predefined image profile.

AN OVERVIEW

This chapter has presented a systematic approach to early stage development. Many of the tools are traditional procedures, hallowed in use, and commonly accepted. Others, such as multidimensional scaling and its allied mapping procedures, represent adoptions of methods commonly used in other fields, such as psychology. Still others, such as category wide modeling and optimization represent new syntheses of methods from different disciplines, designed to make the early development stage easier, more rigorous, and more consumer driven. Finally DesignLab combines these early stages of development into one single coherent, time-efficient process. The objective is to test concepts and products, develop a coherent model, and then provide the developer with straightforward, actionable direction, cost effectively and rapidly.

BIBLIOGRAPHY

Amerine, M. A., R. M. Pangborn, and E. B. Roessler. 1965. *Principles of Sensory Evaluation of Food.* New York: Academic Press.

Boring, E. G., H. S. Langfeld, and H. P. Weld. 1935. *Psychology: A Factual Textbook.* New York: John Wiley & Sons.

Cairncross, S. E. and L. B. Sjöstrom. 1950. "Flavor Profiles—A New Approach to Flavor Problems," *Food Technology,* 4:308–311.

Caul, J. F. 1957. "The Profile Method of Flavor Analysis," *Advances in Food Research,* pp. 1–40.

Clapperton, J., C. E. Dagliesh, and M. C. Meilgaard. 1975. "Progress Towards an International System of Beer Flavor Terminology," *Master Brewers Association of America Technical Journal,* 12:273–280.

Dravnieks, A. 1974. Personal communication.

Meilgaard, M., G. V. Civille, and B. T. Carr. 1987. *Sensory Evaluation Techniques.* Boca Raton: CRC Press, Chapter 8, pp. 119–142.

Meiselman, H. L. 1978. "Scales for Measuring Food Preference," in: *Encyclopedia of Food Science,* M. S. Petersen and A. H. Johnson, eds. Westport, CT: Avi, pp. 675–678.

Moskowitz, H. R., 1981. "Sensory Intensity versus Hedonic Functions: Classical Psychophysical Approaches," *Journal of Food Quality,* 5:109–138.

Moskowitz, H. R. 1983. *Product Testing and Sensory Evaluation of Foods: Marketing and R&D Approaches.* Westport, CT: Food & Nutrition Press.

Moskowitz, H. R. 1985. *New Directions in Product Testing and Sensory Evaluation of Foods.* Westport, CT: Food & Nutrition Press.

Moskowitz, H. R. 1994. *Food Concepts & Products: Just in Time Development.* Trumbull, CT: Food & Nutrition Press.

Moskowitz, H. R. 1996a. "Experts versus Consumers: A Comparison," submitted to *Journal of Sensory Studies.*

Moskowitz, H. R. 1996b. "The DesignLab Paradigm—Sanitary Napkins: How Consumer Research Accelerates the Development Process," *Marketing Research Pays Off,* L. Percy and H. Brenner, eds. In press.

Moskowitz, H. R. 1996c. *Consumer Evaluation and Testing of Personal Care Products.* New York: Marcel Dekker Inc.

Moskowitz, H. R. and A. Gofman. 1996. "The DesignLab Paradigm for Product Creation: Merging Consumers, Creatives, Computers for the Design of a Children's Cereal Product," to appear in an edited volume (S. Poretta ed.). In press.

Moskowitz, H. R., J. G. Kapsalis, A. Cardello, D. Fishken, G. Maller, and R. Segars. 1979. "Determining Relationships among Objective, Expert and Consumer Measures of Texture," *Food Technology,* 23:74–88.

Peryam, D. R. and F. J. Pilgrim. 1957. "Hedonic Scale Method of Measuring Food Preferences," *Food Technology,* 11:9–14.

Pfaffmann, C. M. 1959. "The Sense of Taste," in: *Handbook of Physiology,* Vol 1, Washington, DC: American Physiological Society, pp. 507–534.

Rolls, B. J. and T. M. McDermott. 1991. "Effects of Age on Sensory-Specific Satiety," *American Journal of Clinical Nutrition,* 54:988–996.

Schiffmann, S. S., M. L. Reynolds, and F. W. Young. 1981. *Introduction to Multidimensional Scaling.* New York: Academic Press.

Schutz, H. G. 1965. "A Food Action Rating Scale for Measuring Food Acceptance," *Journal of Food Science,* 30:365–374.

Stevens, S. S. 1946. "On the Theory of Scales of Measurement," *Science,* 103:677–680.

Stevens, S. S. 1953. "On the Brightness of Lights and the Loudness of Sounds," *Science,* 118:576.

Stevens, S. S. 1966. "Matching Functions between Loudness and Ten Other Continua," *Perception & Psychophysics,* pp. 1, 5–8.

Stevens, S. S. 1975. *Psychophysics: An Introduction to Its Sensory, Neural and Social Prospects.* New York: John Wiley.

Stone, H., J. L. Sidel, S. Oliver, A. Woolsey, and R. Singleton. 1974. "Sensory Evaluation by Quantitative Descriptive Analysis," *Food Technology,* 28:28–34.

Szczesniak, A. S., J. A. Brandt, and H. H. Friedman. 1963. "Development of Standard Rating Scales for Mechanical Parameters of Texture and Correlation between the Objective and Sensory Methods of Texture Evaluation," *Journal of Food Science,* 28:39–403.

Vickers, Z. 1988. "Sensory Specific Satiety in Lemonade Using a Just Right Scale for Sweetness," *Journal of Sensory Studies,* 3:1–8.

CHAPTER 13

Shelf Life of Packaged Foods, Its Measurement and Prediction

GORDON L. ROBERTSON

Because all foods change during distribution, usually deteriorating, an understanding of and ability to predict these processes are indispensable components of food product development. It is the food product developer/processor's responsibility to comprehend the changes that take place in distribution. Shelf life prediction consists of two-parts: quantifying the inherent product deterioration characteristics and coupling those with the properties of the packaging/distribution system. Mathematical models have been developed and used with some effectiveness as a guide to shelf life. To reduce the probability of surprises, however, actual shelf life testing is essential. Whether accelerated shelf life testing may be applicable depends on the product and its packaging.

T HE quality of most foods and beverages decreases with time. Exceptions include distilled spirits that develop desirable flavors during storage in wooden barrels, some wines that increase in flavor complexity in bottles, and many cheese varieties where aging leads to desirable flavors and textures.

For the majority of foods and beverages, however, a finite time occurs before the product becomes unacceptable. This time from production to unacceptability is usually designated shelf life.

No simple, generally accepted definition of shelf life exists. The Institute of Food Technologists (Anonymous, 1974) has defined shelf life as the period between the manufacture and the retail purchase of a food product, during which time the product is of satisfactory quality in terms of nutritional value, flavor, texture, and appearance. An alternative definition is that shelf life is that period between the packing of a product and its use, for which the quality of the product remains acceptable to the product user.

Shelf life must be determined for each product by the processor. Storage studies are an indispensable element of food product development, with the processor attempting to provide the longest shelf life practicable consistent with economics and distribution. Inadequate shelf life will lead to consumer dissatisfaction and complaints, and eventually adversely affect the acceptance and sales of branded products.

Since the advent of the consumer movement, many different types of open dating systems have been proposed as part of the consumer's "right to know." An open date on a food product is a legible, easily read date which is displayed on the package with the purpose of informing the consumer about the shelf life of the product. Several types of dates can be used (Dethmers, 1979; Labuza, 1982):

- *pack* date: the date on which the product was packed into its primary package (it does not provide any specific information as to the quality of the product when purchased or how long it might retain its quality after purchase.)
- *display* date: the date on which the product was placed on the shelf by the retailer
- *pull* or *sell by* date: the last date on which the product should be sold in order to allow the consumer a reasonable length of time in which to use it
- *best before* or *best if used by* date: the last date of maximum high quality
- *use by* date or *expiration* date: the date after the which the food should no longer be at an acceptable level of quality

These forms of dating are used infrequently because quality changes generally occur slowly and it is really not possible to state that a food will be acceptable one day and unacceptable the next.

FACTORS CONTROLLING SHELF LIFE

Product shelf life is controlled by three factors:

- product characteristics
- the environment to which the packaged product is exposed during distribution
- the properties of the package

Product shelf life may be altered by changing its composition and form, the environment to which it is exposed, or its packaging system (Harte and Gray, 1987).

PRODUCT CHARACTERISTICS

Perishability

Based on the nature of the changes that can occur during storage, foods may be divided into three categories: perishable, semiperishable, or ambient temperature shelf stable, which translate into very short shelf life products, short to medium shelf life products, and medium to long shelf life products.

Perishable foods are those subject to microbiological and/or enzymatic deterioration and so must be held at chill or freezer temperatures. Examples of such foods would include milk; fresh meat, poultry, and fish; and many fresh fruits and vegetables.

Semiperishable foods are those that contain natural inhibitors (e.g., some cheeses, eggs, etc.) or those that have received minimal preservation treatment (e.g., milk pasteurization, ham smoking, and vegetable fermentation) that delivers greater tolerance to environmental conditions and abuse during distribution.

Ambient temperature shelf stable foods are often regarded as "nonperishable" at room temperatures. Some "natural" foods fall into this category (e.g., cereal grains and nuts, and some confectionery products). Processed food products can be shelf stable if they are preserved by thermal sterilization (e.g., canned foods), contain preservatives (e.g., soft drinks), are formulated as dry mixes (e.g., cake mixes), or processed to reduce their water content (e.g., raisins or crackers). However, ambient temperature shelf stable foods only retain this status if the integrity of the package which contains them remains intact. Even then, their shelf life is finite due to deteriorative chemical reactions that proceed at ambient temperature and the permeation through packages of gases, odors, and water vapor.

Bulk Density

For packages of similar shape, equal weights of products of different bulk densities have different free space volumes and, as a consequence, package areas and package behavior differ. This has important implications when changes are made in package size for the same product, or process alterations are made, resulting in changes to the product bulk density.

The bulk density of food powders can be affected by processing and packaging. Some food powders (e.g., milk and coffee) are instantized by treating individual particles so that they form free-flowing agglomerates or aggregates in which there are relatively few points of contact; the surface of each particle is thus more easily wetted when the powder is rehydrated.

The free space volume has an important influence on the rate of oxidation of foods, because if a food is packaged in air, a large free space volume is undesirable since it constitutes a large oxygen reservoir. Conversely, if the product is packaged in an inert gas, a large free space volume acts as a large "sink" to minimize the effects of oxygen transferred through the film. It follows that a large package area and a low bulk density result in greater oxygen transmission.

Concentration Effects

The progress of a deteriorative reaction in a packaged food can be monitored by following the changes in concentrations of some key components. In many foods, however, the concentration varies from point to point, even at zero time. Because most of these compounds have little opportunity to move, the concentration differences increase as the reactions proceed out from isolated initial foci.

Further, several different deteriorative reactions may proceed simultaneously, and different stages may have different dependence on concentration and temperature. Such a situation is frequently the case for chain reactions and microbiological growth which have both a lag and a log phase with very different rate constants.

Thus, for many foods it may be difficult to obtain kinetic data useful for predictive purposes. Sensory panels to determine the acceptability of the food are therefore the recommended procedure.

DISTRIBUTION ENVIRONMENT

Climatic

The deterioration in product quality of packaged foods is often closely related to the transfer of mass and heat through the package. Packaged foods may lose or gain moisture; they will also reflect the temperature of their environment because very few food packages are good thermal insulators. Thus, the distribution environment has an important influence on the rate of deterioration of packaged foods.

Mass Transfer

With mass transfer, the exchange of vapors and gases with the surrounding atmosphere is of primary concern. Water vapor and oxygen are generally of most importance, although the exchange of volatile aromas from or to the product from the surroundings can be important. Transmission of nitrogen and carbon dioxide may have to be taken into account in some packages.

Generally, the difference in partial pressure of the vapor or gas across the package barrier controls the rate and extent of permeation, although transfer can also occur due to the presence of pinholes in the material, channels in seals and closures, or cracks that result from flexing of the package material during filling and subsequent handling. Because the gaseous composition of the atmosphere is constant at sea level, the partial pressure difference of gases across the package material depends on the internal atmosphere of the package at the time the package was sealed.

In contrast to the common gases, the partial pressure of water vapor in the atmosphere varies continuously, although the variation is generally much less in controlled climate environments (Porter, 1981). Thus, mass transfer depends on the partial pressure difference across the package barrier, and on the nature of the barrier itself.

Heat Transfer

One of the major determinants of product shelf life is the temperature to which the product is exposed during its lifetime. Without exception, food products are exposed to fluctuating temperature environments, and it is important, if an accurate prediction of shelf life is to be made, that the nature and extent of these temperature variations are known. There is little point in carefully controlling the processing conditions inside the factory and then releasing the product into the distribution and retail system without some knowledge of the conditions it will experience in that system. The storage climates inside warehouses and supermarkets are only broadly related to the external climate.

If the major deteriorative reactions causing end of shelf life are known, expressions can be derived to predict the extent of deterioration as a function of available time-temperature storage conditions.

Fundamental to a predictive analysis is that the particular food under consideration follows the "laws" of "additivity" and "commutativity." *Additivity* implies that the total extent of the degradation reaction in the food produced by a succession of exposures at various temperatures is the simple sum of the separate amounts of degradation, regardless of the number or spacing of each time-temperature combination. *Commutativity* means that the total extent of the degradation reaction in the food is independent of the order of presentation of the various time-temperature experiences.

Shelf Life Plots

One useful approach to quantifying the effect of temperature on food quality is to construct shelf life plots (Labuza and Kamman, 1983). Several models are in use to represent the relationship between the rate of a reaction (or the reciprocal of rate which can be time for a specified loss in quality or shelf life) and temperature. The two most-used models are the Arrhenius and linear, and these are shown in Figure 13.1.

The equations for these two plots are

$$\theta_s = \theta_o \, \exp \frac{E_A}{R} \left[\frac{1}{T_s} - \frac{1}{T_o} \right] \qquad (13.1)$$

and

$$\theta_s = \theta_o \, e^{-b(T_s - T_o)} \qquad (13.2)$$

where θ_s = shelf life at temperature T_s and θ_o = shelf life at temperature T_o.

If only a small temperature range is used (less than $\pm 40°F$), there is little error in using the linear plot rather than the Arrhenius plot.

Most deteriorative reactions in foods can be classified as either zero or first order. The way in which these two reaction orders can be used to predict the extent of deterioration as a function of temperature is outlined.

Zero-Order Reaction Prediction

The change in a quality factor A when all extrinsic factors are held constant is expressed in Equation (13.3):

$$A_c = A_o - k_z \theta_s \qquad (13.3)$$

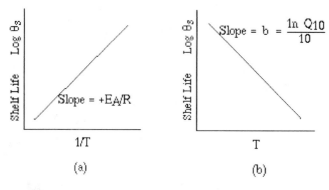

FIGURE 13.1 (a) Arrhenius plot of log shelf life (θ_s) versus reciprocal of the absolute temperature (K) showing a slope of E_A/R, and (b) linear plot of log shelf life versus temperature (°C) showing a slope of b. Reprinted from Robertson (1993), p. 345, by courtesy of Marcel Dekker, Inc.

and

$$A_o = A_c = k_z \theta_s \qquad (13.4)$$

where A_c = value of A at end of shelf life, A_o = value of A initially, k_z = zero order rate constant (time^{-1}), and θ_s = shelf life in days, months, years, etc.

For variable time-temperature storage conditions, Equation (13.3) can be modified as follows:

$$A_c = A_o - \Sigma(K_i \theta_i) \qquad (13.5)$$

where $\Sigma k_i \theta_i$ = the sum of the product of the rate constant k_i at each temperature and T_i times the time interval θ_i at the average temperature T_i for the given time period $\Delta\theta$.

To apply this method, the time-temperature history is divided into suitable time periods and the average temperature in each time period is determined. The rate constant for each period is then calculated from the shelf life plot using a zero-order reaction. The rate constant is multiplied by the time interval θ_i, and the sum of the increments of $k_i \theta_i$ gives the total amount lost at any time.

Alternatively, instead of calculating actual rate constants, the time for the product to become unacceptable (i.e., for A to become A_c) can be measured, and Equation (13.5) modified to give

f_c = fraction of shelf life consumed

= change in A divided by total possible change in A

$$= \frac{A_o - A}{A_o - A_c} \qquad (13.6)$$

$$= \frac{\Sigma(k_i \theta_i)}{\Sigma(k_i \theta_s)} \qquad (13.7)$$

$$= \Sigma \left[\frac{\theta_i}{\theta_s} \right] T_i \qquad (13.8)$$

The temperature history is divided into suitable time periods and the average temperature T_i at each time period evaluated. The time held at that temperature θ_i is then divided by the shelf life θ_s for that particular temperature, and the fractional values summed to give the fraction of shelf life consumed.

The shelf life can also be expressed in terms of the fraction of shelf life remaining, f_r:

$$f_r = 1 - f_c \tag{13.9}$$

Thus for any temperature T_s:

$$f_r \theta_s = (1 - f_c)\theta_s = \text{shelf life at temperature } T_s \tag{13.10}$$

In other words, the shelf life at any temperature is the fraction of shelf life remaining times the shelf life at that temperature.

The above method is referred to as the TTT or time/temperature tolerance approach (Van Arsdel, 1969). To use this method, the period of time (designated as the "high quality life" or HQL) for 70 to 80% of a trained sensory panel to correctly identify the control samples from samples stored at various other temperatures using the triangle or duo-trio difference test is determined. The change in quality at this stage has been designated the "just noticeable difference" (JND). The HQL has no real commercial significance and is quite different from the "practical storage life" (PSL), which is of interest to food processors and consumers. The ratio between PSL and HQL is often referred to as the "acceptability factor" and can range from 2:1 up to 6:1.

Generally, the HQL varies exponentially with temperature. When overall quality rather than just one single quality factor is measured, however, a semilogarithmic plot results in curved rather than straight lines.

Time/temperature tolerance relationships are not strict mathematical functions but empirical data subject to large variability, particularly because of variations in product, processing methods, and packaging (the PPP factors). Therefore, any shelf life prediction made will be specific for a particular product processed, packaged, and stored under specific conditions. Predictions cannot be made with any precision on the quality or quality change in a food from knowledge of its time-temperature history and TTT literature data only. Therefore, in determining the shelf life of foods, the PPP factors in addition to the TTT relationships must be taken into account.

Rather than follow the TTT approach and use a linear model to relate the quality loss with temperature, a distribution system using the Arrhenius model and converting a variable temperature history to equivalent time at a standard temperature may be used (Rosenfeld, 1984).

The distribution system is divided into four stages and tables developed describing the distribution system as equivalent time for a range of Q_{10} values. Data are also collected that enable calculation of the mean and standard deviation days that a product spends at each stage of the distribution system. If the failure point of the product is beyond the 90 or 95 percentile, then the product is considered to have sufficient shelf life to survive the given distribution system. This information could also be used as a guide for stability limits for new product development.

First-Order Reaction Prediction

The expression for a first-order reaction for the case in which all extrinsic factors are held constant is shown in Equation (13.11).

$$A_c = A_o \exp(-k\theta_s) \tag{13.11}$$

From this an expression can be developed to predict the amount of shelf life used up as a function of variable temperature storage for a first-order reaction in the form:

$$A = A_o \exp(-\Sigma k_i \theta_i) \tag{13.12}$$

where A = the amount of some quality factor remaining at the end of the time-temperature distribution, and $\Sigma k_i \theta_i$ has the same meaning as in Equation (13.5).

If the shelf life is based simply on some time to reach unacceptability, Equation (13.12) can be modified to give an analogous expression to that derived for the TTT method. Note that because of the exponential loss of quality, A_c will never be zero. Thus,

$$\ln \frac{A}{A_o} = \Sigma k_i \theta_i \tag{13.13}$$

and

$$k_i = \frac{\ln A_c/A_o}{\theta_s} \tag{13.14}$$

where $\ln A/A_o$ = fraction of shelf life consumed at time θ and $\ln A_c/A_o$ = fraction of shelf life consumed at time θ_s.

The fraction of shelf life remaining, f_r, is

$$f_r = 1 - \frac{\ln A_o/A}{\ln A_o/A_c} = 1 - \Sigma \left[\frac{\theta_i}{\theta_s} \right] T_i \tag{13.15}$$

Sequential Fluctuating Temperatures

Although the above analysis can be applied to any random time/temperature storage regime, in practice many products are exposed to a sequential regular fluctuating temperature profile, especially if held in trucks, rail cars and warehouses. This is because of the daily day-night pattern resulting from exposure to solar radiation. Many of these patterns can be assumed to follow either a square or sine wave form as shown in Figure 13.2.

Equations have been developed (Labuza, 1979) for both zero and first-order reactions that enable calculation of the extent of a degradative reaction for a food subjected to either square wave or sine wave temperature functions. The extent of reaction after a time period is the same as it would have been if the food had been held at a certain steady "effective" temperature for the same length of time. This effective temperature is higher than the arithmetic mean temperature. Comparisons for losses in a theoretical temperature distribution show that for less than 50% degradation the losses are about the same for zero and first order at any time, and thus determination of the reaction order is not critical. However, the temperature sensitivity (Q_{10}) of the reaction is very important in making predictions.

Simultaneous Mass and Heat Transfer

In the majority of distribution environments, many packaged foods undergo changes in both moisture content and temperature during storage as a result of varying temperature and relative humidity conditions in the environment. This has the effect of complicating the calculations for prediction of the shelf life of packaged foods. It is unlikely that a package would be totally

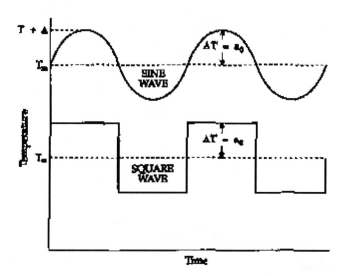

FIGURE 13.2 Square and sine wave temperature fluctuations of packaged foods where a_o is the amplitude. Reprinted from Robertson (1993), p. 352, by courtesy of Marcel Dekker, Inc.

impermeable to water vapor, and therefore the a_w would change with time. This complicates the calculation of quality loss, since the rate is now dependent on both temperature and a_w.

A further complication is that data on the relative humidity distribution of environments in which foods are stored are scarce and not as easily predicted as the external temperature distribution. Therefore, prediction of the actual shelf life loss of packaged foods will only be approximate. More complete data are required about the humidity distribution of food storage environments so that shelf life predictions can be further refined.

PACKAGE PROPERTIES

Foods can be classified according to the amount of protection required, as shown in Table 13.1. The advantage of this sort of analysis is that attention can be focused on the key requirements of the package such as maximum moisture gain or oxygen uptake. This then enables calculations to be made to determine whether or not a particular package structure would provide the necessary barrier required to give the desired product shelf life. Metal cans and glass bottles or jars can be regarded as essentially impermeable, while paper-based packaging materials can be regarded as permeable. Plastic-based packaging materials provide varying degrees of protection, depending largely on the nature of the polymers and their package structures.

The expression for the steady state permeation of a gas or vapor through a thermoplastic material can be written as (Robertson, 1993):

$$\frac{\delta w}{\delta t} = \frac{P}{X} \cdot A \cdot (p_1 - p_2) \qquad (13.16)$$

where P/X is the permeance (the permeability constant P divided by the thickness of the film X), A is the surface area of the package, p_1 and p_2 are the partial pressures of water vapor outside and inside the package, and $\delta w/\delta t$ is the rate of gas or vapor transport across the film, the latter term corresponding to Q/t in the integrated form of the expression.

Water Vapor Transfer

The prediction of the moisture transfer either to or from a packaged food basically requires analysis of Equation (13.16) given certain boundary conditions. The simplest analysis requires the assumptions that P/X is constant, that the external environment is at constant temperature and humidity, and that p_2, the vapor pressure of the water in the food, follows some simple function of the moisture content.

TABLE 13.1. Degree of Protection Required by Various Foods and Beverages [Assuming One Year Shelf Life at 25°C (79°F)].

Food/ Beverage	Maximum Amount of O_2 Gain (ppm)	Other Gas Protection Needed	Maximum Water Gain or Loss	Requires High Oil Resistance	Requires Good Barrier to Volatile Organics
Canned milk and meats	1–5	No	3% Loss	Yes	No
Baby foods	1–5	No	3% Loss	Yes	Yes
Beers and wine	1–5	< 20% CO_2 Loss	3% Loss	No	Yes
Instant coffee	1–5	No	2% Gain	Yes	Yes
Canned soups, vegetables, and sauces	1–5	No	3% Loss	No	No
Canned fruits	5–15	No	3% Loss	No	Yes
Nuts, snacks	5–15	No	5% Gain	Yes	No
Dried foods	5–15	No	1% Gain	No	No
Fruit juices and drinks	10–40	No	3% Loss	No	Yes
Carbonated soft drinks	10–40	< 20% CO_2 Loss	3% Loss	No	Yes
Oils and shortenings	50–200	No	10% Gain	Yes	No
Salad dressings	50–200	No	10% Gain	Yes	Yes
Jams, jellies, syrups, pickles, olives, vinegars	50–200	No	3% Loss	No	Yes
Liquors	50–200	No	3% Loss	No	Yes
Condiments	50–200	No	1% Gain	No	Yes
Peanut butter	50–200	No	10% gain	Yes	No

Reprinted from Robertson (1993), p. 355, by courtesy of Marcel Dekker, Inc.

External conditions do not remain constant during storage, distribution, and retailing of a packaged food. Therefore, P/X will not be constant. However, using WVTRs determined at 98°F (37°C)/90% RH gives a "worst-case" analysis, but if the food is being sold in markets in temperate climates, use of WVTRs determined at 77°F (25°C)/75% RH would be more appropriate. WVTRs can be converted to permeances by dividing by Δp.

A further assumption is that the moisture gradient inside the package is negligible, i.e., the package should be the major resistance to water vapor transport. This is the case whenever P/X is less than about 10 g m^{-2} day^{-1} (cm Hg)$^{-1}$, which is the case for most films but not paperboard under high humidity conditions.

The critical point about Equation (13.16) is that the internal water vapor pressure is not constant but varies with the moisture content of the food at any time. Thus the rate of gain or loss of moisture is not constant but falls as ·p gets smaller. Therefore some function of p_2, the internal vapor pressure, as a function of the moisture content, must be inserted into the equation to be able to make proper predictions. If a constant rate is assumed, the product will be overprotected.

In low and intermediate moisture foods, the internal vapor pressure is determined solely by the water sorption isotherm of the food. Several functions can be applied to describe a sorption isotherm, although the preferred one is the G.A.B. (from Guggenheim-Anderson-de Boer) model (Van Den Berg and Bruin, 1981). If a linear model is used, the result is directly integratable, but if the G.A.B. model is used, it must be numerically evaluated using computational techniques.

In the simplest case, the isotherm is treated as a linear function:

$$m = b \, a_w + c \tag{13.17}$$

where m = moisture content in g H_2O per g solids, a_w = water activity, b = slope of curve, and c = constant.

The moisture content can be substituted for water gain using the relationship:

$$m = \frac{W \text{ (weight of water transported}}{W_s \text{ (weight of dry solids enclosed)}} \tag{13.18}$$

$$\therefore W = mW_s \tag{13.19}$$

and

$$\delta W = \delta m W_s \tag{13.20}$$

By substitution:

$$\frac{\delta W}{\delta t} = \frac{\delta m W_s}{\delta t} = \frac{P}{X} \cdot A \cdot \left[\frac{p_o m_e}{b} - \frac{p_o m}{b} \right] \tag{13.21}$$

which on rearranging gives

$$\frac{\delta m}{m_e - m} = \frac{P}{X} \cdot \frac{A}{W_s} \cdot \frac{p_o}{b} \cdot \delta t \tag{13.22}$$

and on integrating

$$\ln \frac{m_e - m_i}{m_e - m} = \frac{P}{X} \cdot \frac{A}{W_s} \cdot \frac{p_o}{b} \cdot t \qquad (13.23)$$

where m_e = equilibrium moisture content of the food if exposed to external package RH, m_i = initial moisture content of the food, m = moisture content of the food at time t, and p_o = water vapor pressure of pure water at the storage temperature (not the actual vapor pressure outside the package).

A plot of the log of the unaccomplished moisture change (the term on the left-hand side of Equation (13.23)) versus time is a straight line with a slope equivalent to the bracketed term on the right-hand side of the equation.

The end of product shelf life is reached when $m = m_c$, the critical moisture content, at which time $t = \theta_s$, the shelf life. Thus Equation (13.23) can be rewritten as

$$\ln \frac{m_e - m_i}{m_e - m_c} = \frac{P}{X} \cdot \frac{A}{W_s} \cdot \frac{p_o}{b} \cdot \theta_s \qquad (13.24)$$

The relationship between the initial, critical, and equilibrium moisture contents is illustrated in Figure 13.3.

To simplify matters, the packaging parameters can be combined into one constant as

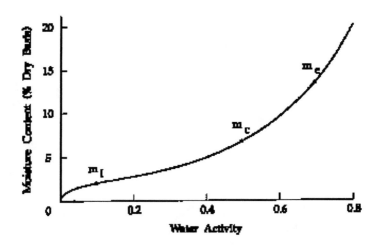

FIGURE 13.3 Typical moisture sorption isotherm for a food product in which m_i = initial moisture content; m_c = critical moisture content of product; m_e = equilibrium moisture content. Reprinted from Robertson (1993), p. 358, by courtesy of Marcel Dekker, Inc.

$$\Omega = \frac{P}{X} \cdot \frac{A}{W_s} \qquad (13.25)$$

Using Equation (13.25), one can calculate a minimum Ω, given a critical moisture content and maximum desired shelf life. Then from Equation (13.24) for a given package size and weight of product, the permeance can be calculated and a package structure to satisfy this condition selected.

Equation (13.23) and the corresponding one for moisture loss:

$$\ln \frac{m_i - m_e}{m - m_e} = \frac{P}{X} \cdot \frac{A}{W_s} \cdot \frac{p_o}{b} \cdot t \qquad (13.26)$$

have been extensively tested for foods and found to give excellent predictions of actual weight gain or loss (Labuza, 1984). These equations are also useful when calculating the effect of changes in the external conditions (e.g., temperature and humidity), the surface area-to-volume ratio of the package, and variations in the initial moisture content of the product.

Given specific external conditions and a critical a_w for moisture gain, the shelf life is

$$\theta_s = \Phi \cdot \frac{W_s}{A} = \Phi' \cdot \frac{V}{A} = \Phi'' \cdot r \qquad (13.27)$$

where Φ, Φ', and Φ'' are constants proportional to

$$\left[\ln \frac{m_e - m_i}{m_e - m}\right] \div \left[\frac{P}{X} \cdot \frac{p_o}{b}\right] \qquad (13.28)$$

where W_s = weight of food solids = $p \times V$, p = density of food, V = volume of food, r = characteristic package thickness, and θ_s = time to end of shelf life.

Because the V/A ratio decreases as package size gets smaller by a factor equivalent to the characteristic thickness of the package, the shelf life using the same film will decrease directly by this thickness. Thus, to ensure adequate shelf life for a food in varying sizes of packages, shelf life tests should be based on the smallest package.

Gas and Odor Transfer

The gas of major importance in packaged foods is oxygen since it plays a crucial role in many reactions which affect the shelf life of foods, e.g., microbial growth, color changes, oxidation of lipids and consequent rancidity, and senescence of fruit and vegetables.

The transfer of gases and odors through packaging materials can be analyzed in an analogous manner to that described for water vapor transfer, provided that values are known for the permeance of the packaging material to the appropriate gas, and the partial pressure of the gas inside and outside the package.

Packaging can control two variables with respect to oxygen, and these can have different effects on the rates of oxidation reactions in foods (Karel, 1974):

- Total amount of oxygen present. This influences the extent of the reaction and, in impermeable packages, where the total amount of oxygen available to react with the food is finite, the extent of the reaction cannot exceed the amount corresponding to the complete exhaustion of the oxygen present inside the package at the time of sealing. This may or may not be sufficient to result in an unacceptable product quality after a period of time, dependent on the rate of the oxidation reaction. Such a rate is, or course, temperature dependent. With permeable packages (e.g., plastic packages), where ingress of oxygen occurs during distribution, two factors are important: sufficient oxygen may be present inside the package to cause product unacceptability when it has all reacted with the food, or there may be sufficient transfer of oxygen into the package over time to result in product unacceptability through oxidation.
- Concentration of oxygen in the food. In many cases, relationships between the oxygen pressure in the space surrounding the food and the rates of oxidation reactions can be established. If the food itself is very resistant to diffusion of oxygen, then it will probably be very difficult to establish a relationship between the oxygen pressure in the space surrounding the food and the concentration of oxygen in the food.

The principal difference between predominantly water vapor–sensitive and oxygen-sensitive foods is in the fact that the latter are generally more sensitive by 2 to 4 orders of magnitude (Heiss, 1980). Thus, the amount of oxygen present in the air-filled headspace of oxygen-sensitive foods must not be neglected when predicting their shelf life. This amount is actually 32 times higher per unit volume of air than per unit volume of oxygen-saturated water. A further complicating factor with oxygen-sensitive foods is that a concentration gradient occurs in them much more frequently than in moisture-sensitive foods.

Prediction of the shelf life of food products which deteriorate by two or more mechanisms simultaneously is more complex. Some general approaches that can be applied to any number of deteriorative reactions have been proposed for food such as chips which deteriorate by two mechanisms simultaneously: oxidation due to ingress of atmospheric oxygen, and loss of crispness due to ingress of moisture (Quast and Karel, 1972).

The simultaneous transfer of water vapor and gases through the package when in an environment with fluctuating temperature and humidity makes quantitative analysis of the deteriorative reactions occurring in the foods (and hence prediction of shelf life) exceedingly complex. Reliance is being placed on the use of accelerated shelf life testing (ASLT) procedures as a more cost-effective and simpler method for the determination of product shelf life.

SHELF LIFE TESTING

Generally the shelf life testing of food products falls into one of three categories (Gacula, 1975):

- experiments designed to determine the shelf life of existing products
- experiments designed to study the effect of specific factors and combinations of factors such as storage temperature, package materials, or food additives on product shelf life
- tests designed to determine the shelf life of prototype or newly developed products

Basic approaches to determining the shelf life of a food product include

- Literature study: the shelf life of an analogous product has been recorded in the published literature or organization files or reports.
- Turnover time: the average length of time which a product spends in distribution is found by monitoring sales and, from this, the required shelf life is estimated. This does not give the "true" shelf life of the product but rather the "required" shelf life, it being implicitly assumed that the product is still acceptable for some time after the average period on the retail shelf.
- End point study: random samples of the product are purchased from distribution channels and tested in the laboratory to determine their quality; from this a reasonable estimate of shelf life can be obtained since the product has been exposed to actual environmental stresses encountered during distribution.
- Accelerated shelf life testing: laboratory studies are undertaken during which environmental conditions are accelerated by a known factor so that the product deteriorates at a faster than normal rate. This method requires that the effect of environmental conditions on product shelf life can be quantified.

Regardless of the method chosen or the reasons for its choice, sensory evaluation of the product is likely to be used either alone or in combination with instrumental analyses to determine the quality of the product. Because human judgment is the ultimate arbiter of food acceptability, it is essential that the results obtain from any instrumental or chemical analysis correlate closely with the sensory judgments for which they are to substitute.

In chemical and physical tests, analytical parameters are isolated so that a single signal is monitored, whereas sensory responses are more complex because of the integration of multiple signals due to the interdependence of appearance, texture, aroma, and flavor of a food. Hedonic responses and, to a lesser extent, intensity judgments are subject to many experimental influences such as past exposure to the product and those created by the actual test protocol.

Difference methods are used to measure whether reference samples are different from stored samples, or "control" samples from test samples. These methods require trained or experienced panelists. Three experimental designs are commonly used for the purpose of shelf life testing (Labuza and Schmidl, 1988): the paired comparison test, the duo-trio test, and the triangle test. Further details about these tests can be found in Chapter 14 of this book.

In shelf life testing there can be one or more criteria which constitute sample failure. One criterion is an increase or decrease by a specified amount in the mean panel score. Another criterion is microbiological deterioration of the food to an extent that renders it unsuitable or unsafe for human consumption. Finally, any physical changes such as changes in color, mouthfeel, flavor, etc., that render the sample unacceptable to either the panel or the consumer are criteria for product failure. Thus sample failure can be defined as the condition of the product that exhibits either physical, chemical, microbiological, or sensory characteristics that are unacceptable to the consumer, and the time required for the product to exhibit such conditions is the shelf life of the product.

However, a fundamental requirement in the analysis of data is knowledge of the statistical distribution of the observations, so that the mean time to failure and its standard deviation can be accurately estimated, and the probability of future failures predicted. The length of shelf life for food products is usually obtained from simple averages of time to failure on the assumption that the failure distribution is symmetrical (Gacula and Kubala, 1975). If the distribution is skewed, estimates of the mean time to failure and its standard deviation will be biased. Further, when the experiment is terminated before all the samples have failed, the mean time to failure based on simple averages will be biased because of the inclusion of unfailed data.

In order to improve the method of estimation of shelf life, knowledge of the statistical distribution of shelf life failures is required, together with an appropriate model for data analysis.

One problem with shelf life testing is to develop experimental designs which minimize the number of samples required, thus minimizing the cost of the testing while still providing reliable and statistically valid answers.

ACCELERATED SHELF LIFE TESTING (ASLT)

BASIC PRINCIPLES

The basic assumption underlying accelerated shelf life testing (ASLT) is that the principles of chemical kinetics can be applied to quantify the effects which extrinsic factors such as temperature, humidity, gas atmosphere, and light have on the rate of deteriorative reactions. By subjecting the food to controlled environments in which one or more of the extrinsic factors is maintained at a higher than normal level, the rates of deterioration will be accelerated, resulting in a shorter than normal time for product failure. Because the effects of extrinsic factors on deterioration can be quantified, the magnitude of the acceleration can be calculated and the "true" shelf life of the product under normal conditions calculated.

The need for ASLT of food products is simple: since many foods have shelf lives of one year, evaluating the effect on shelf life of a change in the product, the process, or the packaging would require shelf life trials lasting at least as long as the required shelf life of the product. Companies cannot afford to wait for such long periods before knowing whether or not the new product/process/packaging will give an adequate shelf life, because other decisions have lead times of months and/or years. The use of ASLT in the food industry is not as widespread as it might be, due in part to the lack of basic data on the effect of extrinsic factors on the rates of deteriorative reactions, in part to ignorance of the methodology required, and in part to a skepticism of the advantages to be gained from using ASLT procedures.

Quality loss for most foods follows either a zero-order or first-order reaction. Figure 13.1 showed the logarithm of shelf life versus temperature and the inverse of absolute temperature. If only a small range of temperature is considered, the former shelf life plot generally fits the data for food products.

For a given extent of deterioration and reaction order, the rate constant is inversely proportional to the time to reach some degree of quality loss. Thus by taking the ratio of the shelf life between any two temperatures 10°C (18°F) apart, the Q_{10} of the reaction can be found. This can be expressed by Equation (13.31):

$$Q_{10} = \frac{k_{T+10}}{k_T} = \frac{\theta_{S_T}}{\theta_{S_{T+10}}} \qquad (13.29)$$

where θ_{s_T} = shelf life at temperature T°C and $\theta_{s_{T+10}}$ = shelf life at temperature $(T + 10)^\circ$C assuming a linear shelf life plot. The effect of Q_{10} on shelf life is shown in Table 13.2, which illustrates the importance of accurate estimates of the Q_{10} value when making shelf life predictions. Typical Q_{10} values for foods have been reported as 1.1 to 4 for canned products, 1.5 to 10 for dehydrated products, and 3 to 40 for frozen products (Labuza, 1982).

A further use for Q_{10} values is illustrated in Figure 13.4, which depicts a shelf life plot for a product that has at least 18 months shelf life at 23°C (73°F). To determine the probable shelf life of the product at 40°C (104°F), lines are drawn from the point corresponding to 18 months at 23°C (73°F) to intersect a vertical line drawn at 40°C (104°F); the slope of each of the straight lines so drawn is dictated by the Q_{10} value. Thus if the Q_{10} of the product were 5, its shelf life at 40°C (104°F) would be 1 month, increasing to 5.4 months if the Q_{10} were 2. Such a plot is helpful in deciding how long an ASLT is likely to run.

ASLT PROCEDURES

The following procedure should be adopted in developing a shelf life test for a food product:

- Determine the microbiological safety and quality parameters for the product.
- Select the key deteriorative reaction(s) that will cause quality loss and thus consumer unacceptability in the product, and decide what tests (sensory and/or instrumental) should be performed on the product during the trial.
- Select the package to be used; often a range of packaging materials will be tested so that the most cost-effective material can be selected.
- Select the extrinsic factors which are to be accelerated. Typical

TABLE 13.2. **Effect of Q_{10} on Shelf Life.**

Temperature		Shelf Life (Weeks)			
°C	°F	$Q_{10} = 2$	$Q_{10} = 2.5$	$Q_{10} = 3$	$Q_{10} = 5$
50	122	2*	2*	2*	2*
40	104	4	5	6	10
30	86	8	12.5	18	50
20	68	16	31.3	54	4.8 years

* Arbitrarily set at 2 weeks at 50°C (122°F). Shelf lives at lower temperatures are calculated on this arbitrary assumption.
Adapted from Robertson (1993), p. 368, by courtesy of Marcel Dekker, Inc.

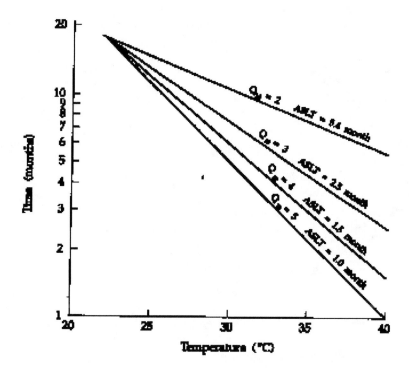

FIGURE 13.4 Hypothetical shelf life plot for various Q_{10}s passing through a shelf life of 18 months at 23°C (73°F). Accelerated shelf life times (ASLT) are those required at 40°C (104°F) for various Q_{10}s. Reprinted from Robertson (1993), p. 369, by courtesy of Marcel Dekker, Inc.

storage conditions used for ASLT procedures are shown in Table 13.3, and it is usually necessary to select at least two.

- Using a plot similar to that shown in Figure 13.4, determine how long the product must be held at each test temperature. If no Q_{10} values are known, then an open-ended ASLT will have to be conducted.
- Determine the frequency of the tests. A good rule of thumb (Labuza, 1985) is that the time interval between tests at any temperature below the highest temperature should be no longer than

$$f_2 = f_1 Q_{10}^{\Delta/10} \tag{13.30}$$

where f_1 = the time between tests (e.g., days, weeks) at the highest test temperature T_1, f_2 = the time between tests at any lower temperature T_2, and Δ = the difference in degrees Celsius between T_1 and T_2.

TABLE 13.3. **Recommended Storage Conditions for ASLT.**

Frozen Foods	Dry and Intermediate Moisture Foods	Canned Foods
−40°C (−40°F) (control)	0°C (32°F) (control)	5°C (−41°F) (control)
−15°C (10°F)	23°C (73°F) (room temp.)	23°C (73°F) (room temp.)
−10°C (14°F)	30°C (86°F)	30°C (86°F)
−5°C (25°F)	35°C (95°F)	35°C (95°F)
	40°C (104°F)	40°C (104°F)
	45°C (113°F)	

Reprinted from Robertson (1993), p. 370, by courtesy of Marcel Dekker, Inc.

Thus if a product is held at 40°C (104°F) and tested once a month, then at 30°C (86°F) with a Q_{10} of 3, the product should be tested at least every

$$f_2 = 1 \times 3^{(10/10)}$$

$$= 3 \text{ months}$$

More frequent testing is desirable, especially if the Q_{10} is not accurately known, and because at least six data points are needed to minimize statistical errors, otherwise the confidence in θ_s is significantly diminished.

- Calculate the number of samples that must be stored at each test condition, including those samples which will be held as controls.
- Begin the ASLTs, plotting the data as it comes to hand so that, if necessary, the frequency of sampling can be increased or decreased as appropriate.
- From each test storage condition, estimate k or θ_s and construct appropriate shelf life plots from which to estimate the potential shelf life of the product under normal storage conditions. Provided that the shelf life plots indicate that the product shelf life is at least as long as that desired by the company, then the product has a chance of performing satisfactorily in distribution.

ASLT PROCEDURES FOR OXYGEN-SENSITIVE PRODUCTS

In all classical ASLT methods, temperature is the dominant acceleration factor used, and its effect on the rate of lipid oxidation is best analyzed in terms of the overall activation energy E_A for lipid oxidation in fatty foods. An inherent assumption in these tests is that E_a is the same in both the presence and absence of antioxidants, although indications are that it is in fact considerably lower in the latter case.

Other acceleration parameters which are used for shelf life are oxygen pressure, reactant contact, and the addition of catalysts. The effect of these factors is generally much less important than that of temperature.

PROBLEMS IN THE USE OF ASLT PROCEDURES

The potential problems and theoretical errors which can arise in the use of ASLT procedures include the following (Labuza and Schmidl, 1985):

- Error in analytical or sensory evaluation. Generally any analytical measure should be done with a variability of less than ±10% to minimize prediction errors.
- As temperature rises, phase changes may occur (e.g., solid fat becomes liquid) which can accelerate reactions, with the result that at the lower temperature the actual shelf life will be shorter than predicted.
- Carbohydrates in the amorphous state may crystallize at higher temperatures, with the result that the predicted shelf life is shorter than the actual shelf life at ambient conditions.
- Freezing ''control'' samples can result in reactants being concentrated in the unfrozen liquid, creating a higher rate at certain temperatures that is accounted for in the measured k value.
- If two reactions with different Q_{10} values cause quality loss in a food, the reaction with the higher Q_{10} may dominate at higher temperatures, whereas at normal storage temperatures the reaction with the lower Q_{10} may dominate, thus confounding the prediction.
- The a_w of dry foods can increase with temperature, causing an increase in the reaction rate for products of low a_w in sealed packages. This results in over-prediction of true shelf life at the lower temperature.
- The solubility of gases (especially oxygen in fat or water) decreases by almost 25% for each 10C° (18F°) rise in temperature. Thus an oxidative reaction such as loss of ascorbic or linoleic acid can decrease in rate if oxygen availability is the limiting factor. Therefore at the higher temperature, the rate will be less than theoretical which in turn will result in an underprediction of true shelf life at the normal storage temperature.
- If the product is not placed in a totally impermeable pouch, storage in high temperature low humidity cabinets generally enhances moisture loss, and this should increase the rate of quality loss compared to no moisture change. This will result in a shorter predicted shelf life at the lower temperature.
- If high enough temperatures are used, the product may actually be cooked.

Therefore, the use of ASLT to predict actual shelf life can be limited except in the case of very simple chemical reactions. Consequently, food packaging technologists should always confirm the ASLT results for a particular food product by conducting shelf life tests under actual environmental conditions. Once a relationship between ASLT and actual shelf life has been established for a particular product, then ASLT can be used for that product when process or package variables are to be evaluated.

BIBLIOGRAPHY

Anonymous. 1974. *Shelf Life of Foods*, Report by the Institute of Food Technologists' Expert Panel on Food Safety and Nutrition and the Committee on Public Information. Chicago, IL: Institute of Food Technologists, August 1974. *J. Food Sci.*, 39:861.

Dethmers, A. E. 1979. *Food Technology*, 33(9):40.

Gacula, M. C. 1975. *J. Food Sci.*, 40:399.

Gacula, M. C. and J. J. Kubala. 1975. *J. Food Sci.*, 40:404.

Harte, B. R. and J. I. Gray. 1987. "The Influence of Packaging on Product Quality," in *Food Product-Package Compatibility Proceedings*. J. I. Gray, B. R. Harte, and J. Miltz, eds. Lancaster, PA: Technomic Publishing Co., Inc., p. 17.

Heiss, R., U. Schrader, and G. R. Reinelt. 1980. "The Influence of Diffusion and Solubility on the Reaction of Oxygen in Compact Food," in *Food Process Engineering*, Vol. I, P. Linka, Y. Malkki, J. Olkku, and J. Larinkari, eds. London, England: Applied Science Publishers Ltd., Chapter 45.

Karel, M. 1974. *Food Technology*, 28(8):50.

Labuza, T. P. 1979. *J. Food Sci.*, 44:1162.

Labuza, T. P. 1982. *Shelf-Life Dating of Foods*. Westport, CT: Food and Nutrition Press Inc.

Labuza, T. P. 1984. *Moisture Sorption: Practical Aspects of Isotherm Measurement and Use*. St. Paul, MN: American Association of Cereal Chemists.

Labuza, T. P. 1985. "An Integrated Approach to Food Chemistry: Illustrative Cases," in *Food Chemistry*. O. R. Fennema, ed. New York: Marcel Dekker Inc., Chapter 16.

Labuza, T. P. and J. F. Kamman. 1983. "Reaction Kinetics and Accelerated Tests Simulation as a Function of Temperature," in *Computer-Aided Techniques in Food Technology*. I. Saguy, ed. New York: Marcel Dekker Inc., Chapter 4.

Labuza, T. P. and M. K. Schmidl. 1985. *Food Technology*, 39(9):57.

Labuza, T. P. and M. K. Schmidl. 1988. *Cereal Foods World*, 33(2):193.

Porter, W. L. 1981. "Storage Life Prediction Under Noncontrolled Environmental Temperatures: Product-Sensitive Environmental Call-Out," in *Shelf-Life: A Key to Sharpening Your Competitive Edge Proceedings*. Washington, DC: Food Processors Institute, p. 1.

Quast, D. G. and M. Karel. 1972. *J. Food Sci.*, 39:679.

Robertson, G. L. 1993. *Food Packaging: Principles and Practice*. New York: Marcel Dekker Inc.

Rosenfeld, P. E. 1984. "Shelf Life Testing: Utilizing the Arrhenius Model to Characterize a Distribution System," in *Engineering and Food: Processing Applications*, Vol. 2, B. M. McKenna, ed. Essex, England: Elsevier Applied Science Publishers, Ltd., Chapter 16.

Samaniego-Esguerra, C. M. L., I. F. Boag, and G. L. Robertson. 1991. *Lebensm.-Wiss. u.-Technol.*, 24:53.

Van Arsdel, W. B. 1969. "Estimating Quality Change from a Known Temperature History," in *Quality and Stability of Frozen Foods.* W. B. Van Arsdel, M. J. Copley, and R. L. Olson, eds. New York: Wiley-Interscience, Chapter 10.

Van Den Berg, C. and S. Bruin. 1981. "Water Activity and Its Estimation in Food Systems: Theoretical Aspects," in *Water Activity: Influences on Food Quality.* L. B. Rockland and G. F. Stewart, eds. New York: Academic Press, Chapter 1.

CHAPTER 14

Toward the Development of an Integrated Packaging Design Methodology: Quality Function Deployment—An Introduction and Example

STEPHEN A. RAPER

Although quality function deployment (QFD) is more typically used for product development, the methodology is equally versatile for use as a packaging design/development tool. QFD is essentially a method of "mapping" the elements, events and activities necessary throughout the development process to achieve customer satisfaction—a techniques-oriented approach using surveys, reviews, analyses, and robust design all centered on the theme of translating the "voice of the customer" into items that can be measured, assessed, and improved. It is a planning tool for translating customer needs and expectations into appropriate organizational requirements, i.e., a system for translating consumer/customer requirements into company requirements at each stage from research and product development to engineering and manufacturing to marketing/sales and distribution. A common theme among each definition of QFD is the customer and his/her requirements and satisfaction. Though the method offers many benefits, there are also challenges in implementing it. QFD can be applied to packaging design and development and serve as an effective communications tool among diverse internal and external customers.

INTRODUCTION

THE design and development of food products is an area well-researched and, for the most part, well understood with a multitude of tools and

Revised, modified, and updated from "Quality Function Deployment: A Tool for Packaging Design," by Dr. Stephen A. Raper, Dr. Henry Wiebe, Martin Topi, and Russell Espinoza, published in September 1998, *Packaging Technology & Engineering*, 7(9): 14–18, 76.

techniques to assist with the product development process. However, tools and techniques for food packaging design and development are relatively few and far in between. Moreover, those that do exist are sometimes directed toward a specific area of packaging. Most individuals involved in packaging are fully aware of its diverse nature, and also recognize it is among the most interdisciplinary of disciplines. The package and packaging system must fulfill many different functions driven by sometimes conflicting demands. A comprehensive design process for packaging must address the total spectrum of these demands and involve both internal and external customers. Furthermore, this design process should include a communications mechanism for the voices of these customers, and should also be able to prioritize the diverse needs of the customers involved.

As noted above, there are few widely used models for packaging design and development. Those that exist usually relate to general areas of mechanics, distribution, cost, and environmental impact (Topi, 1997). For instance, methodologies such as the 4-step and 6-step methods (Bresk, 1992) provide good insight and a logical approach for protective packaging design. Software-driven approaches, such as those espoused by CAPE Systems, Inc. (CAPE) and TOPS Engineering Corporation (TOPS), also help in the design of all levels of packaging. More recently an approach proposed by Sun (Sun, 1991) provides a framework which integrates packaging design with product design, manufacturing systems design, and distribution systems design. However, none of these methods or approaches provides a simple, efficient way to accommodate the voice of the customers and prioritize the various needs expressed.

One method traditionally used in the area of product design shows strong potential for use as an integrated packaging design methodology. This method, referred to as quality function deployment (QFD), has been used to solve packaging design problems (also see Chapter 5 for the use of this technique for food product development). QFD was used because of its ability to incorporate internal and external customer needs into the packaging design process. The remainder of this chapter will provide a brief overview of QFD, present one representative application, and provide conclusions with regard to its merit as a packaging design methodology.

QUALITY FUNCTION DEPLOYMENT

Pioneered and developed in Japan in the early 1970s at the Mitsubishi Kobe shipyard, QFD's major thrust is to make sure that the voice of the customer is heard throughout the organization. This is accomplished by providing a framework for communications with emphasis on the functions responsible for designing, manufacturing (packaging included), and marketing the product or service.

QFD has been successfully used for a wide range of applications in service, manufacturing, and government organizations. Some of the companies include Ford, General Motors, Boeing, Hewlett Packard, Westinghouse, and 3M to name just a few (Bahill, 1993; Jacobs, 1995).

QFD has been defined in several ways. Three such definitions are as follows:

(1) "A method of mapping the elements, events and activities necessary throughout the development process to achieve customer satisfaction. A techniques oriented approach using surveys, reviews, analyses, and robust design all centered on the theme of translating the Voice of the Customer into items that can be measured, assessed, and improved" (Hauser, 1988).

(2) QFD is defined at the Ford Motor Company as "a planning tool for translating customer needs and expectations into appropriate company requirements" (Ford Motor Company, 1989).

(3) The American Supplier Institute defines QFD as "A system for translating consumer/customer requirements into company requirements at each stage from research and product development to engineering and manufacturing to marketing/sales and distribution" (American Supplier Institute, 1991).

A common theme among each definition is the customer. This item includes the voice of the customer, customer requirements, and customer satisfaction. A basic description of QFD is presented here.

Although QFD takes different forms, all use some type of a matrix representation to portray the information being used and the results of the decisions made. This matrix representation is often called the house of quality. The name derives from the basic shape of the matrix shown in Figure 14.1. The names of the various matrix components are also shown in Figure 14.1 and indicate the type of information contained. A more detailed representation of the house of quality is shown in Figure 14.2. The essence of this representation is its focus on the needs of the customer and its ability to provide a simple and useful tool for planning, communication, and coordination during the design process (Hauser, 1988).

The development of the house of quality begins with determining the customer needs. This is reflected in block 1 and is often termed customer requirements or customer attributes, or simply "Wants." The requirements are stated in the customer's terms, and may be bundled into groups of requirements. The customer requirements may be determined from focus groups, in-depth interviews, or similar techniques. It should also be noted that customers include a wide range of interested parties such as functional areas within an organization, end users, regulatory agencies, and distributors. Block 1A allows the requirements to be prioritized. The weights used to establish priorities may be derived from customer input, or from direct experience of those involved in developing the matrix.

Block 2 of the house helps to determine whether satisfying perceived customer needs will yield a competitive advantage. If a company wants to match

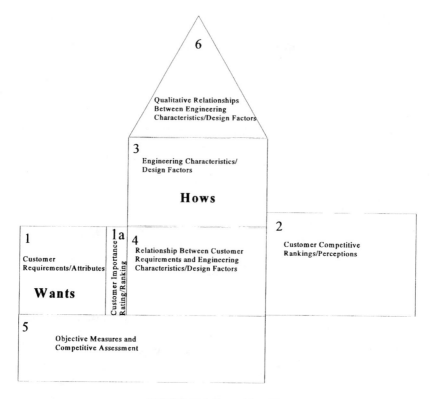

FIGURE 14.1 House of quality.

or exceed the competition, they must first determine their relative standing among competitors. The relative standing can be determined in a number of ways, including customer evaluations and benchmaking. A twofold by-product of this exercise is the identification of the organization's areas of strength and areas of weakness, as compared to the competition.

Block 3, engineering characteristics/design factors describes "How" the engineer can meet the customer's needs. Thus, the voice of the customer is translated into engineering terminology by highlighting engineering characteristics, which affect the customer attributes. Engineering characteristics should describe the product in measurable terms, and should directly affect customer perceptions (Hauser, 1988). It should also be noted that a single engineering characteristic may affect more than one customer attribute.

The next step in building the house of quality is shown in block 4. This is a relationship matrix which indicates the degree to which each engineering characteristic impacts each customer attribute. These relationships are usually stated symbolically in some manner, as shown in Figure 14.1, but the symbols may be assigned a numerical value to indicate relative importance. For example, referring to Figure 14.2, the three customer ''Wants'' for a package have been determined to be a package that is easy to open, yet resealable, and also one with appealing graphics. The engineering requirements, or ''Hows'' may be translated as opening and closing force, and pleasing appearance. The symbols shown within the matrix indicate the relationship between the ''Wants'' and ''Hows.'' For instance, appealing graphics and opening/closing force show a strong negative relationship (as usually expected), and a strong positive relationship for pleasing experience. When the matrix is completed, objective measures are included in block 5 of the house. These measures help to establish target values for the engineering characteristics, and also compare the organization to the competition.

The top portion of the house of quality is its ''roof'' and is shown as block 6. The roof is also often referred to as the correlation matrix (Ford Motor Company, 1989). The correlation matrix serves to identify the qualitative correlations between the various engineering characteristics and is accomplished by use of the relationships symbols. These correlations may be either positive or negative and may range from weak to strong. Too many positive

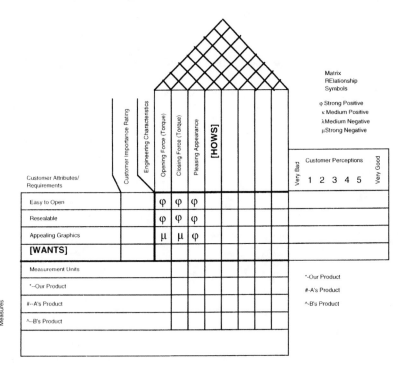

FIGURE 14.2 Generic house of quality.

interactions may indicate redundancy in critical product requirements or technical characteristics. Negative interactions point to the need to consider engineering trade-offs to address customer requirements. These trade-offs can be considered based upon company priorities, competitive strategies, etc. In particular, this segment of the house of quality enables the team involved in the process to note how one engineering change may affect other characteristics.

The completed house provides a number of benefits such as measurable target values (block 5), competitive assessment and competitive position (blocks 2 and 5), relationships between customer requirements and engineering attributes (block 4), and the relationship between engineering characteristics (block 6). If numerical emphasis is used, the house can readily identify the most important customer attributes to pursue through engineering design characteristics. A major benefit from the use of QFD is the communication that must occur in the development of the house. Clearly a team approach must be taken in applying the QFD methodology. This point should not be under emphasized. That is, the use of teams in this process can be critical to the overall success of the entire methodology. This process is oriented toward the use of multifunctional or cross-functional teams who represent various functional elements of the organization, and who are involved in product (packaging included) design and development. The team approach ensures that the ''voice of the customer'' can be understood from the perspective of each functional area represented, and that the ''voice'' is then shared among the team members so that the customer requirements are ''harmonized'' across the organization.

The methodology does not have to end with a single house. At the most basic level, block 1 of the house can be described as ''Wants,'' and block 3 of the house can be described as ''Hows.'' Using this convention, important ''Hows'' from the first house become ''Wants'' in the next house. This creates a cascade effect as shown in Figure 14.3 and indicates how the voice of the

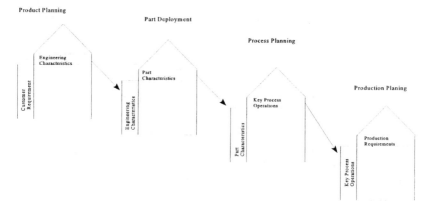

FIGURE 14.3 Cascade effect (modified from Hauser and Clausing, 1988).

customer cascades through design to manufacturing and the ultimate end user. The deployment may occur as shown, but the cascade flow may occur in general terms as follows: customer requirements, design requirements, engineering design, product characteristics, manufacturing characteristics, and quality characteristics. This logic, or deployment process, could easily include separate houses for the deployment of packaging processes, packaging machinery systems, packaging logistics systems, and packaging environmental systems. Moreover, where product and package synergy exists, such as the case for consumer foods and consumer products, the QFD methodology allows for the full integration of packaging systems in the process. In other words, the QFD methodology could be used for product design and for package design, then in an integrative fashion for the deployment of the product-package system. The example presented in this chapter illustrates the case of package design using only the initial house of quality. Examples actually using a package as an illustration of a product can be found in some of the references.

ENHANCED QUALITY FUNCTION DEPLOYMENT

QFD can be used as the central mechanism for integrated product/packaging development with the addition of other product design concepts and philosophies. Some researchers and practitioners refer to this as enhanced quality function deployment (EQFD). This approach consists of five parts as follows:

(1) Concept selection
(2) Deployment through the levels
(3) Contextual analysis
(4) Structured specifications
(5) Static/Dynamic status evaluation (Clausings)

With further study and perhaps refinement, EQFD may well serve as an integrated packaging or product-packaging design model, thus overcoming the deficiency initially noted in this paper. The five parts of EQFD are briefly described below.

Concept selection is another matrix-based technique developed by Stuart Pugh in the 1980s. In this process/method, a multifunctional team, usually made up of the same team that develops the basic QFD house of quality, convenes to generate ideas or concepts used to initially address customer wants. Further deployment of concepts or solutions may occur as QFD is cascaded through the levels. A list of criteria is developed by the team to be rated against various concepts. A datum concept is selected to compare against all others. Each alternative is then evaluated against the criteria list, usually as worse than (−), better than (+), or the same as (S). An initial tally of the alternative concepts are made which often provides an initial view of the concepts versus the criteria. The team then selects the best of the concepts and looks for potential opportunities to create hybrid solutions. These may then be put in the matrix and evaluated again against the datum and other concepts. In this way, bad points or potential bad points of alternatives may be eliminated or minimized. More advanced analysis of the concepts may occur in the form of finite element analysis, failure analysis, design of experiments, simulations, etc. The success of the Pugh selection concept, much like basic QFD, depends on the input and participation of the team members and their ability to communicate their ideas and understand the ideas of others leading to common understanding, consensus, and commitment (Ulrich, 1995; Biren, 1996). The Pugh concept selection technique offers tremendous potential for the evaluation of various packaging concepts, and perhaps development of new or hybrid concepts.

Contextual analysis refers to conducting an analysis to better determine the context of the product or package in a new market. Structural specifications is a detailed process for developing specifications in a life-cycle concept, and the results may be integrated into the house of quality. Structured specifications also are used in the overall guide of the design of the product or package. Static/dynamic status refers to evaluating designs and concepts in terms of the competitive market, and the technology streams present in the marketplace and those under development by the organization. In other words, with regard to design, what is the correct balance between current (static) and new (dynamic) technologies.

PACKAGE DESIGN USING QFD

CASE APPLICATION

To serve as an example, a packaging design problem for a major manufacturer of remanufactured diesel engine components (obviously, not a food product, but used here to illustrate the technique) was addressed with the basic QFD methodology. One plant within the firm was receiving complaints from marketing concerning a single part package used for some fuel systems products. A marketing director within the firm had attempted to develop a survey based on the complaints received in order to determine the extent of the problem and develop solutions. This effort did not solve the problems. QFD was then applied in order to address the intrafunctional communication problems as well as the packaging design problem. Internal and external customers were involved and included fleet users, dealers, distributors, warehousing and logistics personnel, and plant personnel from the production, operations, purchasing, and marketing areas. Dealers, distributors, and warehousing personnel were chosen from several geographic locations to insure that any data gathered was of national scope, and not limited to any particular region.

Customer requirements (''Wants'') were determined through personal interviews. Individuals interviewed were asked to elaborate on several major issues and sub-issues within each category as shown in Table 14.1. Consolidated customer ''Wants'' were developed for the five major categories of packaging design issues related to unitization, communication, protection, shipping, and operations as shown in Table 14.2. The customers were also asked to rank the ''Wants'' on a five point scale, where a score of five was high and a score of one was low. A customer analysis was conducted which compared the package currently used to three competitors. Two competitors were external to the firm, and one was a multiple pack used internally by the firm.

A list of technical requirements or ''Hows'' necessary to satisfy the ''Wants'' was developed. As a matter of note, this was the most difficult part of the model application. Common mistakes included mixing of general and specific solutions and restating the desired ''Wants.'' The customer requirements, their ranking of importance, competitive assessment information, and technical requirements were then used to construct a house of quality. This

TABLE 14.1. **Package Design Issues.**

I. Unitization	A. Number used per order
	B. Ease of handling
	C. Ease of palletizing
	D. Determination of number remaining in package
	E. Optimum number per unit: 1, 2, 6, 8, 12
	F. Other unitization problems
II. Communication	A. Brand identification consistent/clear
	B. Part identification consistent/clear
	C. Product information
	D. Warranty information
	E. Core return instructions
	F. Bar code
III. Protection	A. Damage
	B. Oil/soil
	C. Core return
	D. Method of closure
IV. Shipping	A. Problems
	B. Distribution chain
V. Manufacturing	A. Ease of assembly
	B. Ease of handling
	C. Method of closure
	D. Cost
	E. Vendor information
VI. Other	A. Miscellaneous complaints
	B. Environmental concerns

TABLE 14.2. **Consolidated Customer Wants.**

U (Unitization)	Sell mostly 6×
	Easy to count
	Easy to handle full
	Standard box
	Easy to palletize
	Fits in wire basket
	Sell singles
	Looks good alone
	Minimal pack parts
	Stackability
P (Protection)	Individual protection
	Oil/soil protection
	Strong box
	Tight closure
	Box protection
	Box core returnable
	Product fits in box
	No axial movement
	No rolling movement
C (Communication)	Clear information
	Label sticks to box
	Easy to read
	Untorn label
	Read when stacked
	Distinct product package
	Dirt doesn't show
	Full box indicator
O (Operations)	Easy to assemble
	Easy to handle
	Filling speed
	Low inventory cost
S (Shipping)	Individual shipping

effort resulted in the matrix shown in Figure 14.4, which indicates the correlation between the respective "Wants" and "Hows." Each of those who participated in determining "Wants" and "Hows" was given a copy of the initial matrix and allowed to fill in the relationships from their viewpoint. Correlation relationship matrix values were assigned as strong, medium, or weak. In this case, strong was equal to 9 points, medium was equal to 5 points, and weak was equal to 1 point. A weighted scoring system was developed to help prioritize or weigh the importance of the "Hows." The scores in Figure 14.4 are the sum of the product of the "Wants" rank and the matrix value. The value of the "Hows" score associated with the person surveyed was determined from

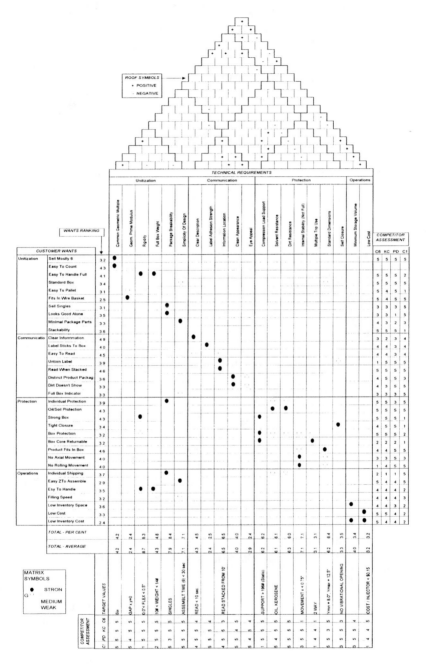

FIGURE 14.4 Packaging design application—house of quality.

their "Wants" ranking and matrix values. The final scores were determined by calculating the percentages of the total and the average of rank/matrix products of all surveyed. Based on those scores, the requirements, in order of importance, were rigidity, package breakability, simplicity of design, internal stability, information location, dirt resistance, compression load support, and standard dimensions.

Target values for the "Hows" were developed based on the survey comments and through interactions with engineering personnel. In some cases, a target value was replaced by an upper or lower threshold. Target values are also included in Figure 14.4. A competitive analysis of the "Hows" was also conducted. Generally, the external competitor's packages were judged to satisfy these demands more completely than the current package, and the internal competitive package.

CASE SUMMARY

The knowledge gained from this particular QFD application was used to ultimately develop a package that met customer requirements ("Wants"). Although the package design was not finalized for this case, the QFD process did identify critical information. The determination of the customer "Wants," technical "Hows," weighted scores, and competitor analyses all provided direction to alternative package development. The customer wanted a package that had clear information, was easy to read from a distance when stacked, protected from the environment, and allowed for easy counting. To meet these requirements the QFD analysis indicated the package should be rigid, simple, and strong while allowing breakability (multipack) and clear labeling.

In this application, the direct survey process used to determine the voice of the customer worked well although extensive customer contact was required. This experience indicated that a positive relationship must exist between those working on the project and those whose opinions are used to solve the problem. The most difficult step in the process involved the determination of the technical "Hows." As mentioned previously, general requirements were often mixed with characteristics of specific alternatives. However, despite these difficulties, preliminary responses to design alternatives derived from this process indicated that the customers voice was successfully translated into satisfactory packages.

The QFD process is surprisingly popular at the plant location where it was applied. It was implemented in all new product development projects. Furthermore, the plant personnel felt that customer input was essential in developing successful products, and QFD helped achieve desired results.

CONCLUSIONS

Quality function deployment can be used as a packaging design strategy.

The example described here, along with successful application at Procter & Gamble (1992), supports that conclusion. Because the development of appropriate packaging involves many internal customers within a firm, as well as many external customers, a general model that incorporates customer needs in the process, and also translates those needs into technical terms, is a positive development. Some might argue that QFD can only be successful for product development. However, a package is in essence a product itself and the application of sound principles of product development will greatly benefit the package design process. Moreover, enhanced QFD, can serve as an integrative packaging design strategy, if not an integrated product and packaging design strategy. Further study of both QFD and enhanced QFD as related to packaging systems design and development is warranted.

Initially, the QFD process might be time consuming, with information hard to find, and the task of creating the house of quality may be somewhat tedious. However, QFD software, readily available from many sources, minimizes some of these disadvantages. QFD, in general, has many benefits including, but not limited to, its systems approach philosophy, its customer driven philosophy, its ability to shorten product/package development cycles, its team/communications driven approach, and its ability to serve as a mechanism for translating knowledge to new teams. (Ford Motor Company, 1989). Clausings (Clausings, 1988) shares a similar view as follows: "QFD is a useful tool for bonding the multifunctional team. It also serves as a training tool because each team member learns from the other's views and limitations. QFD becomes an asset to the corporation . . . it is a historical record if you will."

BIBLIOGRAPHY

Anonymous. 1991. "Quality Function Deployment: Excerpts from the Implementation Manual for Three-Day QFD Workshop," *The Third Symposium on Quality Function Deployment*, American Supplier Institute, pp. 19–40.

Bahill, A. T. and W. L. Chapman. 1993. "A Tutorial on Quality Function Deployment," *Engineering Management Journal*, Vol. 5, No. 3, Sept., pp. 24–35.

Biren, Prasad. 1996. *Concurrent Engineering Fundamentals: Integrated Product and Process Organization*, Vol. 1, Englewood Cliffs, NJ: Prentice Hall PTR.

Bresk, F. C. 1992. "Using a Transport Test Lab to Design Intelligent Packaging for Distribution," *TEST Engineering & Management*, Oct./Nov., pp. 10–17.

CAPE: Computer Assisted Packaging Evaluation, CAPE Systems Inc., Plano, Texas.

Clausings, Don P. 1988. *Enhanced Quality Function Deployment: Video Series*, MIT Center for Advanced Engineering Study.

Ford Motor Company. 1989. *Quality Function Deployment Awareness Seminar, Reference Guide*.

Hauser, J. R. and D. Clausing. 1988. "The House of Quality," *Harvard Business Review*, May–June, pp. 63–73.

Jacobs, D. A., S. R. Luke, and B. M. Reed. 1995. "Using Quality Function Deployment as a Framework for Process Measurement," *Engineering Management Journal*, Vol. 7, No. 2, June, pp. 5–9.

Procter & Gamble, Inc., Guest Lecturer, Associate Director of Packaging, Cincinnati, Ohio, from lecture and notes on package design, The University of Missouri-Rolla, March 20, 1992.

Raper, S. A. and M. R. Sun. 1994. ''Understanding the Role of Packaging in Manufacturing and Manufacturing Systems,'' *Handbook of Design, Manufacturing and Automation,* Richard C. Dorf and Andrew Kusiak, eds. New York: Wiley-Interscience Publications, pp. 331–342.

Sun, M. R. 1991. ''Integrating Product and Packaging Design for Manufacturing and Distribution: A Survey and Cases,'' Ph.D. Dissertation, The University of Missouri-Rolla.

Topi, M. A. 1997. ''Using Quality Function Deployment to Establish Package Design Requirements,'' M.S. Thesis, The University of Missouri-Rolla.

TOPS: Total Optimization Packaging Software, TOPS Engineering Corporation, Plano, Texas.

Ulrich, Karl T. and Steven D. Eppinger. 1995. *Product Design and Development.* New York: McGraw-Hill, Inc.

CHAPTER 15

Innovative Food Packaging Graphics and Testing: A Packaging Decision Can Change the Course of a Company

J. ROY PARCELS

Using the principle that, to the consumer in a purchase decision-making mode, the package is the product, the food package design has enormous impact on the success of any new food product development. No less than the entire food product development process itself is the complex of developing the brand, trademark, trade dress, copy, and design of both primary and corollary packaging. Professional packaging designers must be involved from the outset in the product objectives, consumer identification, package function, retail environment, and competition. If possible, the design elements should strive for a proprietary differentiation without confusing the consumer. Package designs should be evaluated as objectively as possible before reducing the design to reproduction. The commercial package design should duplicate the designs that have been approved and accepted by consumers.

INTRODUCTION

THIS chapter focuses only on the graphic design and brand creation aspects of package design. The package physical structural design is not discussed as it is assumed that the correct physical package for the specific food to be contained has already been determined (Chapters 8 and 14). It is important, however, to remember that the graphic design of a package is only one part of the total packaging concept. The materials selected and the physical configuration of the package are equally important to its marketing success, and therefore should be determined prior to or concurrent with the development of the packaging graphics.

In this chapter, the different aspects of food packaging graphics creation and design are reviewed, with in-depth analysis of the predesign, design, and

369

postdesign phases, including brand name creation and trademark design as well as the legal implications of package design.

There is a myriad of ways to tackle this phase of the product planning process. However, in over 30 years of creating and developing brands for a wide variety of products, Dixon and Parcels have found that the approach outlined in this chapter has proven the most successful for us and our clients.

The following deals with food packaging graphics in three major areas:

(1) Prepackaged Retail—supermarkets, warehouse/club stores, super drug stores, convenience stores, fast food restaurants, vending machines, and web sites on the Internet

(2) Ready-to-Eat Retail (Home Meal Replacement)—perhaps the fastest growing segment in both supermarkets and convenience stores

(3) Food Service [Hotel/Restaurant/Institutional (HRI)]—restaurants, institutions, fast food (This area is equally important as prepackaged retail because these products are purchased by buyers who are retail consumers themselves.)

THE PACKAGE *IS* THE PRODUCT

For the consumer, the package has often become the only link to the product marketer. For the marketer, it is often the sole direct contact with the consumer. As several food marketing executives have remarked, we taste with our eyes. Therefore, the packaging graphics must portray the product in the best manner possible, without overpromising.

One of the first principles we learned in designing packages was that, to the consumer, the *package is the product* until purchased, and the product must meet the consumer's expectations or there will not be repeat purchases.

In the 19th century, when food packaging was in its infancy, package labels were simply a means to identify what was inside the package. Food was packaged for easiest transportation to the marketplace and for the convenience of the end user. Packaging labels simply stated the generic name of the product, its source of origin, and the brand name (Figure 15.1).

Over the years, the importance of packaging graphics has grown tremendously—to the point where the graphics have become integral to the success of the product. Now, packaging features not only the brand name and product identity, but the reason(s) to select that particular product over the competitor's. The package has become the ultimate point-of-purchase medium by acting as the final sales tool—by distinguishing the product from its competitors through innovative use of design, copy, color, graphics, product presentation materials, etc. (Figure 15.2).

Thus, a well-designed package represents a critical, one-time opportunity for both consumer and marketer; for the marketer to sell the product, and for the consumer to learn enough about the product to want to purchase it.

FIGURE 15.1 Nineteenth century packaging was elemental—only the basic facts.

FIGURE 15.2 Late 20th century packaging attempts to distinguish and entice, as well as to identify.

Consider the enormous job packaging does for a food product.

It makes the product possible in commercial form; without a package it cannot be distributed or sold. It enhances the quality appeal of the product, protects it, distributes it, displays it, names and brands it, describes it, prices it, stores it, dispenses it, legalizes it, coupons it, and differentiates it from all other products.

Packaging positions the food product, shapes it, colors it, illustrates it, glamorizes it, personalizes it, and advertises it.

Most important, packaging sells it, and then resells it.

Of course, if the package promises more than the package can deliver, there will be no repeat sales. However, that is just another argument for getting the right package for the right product—and from the very beginning of its life.

The real challenge then is to design the graphics in such a way as to convince the consumer, that

- a necessary or desirable product is being made available
- in a package that combines freshness, satisfaction and convenience
- with minimal waste
- at a reasonable cost
- that encourages repeat purchase

And, as summed up once by Sam Gardner, former Product Director of M&M/Mars, Inc., "If you can't find it, you can't buy it!" (Gardner, 1963). This is as true today as ever, with ever increasing numbers and types of outlets in which to purchase food products.

It is up to those responsible for designing packages to fulfill the marketing needs of food manufacturers and marketers. To do so, designers, design managers and other decision makers must have an understanding of the public's current eating trends; an appreciation for the history of containing, protecting, informing, identifying, and selling packaged food; a clear idea of what motivates the consumer to buy; and, of course, knowledge of the many different relevant food packaging regulations.

IMPACT OF PACKAGE DESIGN ON PRODUCT SUCCESS

The significance of effective package design is underscored by some of the nation's leading food marketing executives. Robert P. Crozier, Vice Chairman of the Board of Flowers Industries, notes that, "With the cost and fragmentation of today's electronic media, a product's point of sale impact is crucial. As a result, good packaging graphics are essential in the product marketing mix," (Crozier, 1998).

Equally forceful is Joseph F. Welch, Chairman of snack foods marketing, The Bachman Company: "The communication that takes place between Bachman and the consumer at the point of sale in every store relies almost totally on our snack food package . . . Packaging in the long run is second only to

Bachman's product quality in the ranking of factors influencing the purchase decision'' (Welch, 1998).

John A. Celauro, President of 4C Foods Corp., observes, ''I see packaging as the introduction of our product line to the consumer. It creates the first impression. As such, it should be one that reflects what 4C Foods is trying to say and accomplish. Our corporate message, strategy and efforts are all represented in our packaging'' (Celauro, 1998).

Most succinctly, William C. Burkhardt, President of Austin Quality Foods, says ''Packaging of consumer products can be your most important ingredient to obtain consumer trials'' (Burkhardt, 1998).

And Wrede Smith, President of American Pop Corn Co., asserts that ''A well designed package containing good colors and good graphics will serve as a billboard on the shelf. A strong design will also look good in advertising—print or television. Consumers expect reliable packaging . . . it is a reflection of the contents . . . and (they) will change brands if the package is not the best'' (Smith, 1998).

In summary, according to Gerry Thomas, former Vice President of Frozen Foods at Campbell's® Soup Company, ''Package design is advertising at its most cost efficient and in its purest form. This advertising takes place at the point of sale, in the store, where we find the moment of truth. Good package design has all of the elements of good advertising: honesty, distinction, originality, conviction, persuasion, perceived value, education, and it must stir the emotions'' (Thomas, 1990).

SELECTING THE INDEPENDENT BRAND AND PACKAGE DESIGN PARTNER

Perception, perspective, and performance are important qualities to look for in choosing the design partner for any food packaging project. The ability to perceive and define the product/brand's true communication challenges is critical, as is the firm's ability to remain the distant observer of the communication interaction between marketer and observer, yet become wholeheartedly involved in the challenge. Look for time-proven successes with food products, both with new and repositioned products.

Prior to undertaking of any of its services, a written proposal is submitted by the design consultancy to the food marketer that states the design team's understanding of the project outlining each of the phases required for it together with time and cost estimates for consideration.

EXAMINING THE PACKAGE DESIGN PROCESS

There are time proven methods for developing food packaging graphics that begin with target audience definition and go all the way through consumer testing to post-marketing adjustments to customize the packaging to a growing consumer base.

Because what goes on to the package has become at least as important as what goes into it, careful planning and structured organization are imperative to the graphic design process. This process is normally divided into predesign, design, and postdesign phases. To make certain that all components of the final package optimally communicate the product's identification and individuality, the design team follows set guidelines throughout these phases, including consumer research and product analysis, so that the final package most accurately fulfills the marketing goals.

The package design process is also separated into parts that deal specifically with design for new products and for the redesign of existing products. In a redesign situation, the goals are actually often more complex than those for new products. They may include enhancing brand recognition, doing a better job of coordinating a product line, more clearly differentiating product flavors, updating the perception of the product, or even a total repositioning to take advantage of changes in the marketplace.

PHASE 1: PREDESIGN STEPS

ORIENTATION MEETINGS

We have found that the degree of success in creating a *new* or *redesigned package* or *graphic packaging system* is directly related to the quality and extent of the preparatory work leading up to it. This includes the marketing strategy, collecting and analyzing information related to the food product(s) under consideration, and such important factors as target audience, demographics, language and cultural considerations, geographic markets, marketing plan, product and corporate orientation, and competitive products and selling environments.

Information gathering typically begins during an orientation meeting with the marketer representatives. This enables the design organization to obtain and discuss any relevant consumer research, as well as review the marketing/ project objectives and brand/packaging copy established to date.

At this meeting, the design team reviews and discusses all marketing requirements for the packaged product, obtaining information from the following departments and individuals:

(1) Marketing
(2) Advertising
(3) Sales
(4) Research
(5) Home Economist

(6) Packaging Coordinator and/or Design Director

(7) Intellectual Property Lawyer

(8) Manufacturing

(9) Purchasing

A thorough review is made of each step of the project proposal at this time, including any established timetables and project budgets. (Normally, time schedules are included in the original proposal and highlight all major deadlines from pre-design to product introduction in the marketplace.)

Another key point of discussion at the orientation meeting is the exploration and definition of target audiences for the product: age, gender, eating habits, ethnic background, etc. Often products are very similar with the same physical packaging and generic names. In fact, some are identical. It is vital to identify what will make your product(s) more desirable to the same target audience as that of your competitors and then to communicate that via the best packaging graphics available. The decision to purchase is most often made at the point of sale. Again, the package is the product until it is tried for the first time. One important result of the orientation phase is an agreed upon strategy for the product.

IN-STORE GRAPHIC PACKAGE DESIGN AND COPY AUDIT OF DISPLAY SURFACES

To best determine the project design objectives, the design team must have a strong knowledge of the environments in which the package is offered, as well as of the competition's packages. It is important to analyze the current and competition's trade dress in its various in-store selling environments, noting the following:

- the design of the brand name
- copy and copy emphasis
- illustrations and/or photography
- colors
- package size and shape
- special package features
- selling price
- shelf location

We have found it most helpful to have one or two marketer representatives along on some of these in-store visits, as they can often offer additional information and answer specific questions that might arise in each of the particular selling environments.

It is also important to know the shelf position of the competition relative to your current package(s), and if working with a new product, to survey the shelves to help determine the most desirable shelf locations.

In addition, there are varying degrees of display opportunities in supermarkets, mass merchandisers, club stores, super drug, stores and convenience stores for outer cartons, shelf shipper displays, aisle shipper displays, and special and seasonal promotions.

Note: when designing packages for club stores, one must be sure to obtain any special requirements that the store/chain headquarters may have for particular products prior to beginning graphic design exploration.

ANALYSIS OF PACKAGE DESIGN AND COPY AUDIT

Once the information is gathered from the in-store graphic package design and copy audit, the design team analyzes the visual findings in relation to the current design and competition. As copy plays such a key role in positioning products, the team also analyzes the packaging copy of the current and competitive packaging.

QUALIFIED CONSUMER EVALUATION OF CURRENT AND COMPETITIVE PACKAGING

Based on the desired objectives for the package design, qualified consumers from the target audience should evaluate the current packaging versus that of the major competition to determine the positive and negative connotations of each. This consumer input, along with the design and copy analysis of the packages from the in-store audit, enables the design team to emphasize the positive connotations and eliminate the negative connotations when designing the new packaging graphics.

IF NEW BRAND OR NEW PRODUCT, CONSUMER EVALUATION OF COMPETITIVE PACKAGING

If the package to be designed is for a new brand or new product and there are similar products are already in the market, it is well to evaluate those current designs relative to the design objectives for the new package. This is done to help obtain qualified consumers' attitudes toward the package designs of the competition as well.

These phases in consumer testing can be accomplished by various research methods. However, we have found the results are most reliable when the testing is performed with the target audience in retail outlets, within the desired demographic locale, in the section of the country in which the product is currently being sold/is projected to be sold. See the section below and Chapters

4 and 6 detailing the various methods of target audience consumer research in use today.

Be sure that qualified store and regional brands are included in the predesign research, on-going research (if required), and postdesign research. In many instances manufacturers/marketers, however, do not seriously consider store brands as competitive brands. However, consumers do not always distinguish a store brand from a regional or national brand. A store brand or trademark on a package that identifies the product they purchase becomes their ''brand''; to them a ''store brand'' is a ''brand.'' Therefore, marketers should always consider store brands as competition for the same consumer's dollar.

ANALYSIS OF RESULTS OF CONSUMER EVALUATION

In order for the redesigned or new product package to attract purchasers to your food product, the graphic design must connote the agreed-upon attributes for the brand/product to the consumer. An in-depth analysis of the consumer responses will tell you whether or not the design is accomplishing this. It can also help to tell you those things that can be improved upon, and those things that can be eliminated.

PRESENTATION OF ANALYSIS OF RESULTS OF THE FIRST SIX STEPS WITH RECOMMENDATIONS

Prior to the start of the graphic package design exploration, a predesign analysis is prepared, along with recommendations, based on the information obtained in the first six steps above. Once formulated, this analysis is presented to the food marketer representatives to inform them of prevailing consumer attitudes, and at the same time, enable all concerned with the project to add input into the marketing/design objectives formulated at the orientation meeting.

REVISE OR ADD TO PROJECT OBJECTIVES AND STRATEGIES FOR APPROVAL

If required, the marketing/design objectives for the project are again reviewed to determine if any revisions or additions are necessary as the result of the information gained in the predesign in-store audit and consumer evaluation.

CONFIRM WORKING COPY AND LEGAL REQUIREMENTS

At this point, the design team should obtain all available package sell and legal copy, as well as discuss and review any special legal requirements for the product.

Of enormous significance in the food packaging design is the multiplicity of regulations with which one must comply. In addition to the restrictions of the Food, Drug & Cosmetics Act, the Fair Packaging & Labeling Act of 1966, the Nutrition Labeling & Education Act of 1990 (NLEA), and the U.S. Department of Agriculture, there is the geographical gridwork of state and city regulations applying to specific categories (e.g., dairy products) (Chapters 16 and 18).

Added to these regulations are those that are issued from time to time by the Food & Drug Administration, the Federal Trade Commission, and the U.S. Department of Agriculture. In addition, food packaging labels must also be in compliance with procedures of the U.S. Patent and Trademark Office and, for products in international distribution, the General Agreement on Tariffs and Trade and the North American Free Trade Agreement. Incidentally, do not forget to copyright the entire label or package graphics—© company name or initials and date. Adding the date protects the copy and design internationally.

PHASE 2: CREATION AND DESIGN OF THE BRAND NAME

If required, the design team must consider the brand name *before* entering the graphic package design phase. Whether it is a new product or a redesign project, the brand name must adhere to the marketing objectives in order to communicate the desired attributes to the target audience. If a brand or corporate name has not yet been determined, the name creation process should begin at this time. The next part of this chapter focuses on that process.

A second source of possible brand names for a product is The Trademark Register of the United States. As well as listing all registered trademarks and helping to avert infringement, the Register is an excellent source of marks that may be available for purchase (those not in general use), and could be the least expensive way to obtain a mark.

CREATION OF THE BRAND NAME

Prior to and/or concurrent with the product research and development, the design consultancy begins the process of brand name development if a brand name has not already been determined.

It is imperative that the brand be perceived to represent the best quality product in its field, even when it is displayed alone. Therefore, the successful trademark design (the logo and symbol that represent the brand) must convey the desired personality of the product and must reinforce the product's attributes—if this fails, the total packaging graphics will not achieve the desired image for the product.

Information Gathering

In the early stages of brand name development, the design consultants collect information on the use and application of the new name. Criteria vary by product, but frequently include product description and history, the demographics of the target audience, language and cultural considerations, product and corporate orientation, and brand and product competition.

Name Generation Platform

Once the above information (and any additional pertinent information) is obtained and evaluated, a name generation platform is prepared for the marketer's approval. This platform defines the objectives, image qualities, and functional attributes for the new name and also provides the criteria for judging the name candidates that will be generated throughout the project.

Generating the Name Candidates

Based on the evaluation criteria outlined in the name generation platform, the name generation process employs creative linguists, computer sources, word libraries, and input from copywriters and marketing specialists until a list of brand names emerges that satisfies the objectives, image qualities, and functional attributes desired—typically several hundred names. Avoid purely descriptive words if the goal is to obtain trademark status for the name.

Name Screening

The design team then evaluates this list of names against the name generation platform, retaining only those names with the best potential, reducing the list as a result of the evaluation.

Preliminary Trademark Register Screening

The list of names is then screened, retaining those demonstrating the ability to meet the objectives most effectively. The remaining name candidates are checked for availability with the U.S. Patent and Trademark Office. (Note: product marketing plans will influence the geographical scope of any additional screening.) For products to be marketed internationally, language, geography, and culture are all important factors.

Considering Word Connotations Interculturally

As more and more companies bring new products to the global marketplace, another critical practice comes into play—connotation clearance. Various cultures and languages often evoke meanings for names that are neither intended nor desired. As a result, it is necessary to conduct negative connotation studies among linguists familiar with current idiomatic and slang usage to anticipate problems that might occur in the various countries where the product will be marketed.

Initial Name Recommendations

Following these procedures, it becomes possible for the design team to propose brand name candidates that best meet the criteria for the product. Generally, between three and six are selected for a thorough availability search of state registrations and common law use. If the brand name is to be used for the company name, then a search of trade names registered by each state must be done for availability.

Search, Refined Evaluation, and Selection of Registrable Names

After a thorough legal availability search in all countries of potential interest, the design consultancy evaluates the registrable name candidates keeping in mind the criteria originally formulated in the name generation platform. Examination of such things as letter forms, readability factors, colors, background shapes, symbols, and inter-relationship of design elements are often necessary in this phase, depending on the platform requirements.

Target Audience Evaluation

The registrable name candidates are subjected to the objectivity of the most important market segment for the product. Using testing procedures designed to uncover target audience preferences or biases, the design team further narrows the list of brand name candidates.

Final Name Recommendation

After reviewing the result of the target audience evaluation and its own analysis, the design team presents its recommended name candidate, along with the results of the target audience testing.

Registration of the Brand Name

The first step in obtaining the registration is submitting an application for

"Intent to Use" status with the U.S. Patent and Trademark Office. The marketer has six months to have the trademark in the marketplace.

DESIGN OF THE BRAND NAME

The Importance of the Trademark—Seven Reasons

Once the brand design is completed, it should be registered as a trademark so that the company has complete control over its future use. Although the trademark system is looked upon as a protection system solely for the manufacturer, the consumer also benefits tremendously. The following are seven reasons trademark rights are so important for the consumer as well as the marketer:

(1) Encourages market competition—competition depends on the ability of competing producers/marketers to identify their goods and services. With trademarks to differentiate the source of one product from another, an important incentive to offer superior quality is stimulated.

(2) Fixes responsibility—trademarks offer the business and public consumer an avenue for recourse when products fail to live up to claims or expected standards of quality. Trademarks force manufacturers/marketers to pay attention to the details if they are to continue in business.

(3) Stimulates innovation—continuous new product development in a free market society requires an efficient and effective trademark system. With it the innovator is assured the recognition and rewards when newly developed products prove successful.

(4) Lowers costs—the economies of mass production and mass distribution depend on trademarks to develop and hold large markets. Without trademarks, cost-saving distribution techniques such as self-service supermarkets, club stores, and convenience stores would be impaired. With such distribution assured, producers/marketers have the confidence of repeat sales and are willing to package products at the rate of millions of units per month.

(5) Saves consumer time—fast product identification saves valuable time for consumers. The average American food shopper today spends only 30 minutes per week selecting 50 different products on a typical trip to the supermarket.

(6) Gives consumers a choice—trademarks make it possible for consumers to distinguish one product from another, choose their favorites and reject the brands they do not like.

(7) Creates foreign markets—American companies' foreign trade has played a major role in building the U.S. economy. Trademarks make it possible for American producers to create worldwide markets for their products.

Once the brand name candidate is agreed upon, the designers explore the graphic possibilities inspired by the name itself, hoping to achieve design excellence and strengthen communication values based on the criteria in the name generation platform. A format is designed to incorporate the brand name and a graphic design in a memorable, attractive fashion. The design is created to enhance the memorability of the brand name and must successfully represent the product in various types of packaging and advertising.

Nine Steps to Successful Trademark Designs

Creative design will include the exploration and development of appropriate symbols. A symbol can add to the memorability of the name as well as reinforcing its desired attributes. Sometimes symbols are an integral part of the logotype itself, whereas other times they are used only in one or two fixed relationship(s) to the name, and often the symbol is used alone to identify the brand of product or service.

After completing and refining the trademark (logotypes and/or symbols design concepts) the design candidates selected as best fulfilling the marketing objectives are presented to the client. Several trademark applications are judged against the previously agreed-upon criteria for the selection of one trademark design. The final brand design will undergo a last examination to attempt to bring it to ''ultimate'' design perfection.

(1) Develop the design criteria—all relevant information on the visual image that should be projected by the new trademark to reinforce the brand name objectives is collected. The mark's intended use in connection with other corporate established trademarks is a serious design consideration as well.

(2) Design exploration—a thorough exploration of design approaches is undertaken. Typically the name is designed both as a ''stand-alone'' and with a graphic symbol to support it. The concept designs that best meet the established criteria are then recommended for trademark design development. Usually there are three to five design candidates selected from the design exploration.

(3) Design development and refinement in reproduction art form—the selected designs are developed and refined. The effectiveness of candidates is then ascertained by applying the designs in dummy form to such everyday items as letterheads, envelopes, mailing labels, products, packages, and whatever media would be appropriate to the project.

(4) Consumer testing—the trademark candidates are evaluated by target audiences. The evaluation is based on the previously agreed upon design criteria. The winning trademark(s) is tested against competitive trademarks to determine the positive and negative attributes of the proposed design so that the positive attributes can be strengthened and the negative ones eliminated.

(5) Legal clearance and registration—it is important that the graphic designs, like the brand name they are based upon, be searched for legal availability.

(6) Design selection and trademark design registration—based upon the marketer's preference and consumer testing and legal availability, a single trademark will be chosen and a registration applied for.

(7) Preparation of reproduction art—when a design is finalized, a master trademark design is prepared in color and black and white art in diskette form for all methods of reproduction.

(8) Transliteration—if the trademark is to be used in export, the transliteration of the trademark will be made for each country that does not use the Latin alphabet. This must be done prior to applying for trademark registration.

(9) Preparation of identification standards manual (electronic and printed)— it is helpful to have the design team prepare rules for its proper use to guide all those who will be working with the trademark. Misuse of the trademark could result in a company losing its rights to that mark.

PHASE 3: GRAPHIC PACKAGE DESIGN

When more than one product is to be packaged for the marketplace under a single brand name, a package design system that clearly distinguishes each production from the other is paramount.

DESIGN TRADEMARK (LOGO AND SYMBOL) FOR NEW BRAND NAME OR UPDATE CURRENT TRADEMARK (IF REQUIRED)

Graphic interpretation of a brand name or trade name (company name) can have a dramatic impact on how effectively the name will project the desired image of a company, product, or service. Symbols or other graphic devices used with the name should reinforce the desired image. In cultures other than those using the Latin alphabet a symbol may be essential to maintain image continuity.

A brand name becomes a trademark when it is registered in the U.S. Patent and Trademark Office. Since trademarks and trade names appear on product packages and promotional literature, as well as in print and television advertising, graphics which command attention and enhance the name will greatly assist marketing efforts.

During this stage, the design team creates, refines, and completes the trademark design to go with the new brand name if it has not already been done. That trademark design (logo and symbol) is then incorporated in the overall package label design.

This is an opportune time for the marketer and design team to consider evaluating and updating the current trademark as well.

THOROUGH GRAPHIC DESIGN EXPLORATION OF PACKAGE'S PRINCIPAL DISPLAY PANEL(S)

During the packaging graphics design phase, initial concepts, often referred to as comprehensive sketches or "comps," are created. Though in every sense resembling a production package, the comps are simply concepts made tangible. The design concepts are studied and critiqued by the design team, marketing staff and target audience consumers to determine which of them most closely satisfies or exceeds the marketing, merchandising, production and other criteria.

The design team undertakes a thorough exploration of principal display panel concept designs for its own internal review.

REVIEW GRAPHIC DESIGN EXPLORATION AND SELECT CONCEPTS FOR EXPERIMENTAL PHOTOGRAPHY

The design team reviews the prospective principal display panels, selecting those design concept(s) that best meet the stated project objectives (agreed upon at the completion of the predesign stage) for experimental product photography.

DEVELOP SELECTED GRAPHIC DESIGN CONCEPTS IN 3-D IRIS COLOR PRINT FORM

Once the number of prospective design concepts has been narrowed down, the team develops those remaining in 3-dimensional color print form in preparation for evaluating them in a retail environment.

PRESENTATION AND REVIEW OF SELECTED GRAPHIC DESIGN CONCEPTS AND VIEW IN RETAIL ENVIRONMENT, OR IN DESIGN SELECTION PLANOGRAM (DSP$_{SM}$) COLOR SLIDE FORM

The selected designs are prepared and presented in "finished" 3-D form or as Design Selection Planogram slides (DSPs) for viewing in a retail environment.

SELECT DESIGN CANDIDATES FOR FURTHER DEVELOPMENT AND REPRODUCTION PHOTOGRAPHY AND DESIGN REMAINING PANELS AND PREPARE ADDITIONAL PHOTOGRAPHY (FINAL PACKAGE COPY NEEDED FOR ALL PANELS TESTED)

The design team finishes designing the remaining package or label areas that are available for graphics including the all important product "sell copy."

For a new product, consumers get a sense of the product from the copy as well as from any photographs or illustrations. Appetizing descriptive copy does much to help convince shoppers to try your food products. Additional package photography is done at this time, if required, suitable for reproduction.

REFINE GRAPHIC DESIGN CANDIDATES IN PRELIMINARY ART FOR REPRODUCTION FORM AND PRESENTATION OF SELECTED DESIGN CANDIDATES IN 3-D IRIS COLOR PRINT FORM FOR APPROVAL FOR CONSUMER EVALUATION

The new package design candidates are refined in preliminary art for reproduction form including any revisions requested by the marketer and/or design team to more closely achieve the project objectives. Typically, two to four design candidates are then selected for evaluation among target audience consumers and 3-D Iris Print "dummy packages" are prepared for use as research stimuli.

PHASE 4: POSTDESIGN TARGET AUDIENCE RESEARCH

PACKAGE DESIGN RESEARCH METHODS

Package graphic design research can be used to measure physical, emotional, and psychological consumer responses to packaging. It is concerned with consumer awareness, opinions, and attitudes. There are numerous techniques that employ qualitative and/or quantitative methods depending on the type of information sought, the time available, and the budget. Several types of research can be employed in combination to measure different factors such as impact, communications, image, and preference (Chapter 12).

The following are some research methods in use today, all using qualified consumers.

Focus Groups

Focus groups are a way to learn, not measure. They are a tool to explore and diagnose, to gain an impression of what is going on. Focus groups are used to probe thoughts, feelings, and emotions of those involved. They do not produce hard facts or projectible data and are not definitive. They can provide clues or insights into the nature of an attitude or reaction but not its extent.

Focus groups consist of consumer panels selected for their appropriateness to the material under consideration and a moderator who, working from a preplanned outline, leads a discussion. Usually the group is observed from behind a one way mirror by the marketing, design, and research people directly involved with the subject under discussion. They are often permitted to put questions to the panel but only through the moderator.

Because the groups do not exceed 10 to 12 participants it is not easy to obtain geographical dispersion. There are locally run interview centers in every good-sized town in the United States. This makes it possible to gain insight into local or regional differences in attitudes or perceptions.

One-on-One Interviews

Direct interviews are qualitative and similar to focus groups. They have the advantage of getting answers that are not influenced by the group dynamics. They suffer conversely from the lack of group interaction sometimes desired, as when probing for new ideas, looking for social bias, or trying to learn consumer language or terminology.

One-on-one interviews are often conducted in an interview facility because they tend to be open ended, and because the respondent is often asked to examine packaging or to look at slides. The interview is conducted by an interviewer who works from an outline, although structured questions can be included.

Depth Interviews

Depth interviews and one-on-ones are similar in that they are both qualitative, use interviewing techniques, and work with one respondent at a time. They differ in purpose, however. Most interviewing techniques are concerned with identifying and recording consumer behavior. The depth interview is more concerned with why consumers behave in a given situation and how these behavioral characteristics are formed. As in the one-on-one, an interviewer works from an outline or question guide but the questioning follows a much broader scope and allows the interviewer to discover and explore areas of interest that may not have been apparent during the planning for the interview.

When the interviews are completed an analyst reviews the reports, summarizes the findings, draws conclusions, and makes recommendations to the client.

One-on-One Intercepts

These are interviews that can be qualitative or quantitative depending on the nature of the information sought and the size of the cells selected. Intercepts are conducted in shopping malls, stores, and other public locations usually close to a shopping outlet. They are normally structured and can involve showing respondents pictures or actual packages. Semantic differential scales are frequently used in this type of research to elicit image and preference responses.

One advantage to this method is that consumers are approached in a shopping situation. This makes it possible to interview people who have just purchased the product or brand under consideration. A disadvantage is the problem of trying to structure the interviews to fit a predetermined demographic breakdown.

Tachistoscope Testing

A Tachistoscope (T-Scope) test is used to measure the impact of elements of a test package. It is a device for measuring visual response and is normally conducted in a central location under tightly controlled conditions.

By exposing a respondent to slides of a test package at predetermined speeds it is possible to determine if, when, and in what order the elements on a package are recognized or read. It is most effective when used to measure elements of a single package even though tests are conducted using shelf displays of competitive products.

The technique is useful in comparing test package alternatives because it produces consistent results in test and retest situations and can discriminate among test alternatives. An advantage of T-scope testing is the ability to identify elements of a package that are being misread or are confusing to the respondents. Reliable results can be obtained with a relatively small test sample of no more than 50 or 60 respondents.

Semantic Differential Tests

These are highly structured quantitative tests which are frequently used in conjunction with other package test techniques such as the T-scope. They are self administered and therefore lend themselves to intercept interviews in malls and stores. The respondent is given a sheet of paper and asked to mark the place on each scale that best describes the package. The test package is on display while the respondent is completing the form.

The semantic differential measures connotations of the package and product. Respondents are given pairs of opposite words with spaces between the opposing words and asked to describe the space on the scale that best describes the package or product. The technique measures evaluative, potency, and activity factors. An evaluative factor might be attractive/unattractive, a potency factor might be strong/weak; and an activity factor might be active/passive.

The test is limited by the inability of many people to think abstractly. Therefore, the interviewer has to explain it carefully and administer it to be certain the respondent is answering properly.

Image and Preference Tests

These are comparative tests in which respondents are shown packages and asked to rate them on a series of scales or multiple choice questions. By their nature they are normally administered in a central location. To produce meaningful results these tests must be quantifiable and drawn from a demographic profile that reflects the marketing objectives.

The objective of the image test is to determine the images and/or impressions communicated by the package designs under test. This is accomplished by asking respondents to rate test packages on a number of attributes or dimensions that would describe the package or product.

The attribute list would cover dimensions related to product efficiency, package esthetics, and personal involvement or referral. In certain cases the physical package can contribute to the imparted image, and the respondents are encouraged to handle the test packages.

In direct comparison preference test respondents are asked to compare test packages against one another. They are shown the packages and read a number of product related statements that represent dimensions or impressions common to each of the packages shown. Respondents are then asked to select the one package alternative preferred for each statement. The list can include statements such as most economical product, most effective product, highest quality product, and best value for the money. The intention of these questions is not to determine a "best choice" but to magnify the differences in consumer perceptions that may exist and determine where the highest level of intensity for each attribute lies.

Eye Movement Tests

Eye movement tests track respondents' reactions to shelf visibility, speed of communication, and involvement in a simulated store shelf environment. They plot the way respondents' eyes move from one display to another and how they evaluate individual packages. They can measure dwell time, which is an indication of interest, but cannot discriminate between positive or negative interest. This is an excellent method to make comparative measurements of the impact and communications of competing brands.

This is a laboratory test in which a single respondent is exposed to slides while a camera records the way the eyeball tracks over the slide and how much time it spends in each location. A computer then takes the imagery from a panel of respondents and translates this into a composite for the group. This technique cannot be used as a substitute for additional studies or to predict purchase intent.

In addition to the basic methodology described above there are a number of research techniques in use which attempt to replicate the shopping experience.

Packtest

Packtest is a system for testing package designs that employs T-scope store slide, store shelf simulation, and interview techniques. These methods are use to obtain consumer responses to package designs as they pertain to in-store recognition, visual impact, shelf impact, image mediation, and overall appeal. Programs are tailored to meet design objectives which range from individual packages to competitive brand analysis.

Miniature Store Testing

In this test technique respondents are asked to shop in a simulated store in which real products are displayed together with the items being tested. In the attempt to recreate an actual shopping experience, respondents are frequently shown commercials, and given seed money to make their purchases. They are not told what is being tested but are given categories of products to shop for which include the test items. Their purchases are then checked to see which brand was selected and interviews conducted on a structured and open-ended basis. This can be a quantifiable technique when large enough panels are selected.

The packaging design research methods reviewed here represent those that are the most basic and commonly used. There are variants to most of them but any attempt at a listing would not add anything significant to this review. In addition there are several esoteric techniques not included, and while no criticism of their validity is implied they are omitted because they do not represent commonly accepted practice among marketing researchers.

It should be emphasized that, of the many research techniques available, there is no single best way to obtain actionable answers. The optimal method supplies answers to real questions within a reasonable schedule and budget. Its impact helps to guide the design organization to optimal results while reducing risk to the marketer.

The following postdesign target audience research steps should be regarded by designers as an essential part of the design process.

QUALIFIED CONSUMER EVALUATION OF GRAPHIC DESIGN CANDIDATES

From the outset, in any comprehensive design program, it is critical for the marketer's management team and the design consultancy to have a clear understanding of the program's objectives. These remain in focus throughout the developmental process, becoming a measuring stick in evaluating the final design candidates.

Final graphic package design candidates are then subjected to in-store consumer evaluation, where they are measured against each other and against the current package design. Qualified consumers are asked to formulate their opinions on attributes in line with the previously agreed-upon project objectives, such as the package's communication of "quality connotation" and "product differentiation." Consumers' preferences and their detailed feedback on the attributes play a major role in the selection of the winning package design candidate.

As these candidates are tested with the target audience, the ease with which the designs are narrowed down often indicates how well the criteria have been met; simply because the objectives have been kept in tight focus.

REVIEW AND ANALYZE RESULTS OF CONSUMER EVALUATION

The design team reviews and analyzes the results of the consumer evaluation in relation to the project objectives.

PRESENT ANALYSIS OF CONSUMER EVALUATION AND RECOMMENDATIONS

The team prepares its analysis for presentation to the marketer, along with its recommendations. These recommendations may include adjustments to the winning design candidate to improve it even further, and may involve such things as color, copy, and/or product portrayal.

REVISE AND REFINE WINNING DESIGN CONCEPT IN PRELIMINARY ART FOR REPRODUCTION FORM FOR TENTATIVE DESIGN APPROVAL (ALL FINAL COPY REQUIRED AT THIS TIME)

With the marketer's approval, any revisions and/or refinements are made to the winning design candidate in preliminary art for reproduction form for final approval. All final copy is needed at this time—ingredients, nutritional information, copyright notices, cooking/preparation instructions, etc.—for all concerned to review and approve it on a dummy 3-D package *prior* to the preparation of final art for reproduction.

ADDITIONAL CONSUMER EVALUATION (IF REQUIRED)

If the results of the postdesign consumer evaluation are such that changes other than minor ones are required, the newly refined design candidate(s) is once again evaluated among qualified consumers to ensure that the changes are, indeed, improvements, and that the package design fulfills the project/marketing objectives.

PHASE 5: ART FOR REPRODUCTION OF THE RETAIL PACKAGE

PACKAGING THE RETAIL PACKAGE—THE "OUTER PACKAGE" AND OTHER PACKAGE SIZE OPPORTUNITIES (CONCURRENT WITH PHASE 5, ART FOR REPRODUCTION OF THE RETAIL PACKAGE)

Decisions made at this juncture of the design process will materially affect the degree of success the product will achieve. The products' distribution channel(s) greatly influences how the package is to be packaged for each customer. Special requirements for each channel will depend upon the type of food, its shelf life (fresh, shelf-stable, refrigerated, or frozen), its individual retail package and the area of the store in which it will be offered for sale.

Supermarkets, mass merchandisers, club stores, chain drug stores, convenience stores, and vending and food service channels all have their own special requirements for regular distribution, and in some channels there are special package display opportunities ranging from display trays to shipper displays.

Once the final packaging graphics design has been determined for the retail package, the design for these outer packages can begin and the reproduction artwork can be prepared concurrent with the preparation of art for reproduction for the retail package.

In addition, multiples may be packaged for special retail environments, also creating a larger billboard area for vastly improved point of sale impact. Conversely, especially in foods, smaller quantities may be considered for single-use consumption.

THE "ORIGINAL" RETAIL PACKAGE

Preproduction Meeting with Color Separator(s) and Converters

This meeting is essential and can be a real time saver, especially if the marketer is under tight time constraints for a new product introduction. It is most helpful if done in person with the converter and color separator together with the design team and marketer. It is possible, however, to accomplish some of the meeting via conference call and/or E-mail. Any special requirements the converter may have can be reviewed and discussed at this time, and the design team and marketer have a chance to ask any questions regarding packaging specifications and/or printing processes and schedules. In addition, at this time preproduction color matching on the substrates to be used in production of the packages is obtained for approval.

PREPARATION OF FINAL ART FOR REPRODUCTION

Reproduction artwork is prepared in final form to the specifications of the selected converter for the type of printing chosen (flexographic, offset, gravure). Reproduction photography and/or illustrations are also completed for approval at this time. Electronic file information is prepared along with any special instructions for the converter, color separator, and/or printer.

SEND HIGH RESOLUTION COLOR PRINT (OR MODEM THE DESIGN) FOR APPROVAL

A high resolution color print (or similarly specified computer color output print) of the completed reproduction artwork is sent to the marketer for final approval and sign-off on design, color, copy, and printing and package specifications. Sometimes, this material can be sent via modem. However, most often the marketer needs this in hard copy format for the company's internal routing and permanent files.

REVISE REPRODUCTION ARTWORK AND SEND NEW PRINT OR MODEM THE REVISED DESIGN FOR APPROVAL

Any requested revisions, corrections, and/or additions are made to the reproduction artwork at this time and a new print of the revised artwork is sent (hard copy or via modem) to the marketer for final approval.

PREPARE MATCHPRINT® AND RELEASE ART FOR REPRODUCTION IN ELECTRONIC FORM

The reproduction artwork is completed in electronic form and furnished to the converter or color separator directly, or via the marketer. The Matchprint is prepared from the reproduction artwork prior to its release to show all concerned with the new package just what can be expected in the specified type and scale of printing.

PHASE 6: POSTDESIGN IMPLEMENTATION

These steps are an essential part of the packaging graphics design process and begin either immediately following, or in some cases because of time constraints, concurrent with preparation of reproduction artwork, Phase 5.

PREPARATION OF PACKAGES FOR TV, PRINT ADS, AND WEB SITE(S)

This is done to the marketer's request and specifications.

DESIGN DISKS AND COLOR PRINTS RELEASED FOR THE PREPARATION OF MATERIALS FOR SALES PROMOTION AND INTRODUCTION TO THE TRADE

This is done to the marketer's request and specifications.

APPROVE INK "DRAW DOWNS," CRITIQUE CONVERTER'S OR SEPARATOR'S 3·M, KODAK, OR OTHER TYPE OF PREPRESS COLOR PRINTS OR PRESS PROOFS

The design team reviews these items prior to the first production run in order to get the marketer's and printer's agreement on what the marketer will accept as quality printed material.

SUPERVISE FIRST PRODUCTION RUN AND ESTABLISH COLOR STANDARDS

A member of the design team often attends the first production run along with the marketer in order to help suggest any needed adjustments prior to the actual press run.

CRITIQUE FINAL PRINTED PACKAGES IN RETAIL ENVIRONMENT

The printed packages are reviewed by the design team and the marketer in several retail environments keeping in mind the original project objectives.

REVIEW CHANGES, IF ANY, IN THE COMPETITIVE PACKAGES SINCE THE INCEPTION OF THIS DESIGN PROGRAM

Changes often occur in the marketplace which can require additional fine tuning of the design before the next press run of packages. The design team relays this information to the marketer for consideration, and together they decide if changes are needed, and/or would be beneficial.

CONSIDERATIONS AND RECOMMENDATIONS FOR THE FUTURE

The design team and the marketer should continually review the printed packages in the marketplace keeping in mind recommendations for future packages (and products!). Sometimes the competition reacts so quickly to the new or repositioned product that the marketer may need to make changes in copy or positioning within a few months.

PACKAGING SHOWCASE

TRADEMARKS

Following is a selection of designs developed for food marketers for their food products. They include food brand marks which have been incorporated into new products' packaging as well as those which served as the basis for repositioning and/or redesigning current products' packaging (Figure 15.3).

CASE STUDIES

Case Study #1: Farm Fresh Egg•Land's Best Eggs/Egg•Land's Best, Inc. (Figure 15.4)

Design Challenge

Create a unique new brand personality in the fresh eggs section of retail stores for the first national brand of fresh eggs. The brand name and overall graphic package design layout had to allow for copy that would clearly distinguish the eggs from the competition.

The marketing plan called for a symbol (derived from the trademark) that could be stamped on each egg for consumer reassurance that the products were genuine Egg•Land's Best brand of eggs. The design was to be printed flexographically on both pulp fiber and plastic containers.

Design Solution

The eye-catching red, white, and blue Egg•Land's Best trademark features the letters "EB" in red which are stamped on each egg for quality reassurance. Printed in 3 colors, red, blue, and yellow on a white background, the overall package design contributes to the quality and freshness image necessary for the sale of the eggs.

Market Impact

The design has proven to be the correct balance of brand impact and product information. Charles Lanktree, president of Egg•Land's Best, states "the color-

FIGURE 15.3 An array of food brand marks (trademarks): 1. Hugs®, 2. Austin Quality®, 3. Jolly Time®, 4. Bahlsen, 5. Borden, 6. Bachman, 7. 4C Foods®, 8. Crackin' Good, 9. Palmer, 10. Florida's Natural Growers' Pride, 11. Mario, 12. Penn Dutch®, 13. Tate & Lyle, 14. Twizzlers, 15. Country Hearth, 16. Amazin' Fruit, 17. M&M/Mars®, 18. Love.

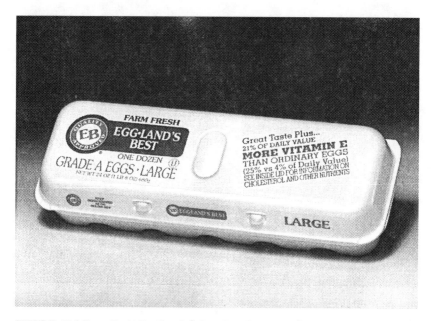

FIGURE 15.4 Farm Fresh Egg•Land's® Best Eggs/Egg•Land's® Best, Inc. Photo by Neil Gershman, NYC.

ful informative packaging graphics played a major role in establishing this new marketing concept, a national brand of fresh eggs, and has continued to play a major role in the success of the brand since its introduction'' (Lanktree, 1998).

Case Study #2: The Bachman Company (Figure 15.5)

Package graphics (trade dress), if unique, can become registered trade-marks. For example, each package in the Bachman Pretzel Packaging Design System is a registered trademark. There are no photographs on the package surface. The product is shown through a window of clear film designed for each "bowl of pretzels" left unprinted in the yellow background. Color-coded copy, napkin, and bowl also help to differentiate each style of pretzel.

Case Study #3: Hershey's Hugs/Hershey Chocolate U.S.A. (Figure 15.6)

Design Challenge

Hershey Chocolate U.S.A. directed that the package design system for Hershey's® Hugs® with Almonds create a unique identity and clearly differenti-

FIGURE 15.5 The Bachman Company. Photo by Neil Gershman, NYC.

FIGURE 15.6 Hershey®'s Hugs®/Hershey Chocolate U.S.A. Photo by Neil Gershman, NYC.

ate the two Hugs products while maintaining a family relationship. The packaging design was also to clearly convey the product's unique appearance and composition, i.e., mini Hershey's Kisses hugged by white chocolate, striped with milk chocolate.

Design Solution

The packages emphasize the Hugs brand identity. *Hershey's* is printed in gold above the white *Hugs* which features a stylized foil-wrapped product between the *u* and *g* in *Hugs*. The product shape is silver with brown stripe for the regular Hugs product and gold and maroon stripes for the Hugs with almonds. The cut-away product illustration immediately relates to the consumer the product's composition as well as adding to the overall taste appeal of the package design. The bottom of the face panel features a clear window, showcasing the gold- or silver-striped foil-wrapped product topped with the paper flag that also identifies Hershey's Kisses®.

Market Impact

The distinctive package design with its memorable Hugs trademark created "instant equity" for this new brand and was instrumental in making Hugs "one of the most successful new-product introductions in Hershey's 99 year history" according to Hershey's 1993 Annual Report.

Case Study #4: Whitewheat® White Bread/Flowers Bakeries, Inc. (Figure 15.7)

Design Challenge

Create a new white bread *product* for entry in a declining market.

Design Strategy

Developed and approved with the client.

Design Solution

The product positioning is a white bread with a Taste of White•Nutrition of Wheat, a Healthy Family White Bread. This required creation of the brand name: "Whitewheat® White Bread" was the winner. The design format is unique and in keeping with the strategy. When introduced, the design included an endorsement from Nature's Own. Nature's Own is America's No. 1 soft variety bread brand.

FIGURE 15.7 Whitewheat® White Bread/Flowers Bakeries, Inc. Photo by Neil Gershman, NYC.

Whitewheat® is featured on the wrapper in white on a medium blue background. A stylized golden wheat sheaf is framed in a red oval containing the copy "Healthy Family White Bread." "Taste of White•Nutrition of Wheat" appears in white against a dark blue background. The color of the Nature's Own trademark is ochre with a deep brown drop shadow along with an orange circle containing wheat and a pot of honey. Dixon & Parcels Associates created the product concept, brand name Whitewheat®, and the packaging graphics.

Market Impact

After successful launch, the brand continues to grow throughout Flowers Bakeries' super-regional distribution in the East, South, and South Central United States.

Case Study #5: Reese Sticks (Figure 15.8)

The Reese Sticks® packaging is an excellent example of packaging for club stores. The skid display features outer display cartons for master cartons containing 36 bars. The interior of the back panel of the outer display carton is printed in brilliant orange and yellow, and features the Reese Sticks yellow and brown trademark. The interior becomes a billboard attracting consumers' attention when empty or nearly empty, thus adding impact to the entire display.

FIGURE 15.8 Reese Sticks®.

Case Study #6: 4C Bread Crumbs/4C Foods Corporation (Figure 15.9)

Design Challenge

The 4C Foods Corporation wished to create a more upscale image for its 4C brand product line through an updating of the 4C trademark and initial development of a new family package design look for the bread crumbs packages. The new bread crumbs package design was also to picture more contemporary product usage alternatives.

Design Solution

Dixon & Parcels Associates leveraged the design equity in the current 4C trademark and bread crumbs packaging. The recognizable 4C logo was enhanced through the addition of a red and green halation complete with a bolder wheat sheaf illustration. The bread crumbs packages were given a more upscale image through simplification of the face panel copy, use of a classic Goudy type, and a stylized photographic treatment of contemporary usage alternatives which also helped create strong appetite appeal.

Market Impact

The redesigned packages contribute powerfully to an expansion of the user franchise in the Northeast and Florida distribution areas.

Case Study #7: Dolphins & Friends/Austin Quality Foods, Inc. (Figure 15.10)

Design Challenge

Austin Quality Foods, Inc., wanted a unique personality for its recently acquired Dolphins & Friends® cheddar cheese crackers which were to be introduced to the market nationally via club stores. The trademark and packaging graphics were to stand out in the snack section to convey the appetite appeal and fun of these uniquely shaped snacks while at the same time maintaining the premium image associated with the Austin Quality brand.

Design Solution

To stand out in the crowded snack cracker and cookie aisles, Dixon & Parcels Associates chose the unique color combination of magenta and orange for the new Dolphins & Friends "pennants" trademark. The halyard on which the pennants "fly" is reminiscent of the previous life preserver trademark while the Austin Quality trademark sits atop the carton's front and back gable. The Dolphins & Friends continue to "swim" across the package but now in a swirl of blue water at the bottom of the face panel out of which "pops" the important copy, "Made with Real Cheddar Cheese."

FIGURE 15.9 4C Bread Crumbs/4C® Foods Corporation. Photo by Neil Gershman, NYC.

FIGURE 15.10 Dolphins & Friends/Austin Quality Foods, Inc. Photo by Neil Gershman, NYC.

Market Impact

Mike Ritchey, Austin Quality Foods Vice President of Marketing, states, "The redesigned packaging graphics did much to help launch the product's introduction on the West Coast which has now been expanded to vending bags and food service bags nationally" (Ritchey, 1999).

Case Study #8: Florida's Natural Orange Pineapple Juice/Citrus World, Inc., a Cooperative of Citrus Growers Based in Lake Wales, Florida (Figure 15.11)

Design Challenge

Create a package design system for over a dozen Florida's Natural Brand Premium Not From Concentrate Juices in keeping with these premium-quality products. Enhance "shelf awareness" and appetite appeal while maintaining brand equity.

Design Solution

Through pre-design consumer research and in-store and media design and copy audits, Dixon & Parcels Associates determined the strengths and weaknesses of Florida's Natural then-current packaging. The result is a more impactful package that includes the upgrading and enhancement of the legibility of the Florida's Natural brand name and a more appetizing presentation of the product illustrations. The total package design clearly differentiates the brand and products from those of competitors while creating a greater visual taste appeal that is clearly preferred by consumers in the postdesign testing.

Market Impact

In less than a year since introduction of the new packages, "Supermarket sales are up eleven percent for Florida's Natural, and we're also seeing double digit growth for the whole premium pasteurized orange juice segment," says Citrus World Director of Marketing Dan McSpadden. Moreover, he adds, "We're projecting that this segment will soon surpass the 'from concentrate' form in dollar sales to become the largest orange juice segment in the supermarket. Apparently, consumers are recognizing quality, and this new packaging is effectively communicating Florida's Natural quality to them. . . . And," Dan McSpadden asserts, "our sales results indicate that the packaging and advertising message which features the new package design prominently is getting through clearly and effectively" (McSpadden, 1998).

FIGURE 15.11 Florida's Natural® Orange Pineapple Juice/Citrus World, Inc., a Cooperative of Citrus Growers Based in Lake Wales, Florida. Photo by Neil Gershman, NYC.

FUTURE PACKAGING TRENDS

HOME MEAL REPLACEMENT

Ready-to-eat retail [home meal replacement (HMR)] is, perhaps, the fastest growing segment in both supermarket and convenience stores. Ready-to-eat retail refers to preprepared foods that are packaged and sold by supermarkets and convenience stores. The concept appeals to busy consumers, people who don't have the time to prepare their own meals or the desire to dine out. According to the USDA Economic Service, in 1996, $0.46 of every dollar spent on food was used to purchase food away from home. This increase in food spending outside the home is attributed in part to the availability of prepared packaged food.

Ready-to-eat retail has introduced a new aspect to package design. The take-home food packaging must not only protect the food until consumed, it must do so in an attractive, tasteful, appealing manner to help ensure repeat purchases. The continued trend toward this kind of consumer food spending will require more and more attention from package designers. This ready-to-eat retail phenomenon not only introduces an entirely alternative packaging need, but also puts an extra burden on "regular" retail food packaging to attract consumers.

A 1997 study by the Food Marketing Institute found that 29% of those who bought prepared meals at supermarkets were aged 18 to 24, compared with 16% in 1996. The only larger age group was shoppers 65 and older, at 32% in 1997, up from 25% (O'Neill, 1998).

HRI

Hotel/restaurant/institutional (HRI) will continue to play a very important role in food packaging. Increased spending on eating out—restaurants, cafeterias, fast food—means more money spent on food services. Marketers must keep in mind that the people who choose the brands served in these locations are consumers themselves. Brand name and packaging are as influential to buyers for restaurants, cafeterias, and agencies responsible for food services as to individual consumers. Food service creates an opportunity for marketers to reach a mass market, not only increasing sales through food service, but also increasing brand recognition for future food service sales. More care will be given to designing portion packs for they are a great trial size for food that will lead to sales of the retail package.

With their specialized packaging requirements, the ready-to-eat HRI segments of the market offer opportunities to create designs that will contribute to the taste appeal and pleasure of consuming the product.

POINT-OF-PURCHASE PACKAGING

Alternate package sizes and styles require a strong awareness of physical considerations in order to meet the demands of the sales environment. Both physical packaging and packaging design must adapt to a variety of point of purchase opportunities.

PRINTING PROCESSES

Six color process printing will become more available. In addition, more six and eight color presses will be available than ever before. This will help to assure top quality reproduction along with tighter color tolerances for food packaging.

Note: the present quality of process printing on most retail packaging substrates exceeds the quality and consistency found in many magazines.

FRAGMENTED CONSUMERS

As food products become increasingly segmented, so will consumers or vice versa. Increased requirements among store buyer committees will require an increased sophistication in product packaging. Designers will be required to find new ways to reach audiences that may not be easily defined.

We believe teen-age purchasing power will increase and teen-age purchasing power in food stores will continue to grow. According to a Rand Youth Poll that surveyed 2,700 teenagers, $48.8 billion were spent by teenagers in grocery stores in 1997. According to a recent *New York Times* article, ''[teen-agers] scoff at product hype, are snack-hungry, brand-loyal and drawn to familiar foods. Most buy and eat on impulse . . . rather than by judicious planning, and are price-conscious'' (O'Neill, 1998). More consideration will also be given to the needs and desires of both the pre-teen and the ''graying population'' segments of the food market.

E-COMMERCE

The package design created for the store shelf may indeed be an excellent shelf package, but it may need some adjusting to be an excellent web site package.

The future is now as far as the web site is concerned. You can sit at home and be taken down a supermarket aisle, make your selection, and have products delivered within the same day you've ordered them.

However we can visualize a different type of web site in the near future where the package will continue to play the major role in selling the product, but it will be a slightly different role.

For example the site will show

- a single brand facing if a unique product
- or two or more single brand facings of similar products
- plus the store brand if the sponsor is a retail marketer

There will be no need for mass displays—every package will be on the "optimal shelf."

There will be no need for a store—just a studio to broadcast from, a warehouse in which to store the products, and a delivery system that will promptly and accurately deliver the groceries.

Approximately 40% of households now have E-mail capability. It is predicted that, by the year 2002, the food industry will sell $6.6 billion worth of products to consumers through the Internet (Cassar, 1998).

INCREASED COMPETITION

The "Battle of the Brands" will become even fiercer. With supermarket chains acquiring other chains, there will be fewer but stronger store brands. However as previously mentioned, to many consumers "a brand is a brand." They do not consciously differentiate a "store brand" from an "advertised brand," one reason being that most stores advertise their major brands in newspapers weekly as well as producing full color circulars. Thus, it is even more important for trademarks to be unique, and to properly connote the desired image quickly and succinctly for the product. Otherwise they are counter productive.

TOTAL PACKAGE GRAPHICS = TRADEMARK

To be even more unique, strive for the total package graphics to become a trademark. Each package in the Bachman Pretzel Packaging Design System (which consists of a yellow background, a clear film silhouette of the pretzels, and a color coded napkin and bowl for each pretzel variety) is an actual registered trademark.

CONCLUSION

The American consumer is changing rapidly, and both marketers and designers must maintain a constant vigilance over trends in merchandising or risk becoming obsolete overnight. You must frequently (every two to three months) review the product *positioning* as well as the product *presentation* against the competition. Successful companies must be strategically a step ahead of the

competition while at the same time examining their product(s) in relation to
the needs and wants of the new and ever-changing consumer. With the package
as the critical communicator, as well as the product until purchased, meticulous
management of the package and its graphics is vital.

BIBLIOGRAPHY

Anonymous. 1998a. "Change In Store," *The Wall Street Journal,* November 20.

Anonymous. 1998b. "New and Improved," *The Wall Street Journal,* November 20.

Anonymous. n.d. "7 Reasons Why," The United States Trademark Association.

Brody, Aaron L. and J. Roy Parcels. 1995. "Designing Packages for a Healthier America and the Nutrition Labeling and Education Act," in *Nutritional Labeling Handbook.* Ralph Shapiro, ed. New York: Marcel Dekker, Inc.

Burkhardt, W. C. 1998. Personal communication.

Cassar, Ken. 1998. "Digital Commerce Strategies/Shopping," *Jupiter Communications.* August 14.

Celauro, John A. 1998. Personal communication.

Crozier, Robert P. 1998. Personal communication.

Dixon & Parcels Associates, Inc. 1960–1996. "The dp Letter," Volumes 1 through 16.

Dixon & Parcels Associates, Inc. 1989–90. "Strategic Marketing Communications," *Business Strategy International.* Cornhill Publications.

Dixon & Parcels Associates, Inc. 1995. "Name*power*® Where Successful Brand Names, Corporate Names & Designs Are Created."

Fraser, Faith S. 1992. "Creating a Brand Name," Visiting Lecture Series at the United States Office of Patent and Trademarks. November 19.

Gardner, Sam. 1963. Personal communication.

Hamel, Gary and Jeff Sampler. 1998. "The e-Corporation," *Fortune,* December 7.

Lanktree, Charles. 1998. Personal communication.

McSpadden, Dan. 1998. Personal communication.

Meyers, Herbert M. and Murray J. Lubliner. 1998. *The Marketer's Guide to Successful Package Design.* Lincolnwood, IL: NTC Business Books.

O'Neill, Molly. 1998. "Feeding the Next Generation; Food Industry Caters to Teen-Age Eating Habits," *The New York Times,* March 14.

Parcels, J. Roy. 1982 "Packaging Often Irks Buyers, but Firms Are Slow to Change," *The Wall Street Journal,* January 28.

Parcels, J. Roy. 1989. "Package Design: Art or Science?" *Remarks . . . Trademark News for Business,* Vol. 2, No. 4. The United States Trademark Association.

Parcels, J. Roy. 1995. "Selling in a Dangerous Environment," *New Product News,* Vol. 31, No. 9, October 12.

Parcels, J. Roy. n.d. "Packaging, The Stepchild in New Product Planning," *Product Management,* Vol. 5, No. 6.

Ritchey, Mike. 1998. Personal communication.

Smith, Wrede. 1998. Personal communication.

Stern, Walter, ed. 1981. *Handbook of Packaging Design Research.* New York: John Wiley & Sons, Inc.

Storlie, Jean. 1996. *Food Label Design, a Regulatory Resource Kit.* Herndon, VA: Institute of Packaging Professionals.

Thomas, G. 1990. Personal communication.

Welch, J. F. 1998. Personal communication.

CHAPTER 16

Mandatory Food Package Labeling in the United States

JEAN STORLIE
AARON L. BRODY

Food package design includes two basic communication elements: those desired by marketers to convey the attributes and benefits of the contained products and the information required by law and regulation. The Food and Drug Administration (FDA) oversees most mandatory food labeling while the U.S. Department of Agriculture (USDA) is responsible for packages containing meat. In general, food package labels must contain a product identity, content volume and weight, a source, an ingredients list in descending order of quantity in the product, and a nutritional facts panel. The two agencies differ to some degree in requirements, but are in large measure very similar so as not to confuse consumers. Food product developers should be cognizant both of the principles and the essential details to guide their total communications message.

REGULATION of food label content in the United States falls mainly under the jurisdiction of two federal agencies: the Food and Drug Administration (FDA) and the United States Department of Agriculture (USDA). USDA's Food Safety and Inspection Service (FSIS) oversees labeling of food products containing meat or poultry. All other food products, including some containing very small quantities of meat or poultry and some products (e.g., traditional sandwiches) not associated by most consumers with meat and poultry products, fall under the jurisdiction of FDA's Center for Food Safety and Applied Nutrition. Although the two federal regulatory agencies approach many aspects of food labeling in a similar manner, sometimes they differ.

This material is excerpted from Storlie, J., *Food Label Design: A Regulatory Resource Kit*. Copyrighted 1996, Institute of Packaging Professionals. Herndon, Virginia. Used with permission.

Other United States government agencies also have requirements that affect food package labeling: The Bureau of Alcohol, Tobacco and Firearms of the United States Treasury Department regulates labeling of alcoholic beverages including beer, wine, and distilled beverages; the Agricultural Marketing Service (AMS) within USDA administers grading of fresh and minimally processed agricultural products—another agency having some labeling jurisdiction.

DIFFERENCES BETWEEN FDA AND USDA PACKAGING REGULATIONS

FDA package label regulations are governed by the Federal Food, Drug, and Cosmetic Act, and the Fair Packaging and Labeling Act (FPLA). The Nutrition Labeling and Education Act of 1990 (NLEA) and the FDA Modernization Act of 1997 are amendments to the Food, Drug, and Cosmetic Act. The Federal Meat Inspection and Poultry Products Inspection Acts define USDA's regulatory role and responsibilities. Because the two agencies are governed by separate laws and therefore by separate regulations, they have different missions, philosophies, and approaches to food package labeling, resulting in differences in regulations as shown in Table 16.1. USDA has a more active role in monitoring meat product production, packaging practices, and package labels through on-site plant inspections and label preapproval.

MISSION AND APPROACH

FDA's regulatory history began with the 1906 passage of the Pure Food and Drugs Acts, which instituted federal oversight of food labeling, replaced in 1938 by the current law. Labeling was intended to protect consumer safety. In 1973, FDA initiated a voluntary nutrition labeling program that represented a regulatory turning point—the inclusion of ''meaningful'' health information on food labels.

Under the Federal Meat Inspection and Poultry Products Inspection Acts, USDA's primary regulatory role has been to reduce the probability of public health hazards arising from improper handling of meat and poultry during production and packaging. USDA has a network of field offices responsible for administering in-plant on-site inspections. A national office administers a label preapproval program.

REGULATORY DOCUMENTS AND GUIDANCE

Both FDA and USDA publish daily regulatory changes in the *Federal Register*. All federal regulations are updated annually and compiled in the

TABLE 16.1. Key Differences between Food and Drug Administration and U.S. Department of Agriculture Labeling Regulations.

	FDA	USDA
Mission and Approach	• Early emphasis on truth in labeling • Broader span of regulatory oversight, limited resources	• Emphasis on prevention of public health hazards from improper handling • Extensive network of field offices available for consulting and monitoring
Regulatory Documents and Guidance	• *Federal Register* • Title 21 of the *Code of Federal Regulations* • Compliance Policy Guides	• *Federal Register* • Title 9 of the *Code of Federal Regulations* • Policy Manual and Memos
Label Preapproval	• No label preview or approval • Mistakes identified through market surveys, and then warning letters issued	• Label designs reviewed in advance and approved before use • Plant inspectors provide greater opportunity to monitor label compliance
Nutrition Labeling	• Mandated by NLEA • Spearheaded nutrition labeling changes	• Exempted from NLEA, but USDA generally follows FDA regulations and so USDA has nutritional labeling regulations
Food Categories Regulated	• All packaged foods containing less than 2% cooked meat or poultry, or 3% or less raw meat or poultry • Traditional sandwiches • Pizzas not containing meat or poultry • Dairy products • Eggs • Fish and seafood • Fruits and vegetables • Soups not containing meat or poultry • Salads not containing meat or poultry • Nuts and legumes	• Food containing 2% or more cooked meat or poultry, or more than 3% raw meat or poultry • Fresh meats and poultry • Processed meats and poultry • Soups with meat or poultry • Salads with meat or poultry • Pizzas containing meat or poultry • Meat- or poultry-based snacks • Mixed dishes containing meat or poultry (e.g., chili, lasagna, TV dinners, stews)

Code of Federal Regulations. (*CFR*); FDA package labeling regulations appear in Title 21 and USDA labeling regulations in Title 9 of the *CFR*. Some USDA regulations cross-reference FDA regulations. FDA has issued Compliance Policy Guides, which reflect that agency's interpretations of the regulations and policies for package label enforcement. USDA publishes a standards and labeling Policy Manual and periodic labeling Policy Memos which address similar issues. Both agencies provide consulting, but meat packers and processors obtain significantly more federal agency interaction and guidance through the label preapproval program and their on-site continuing plant inspections.

LABEL PREAPPROVAL

In the past, USDA administered a label preapproval program in which all food label designs were reviewed and approved by the agency before they were printed. If the processors made mistakes, USDA told them what was wrong, and so processors were able to learn the regulations empirically and rely on the agency's guidance. This program has been restructured to place greater responsibility on the manufacturer to understand and correctly implement the regulations. Some labels are reviewed in generic (sketch) form, and it is not necessary to resubmit a label for additional review when changes are made. Plant inspectors no longer closely monitor labeling and maintain records on label review.

Unlike USDA, FDA does not preview package labels or offer an approval process, but rather places the responsibility on the food processor/packager to understand the regulations and correctly apply the rules. The agency typically does not find mistakes until labels are printed and the packaged products are in the marketplace.

NUTRITION LABELING

FDA's voluntary nutrition labeling program was in existence for more than 20 years before the passage of the Nutrition Labeling and Education Act (NLEA). In 1990, Congress passed the NLEA, which made nutrition labeling mandatory on most packaged food product packages and mandated that FDA initiate uniformity in the content and format of the nutrition label on the package.

Nutrition labeling of USDA products was not included in the congressional mandate, but USDA immediately opted to follow FDA's regulations to avoid consumer confusion from two different nutrition labels. In most significant aspects of the nutrition labeling regulations, the two agencies have the same requirements, but some differences exist.

USDA SPECIAL REQUIREMENTS

USDA product package labels are subject to additional requirements, some of which also pertain to certain FDA products.

Handling Instructions

Some FDA food product package labels must carry warning statements or handling instructions. USDA product labels usually carry special instructions to ensure the safety and quality of the product (e.g., "Keep Refrigerated," "Keep Frozen"). Further, raw meat and poultry product package labels must provide "Safe Handling Instructions" with specified statements and graphics.

Inspection Legends

All USDA product package labels, also must bear an official Inspection Legend in accordance with established inspection procedures.

VOLUNTARY GRADING SERVICES

Voluntary grading services are available for some food products under both FDA and USDA jurisdiction. USDA's Agricultural Marketing Service (AMS) performs grading services for some fresh and minimally processed products (e.g., dairy, egg, fruit, vegetable, meat, poultry) subject to FDA or USDA regulation. To use approved grade designations (e.g., U.S. Grade A, U.S. No. 1, U.S. Fancy), a product must have been inspected by USDA staff and determined to comply with grade standards established by AMS.

PRINCIPAL DISPLAY PANEL

A Principal Display Panel (PDP) is a part of almost every package of all packaged food. The PDP is the part of the package graphics most likely to be displayed to the consumer for retail sale. Mandatory label elements that must be included on the package PDP are: Product Identity statement and Net Quantity of Contents declaration. Some other required statements must also be displayed on the PDP.

Product Identity Statement

The Product Identity statement is intended to identify the contents, i.e., to communicate to the consumer the type and form of food contained in the product package. This information should be incorporated into the PDP with sufficient prominence that the consumer can readily view it on a retail shelf. The terminology as well as placement and type size of the identity statement are decreed by the regulations.

Net Quantity of Contents

The package's Net Quantity of Contents declaration informs the consumer about the quantity or amount of food product contained in the package.

INFORMATION PANEL

The package Information Panel typically is the package panel contiguous to and immediately to the right of the PDP as viewed by the consumer. The Information Panel includes detailed information about the contained food product that is intended to help the consumer make informed purchasing decisions. The information required on the Information Panel includes: the Ingredient List, Nutrition Facts, and Manufacturer Identity statement. At times, other information, such as warning statements, a Percent Juice declaration, and claims-related statements, may be required. Mandatory information may be displayed on the PDP instead of on the Information Panel.

INGREDIENTS LIST

A list of the ingredients of a multi-ingredient product must be included on the food package surface, typically on the Information Panel.

NUTRITION FACTS INFORMATION

With the introduction of the NLEA, the format of nutrition information has become strictly defined. A number of Nutrition Facts layouts are available, and specific criteria dictate their use, as described below.

MANUFACTURER IDENTITY STATEMENT

The Manufacturer Identity statement indicates the name and place of business of the manufacturer, packer, or distributor—a source the consumer may contact with questions or problems. This information usually is placed on the Information Panel, or on the PDP if no Information Panel is present.

OTHER INFORMATION

Additional statements may be required on the Information Panel and/or the PDP. For example, if a PDP bears a comparative claim, such as "reduced," quantitative information comparing relevant nutrients in the product to those in a reference food must be included on the Information Panel or the PDP. Some products may require a warning statement, Percent Juice declaration, or special handling instructions.

EXEMPTIONS

The package labeling regulations apply to primary packages for retail sale, but certain containers and wrappings may be exempt from the labeling require-ments. The regulations define the term "package" as any container or wrap-ping in which a food is enclosed for retail sale. Packages used for shipping, distribution, display, or other functions are exempt. Further, generally, very small packages, e.g., unit portion, are exempt.

INNER WRAPS

When multiple primary packages of food are not intended to be sold individ-ually, their inner packaging that bears no information is exempt from labeling requirements, e.g., individually packaged unit portions of candy dump-filled in pouches or cartons.

SHIPPING CASES

Shipping cases for distribution of food products are exempt from labeling requirements. If the shipping cases are sold as retail units (e.g., in a warehouse club), however, all mandatory package label requirements apply.

SMALL CONFECTIONS

Small confections (e.g., "penny candy") weighing less than $\frac{1}{2}$ ounce (or 15 grams) are exempt from labeling requirements when the shipping container or retail package meets the labeling requirements.

PRINCIPAL DISPLAY PANEL

As indicated previously, retail food packages must have a Principal Display Panel (PDP) that bears a Product Identity statement and a declaration of Net Quantity of Contents. If no Information Panel exists, an Ingredient List, a Nutrition Facts panel, and a Manufacturer Identity statement must appear on the PDP. On some product packages, additional information may be required, such as claims-related information, a warning statement, or special handling instructions.

LOCATION

The PDP is considered the area of the package that is most likely to be displayed, presented, or read by the consumer at the point of sale. The PDP may be a label on a package surface or the entire surface. On glass or plastic jars and bottles, a label on the closure may be used for the PDP. A printed header strip attached across the top of a flexible pouch that contains no other printed or graphic material may serve as the PDP.

More than one PDP may be incorporated into a package design but, when this approach is used, all of the mandatory information must be wholly duplicated.

SIZE

The package PDP must be large enough to accommodate all mandatory label information clearly and conspicuously, without obscuring or crowding by designs, pictorials, or vignettes. The size of the PDP determines the minimum letter height for much of the mandatory information contained in the PDP. The area of the PDP is the same in similar-size packages, regardless of the label size. The container size and not the label size determines the area of the PDP. For example, if a package bears a spot label, the area of the surface displayed to the consumer and not the size of the label determines the area of the PDP.

Rectangular Solid Shaped Packages

For rectangular solid shaped packages, one entire surface of the package, typically the largest surface, is considered the PDP. The area is determined by multiplying the height times the width of that surface: area of PDP = height × width.

Cylindrical Containers

For cylindrical (or approximately cylindrical) containers, such as jars, bottles, and cans, the PDP is considered 40% of the height times the circumference of the container. If a spot label is used, the area of the PDP is still determined by: (height × circumference) × 0.40.

Conical-Shaped Containers

For cone-shaped containers, such as jars, bottles, cans, and tubs, the PDP is considered 40% of the height times the circumference of the container. To determine the circumference, measure both the widest and narrowest circumferences and average the two numbers:

$$\text{PDP area} = \text{height} \times \left(\frac{\text{widest circumference} + \text{narrowest circumference}}{2} \right) \times .40$$

Irregularly Shaped Packages

For containers of other shapes, 40% of the total surface of the container is considered the PDP. If an obvious package surface is present to serve as the PDP (e.g., top of a triangular or cylindrical package of cheese), this entire surface should be measured. For extremely irregular containers, the packager may substitute an easily measured container of the same capacity: PDP area = total surface area × 0.40.

PRODUCT IDENTITY STATEMENT

The Product Identity statement describes the product contents. It includes the name of the food as well as other descriptive characteristics, such as the form of the food, if it is an imitation food, or if it contains specific ingredients.

TERMINOLOGY OF THE PRODUCT IDENTITY STATEMENT

The product name, typically a standard name or a common or usual name of the food, is used for the Product Identity statement (orange juice). Sometimes an appropriately descriptive term (e.g., tomato soup) or a fanciful name (e.g., Coca-Cola®) commonly used by consumers to identify the food is permitted.

Form of Food

If a food is marketed in different forms (e.g., whole, sliced, or diced pickles), the form of the food is a necessary element of the Product Identity statement. If the form of the food is visible through a transparent container (e.g., whole pickles in a clear glass jar), or it is depicted by an appropriate pictorial or vignette, the particular form need not be stated.

Imitation Foods

If a food is an imitation of a "real" food, the word "imitation" must be used as part of the package surface Product Identity statement, immediately preceding the name of the food, and in a readable size.

Juice-Containing Beverages

Special rules affect how beverages containing fruit or vegetable juices may be named and described on the package:

- If a beverage contains less than 100% but more than 0% juice *and* the product name includes the word "juice," the name must also include a defining term such as "beverage," "cocktail," or "drink."

- If a product name specifically identifies a juice that has been reconstituted from a concentrate, the name must also include a qualifying term such as "from concentrate" or "reconstituted."
- On a 100% juice *or* a diluted juice product, if specific juices are identified as part of the product name (or elsewhere outside of the Ingredient List), either the juices must be listed in descending order of quantity by volume, or their relative quantity must be shown by other means. For example, a juice name could be combined with "flavored" to indicate a nondominant juice.
- When a label for a multiple-juice beverage indicates that specific juices are present, but does not identify all juices in the product, the label must state that other juices are present (e.g., ". . . in a blend of two other juices").

In addition to the product name requirements, beverages that contain, or appear to contain, fruit or vegetable juice are subject to requirements for percent-juice labeling.

Characterizing Ingredients

Package labels on some products may be required to include information about characterizing ingredients in the Product Identity statement. An ingredient is considered a characterizing ingredient when the proportion of an ingredient present in a food has an influence on the price or consumer acceptance of the packaged product, or when labeling may create an erroneous perception (e.g., pictorials or vignettes depicting an ingredient not present).

The required information may be

- the presence or absence of a characterizing ingredient
- the percentage of a characterizing ingredient
- the need for the consumer or user to add a characterizing ingredient

For example, foods packaged for use in consumer preparation of main dishes or dinners (e.g., macaroni and cheese dinner kit) require information about the need to add ingredients such as milk and butter.

Characterizing Flavors

If the product package labeling communication and/or advertising makes a direct or an indirect representation of any recognizable flavors through words or pictorial, the flavor is considered a characterizing flavor. A characterizing flavor must be incorporated into the Product Identity statement. For example, depiction of vanilla beans on an ice cream carton would trigger the inclusion of the word "vanilla" in the product name (e.g., vanilla ice cream). The word "flavored" or "artificial" may have to accompany the name of the flavor in certain circumstances, e.g., if the flavor is not natural.

DESIGN ELEMENTS

In addition to dictating the terminology, the regulations describe package surface location, placement, and type-size requirements for the Product Identity statement. When other required information is incorporated into the Product Identity statement (e.g., form of food, "imitation," characterizing ingredient, characterizing flavor, "reconstituted"), additional design requirements apply. USDA defines an additional criterion: no letters can be less than one-third the size of the largest type in the product name.

Product Identity Statement

The product name must be in conspicuous, contrasting type, located prominently on the PDP. The type size should be reasonably related to the largest print face on the PDP.

Form of Food

When language describing the form of the packaged food is required, the letters should be in a type size bearing a reasonable relation to the type size of the other components of the Product Identity statement.

Imitation

If the descriptor word "imitation" is required, it must be in the same type size and prominence as the name of the food in the package.

From Concentrate

When a term such as "from concentrate" or "reconstituted" is required as part of a fruit/fruit-flavored beverage name, the letters may be no smaller than half the height of the letters in the name of the juice.

Characterizing Ingredients

Percent ingredient declarations must be printed in an easily legible, prominent typeface that distinctly contrasts with other printed or graphic material.

Characterizing Flavors

When a characterizing flavor must be included in the product name, the letters must be no smaller than half the height of the letters in the product name. If the word "natural," "artificial," or "flavored" is included, it must be no smaller than half the height of the letters in the characterizing flavor.

FOOD AND DRUG ADMINISTRATION DECLARATION OF NET QUANTITY OF CONTENTS

A declaration of Net Quantity of Contents, which generally must appear on all PDPs, provides quantitative information about the quantity or amount of product contained in the package. If alternate PDPs are present, this declaration must appear on each PDP.

METRIC LABELING

The current regulations are based on the inch-pound system of measures and may not appear to be necessarily consistent with the national movement to convert the United States to metric measures. The regulations pertaining to Net Quantity of Contents that appear below include the metric labeling regulations.

Selecting the Proper Quantity/Amount Unit

The Net Quantity of Contents may be expressed as weight, a fluid volume or a dry measure, or a numerical count. Units used in declaring the Net Quantity vary depending on the form of the food. If the food is in liquid form, a fluid measure must be used. If the food is solid, semisolid, viscous, or a mixture of solid and liquid, a weight measure must be used. Fresh fruits and vegetables and dry commodities may be labeled with a dry measure. If a firmly established consumer usage and/or trade practice exists for declaring the contents of a liquid product by weight—or solid, semisolid, or viscous product by fluid measure—it may be used. Similarly, if a firmly established consumer usage and trade practice of declaring net quantity by numerical count, linear measure, or area measure is employed, it may be used and possibly augmented by a weight or fluid measure. It is *never* acceptable to use any adjective qualifying the unit size or volume (e.g., jumbo quart, full gallon).

Using the Correct Term

When the Net Quantity is declared as a weight measure, the terms "Net Weight," and "Net Mass" are used. Net Weight is used when the inch-pound declaration is first; "Net Mass" is used when the metric appears first. If a fluid measure is used, then either Net or Net Contents is the modifier.

Largest Whole Unit

The Net Quantity of Contents should be declared in the largest whole unit, with any remainder expressed as a decimal or a common fraction of the unit. Alternatively, the remainder may be expressed in terms of the next smaller whole unit and any decimal or common fraction of that unit. Common fractions are halves, quarters, eighths, sixteenths, or thirty-seconds. Common fractions must be reduced to the lowest (simplest) terms. Decimal fractions should be carried to no more than three places.

Dual Inch-Pound Declaration

The current Net Quantity of Contents regulations require a dual inch-pound declaration on packages weighing between one and four pounds or containing between one pint and one gallon. This dual declaration requirement results in a statement of ounces followed by a parenthetical pound-ounce declaration, e.g., 18 oz. (1 lb 2 oz) or a statement of fluid ounces followed by a parenthetical declaration in largest whole fluid unit with the remainder in fluid ounces, e.g., 36 fl oz (1 qt 4 fl oz). Under the metric labeling regulations, this dual inch-pound declaration requirement would be eliminated. Including a dual inch-pound declaration would be optional; if used, however, the inch-pound designation must appear on one line and may precede or follow the metric declaration.

Abbreviations

The regulations list the only abbreviations that may be used in the declaration of Net Quantity of Contents (see Table 16.2). Generally, periods and plural forms are optional.

DESIGN ELEMENTS

The regulations specify design requirements for the Net Quantity of Contents declaration. It must appear as a distinct item in the lower portion of the Principal Display Panel, be surrounded by blank space, and meet a minimum type-size requirement based on the area of the PDP.

Location

The Net Quantity of Contents declaration must appear within the bottom

TABLE 16.2. **Abbreviations Permitted by FDA for Net Quantity of Contents.**

Quantity Unit	Abbreviation
weight	wt
pint	pt
ounce	oz
quart	qt
pound	lb
fluid	fl
gallon	gal
kilogram	kg
meter	m
gram	g
centimeter	cm
milligram	mg
square meter	m^2
cubic meter	m^3
cubic centimeter	cm^3
liter	L or l
milliliter	mL or ml

30% of the PDP. If the PDP is five square inches or less area, any available space on the PDP may be used.

The Net Quantity of Contents declaration must be placed in lines parallel to the base on which the package rests as it is displayed and may appear on more than one line.

Copy-Free Area

The Net Quantity of Contents declaration must be separated from other printed matter. Spacing requirements do not apply to pictorials or other graphics, provided they do not render the quantity declaration inconspicuous. The declaration may be placed closer to the extreme lower border than the space prescribed below the statement.

Typestyle

The Net Quantity of Contents declaration must be in conspicuous, prominent, and easily legible type. The declaration must appear in distinct contrast (by typography, layout, color, embossing, or molding) to other graphics on the package, *unless* it is blown, embossed, or molded on a glass or plastic surface, in which case the contrast is not mandatory.

Type Size

To ensure that all packages of the same size have the same size Net Quantity of Contents, the regulations define minimum type-size requirements based on the area of the PDP (see Table 16.3). To determine the minimum type size for the Net Quantity of Contents, measure the area of the PDP.

SPECIAL LABELING PROVISIONS

The regulations allow for modification of the Net Quantity of Contents declaration in special circumstances. In some cases, a modified declaration is required (e.g., multiple primary packages). In many other situations, the requirements may be relaxed to offer more flexibility in package design.

Secondary Packages

Secondary packages containing two or more individually primary-packaged, identical units that may also be sold individually have special labeling requirements. The Net Quantity of Contents declaration must be on the exterior of the secondary package if the individual primary package labeling is obscured by the exterior secondary packaging. The Net Quantity declaration must include the number of individual primary packages, the quantity of each individual unit, and, in parentheses, the total weight or volume quantity of the multiple package. The total quantity optionally may be preceded by "total" or "total contents."

Bulk Foods

Food sold from bulk containers at the retail level is exempt from declaring the Net Quantity of Contents, provided it is accurately weighed, measured, or counted within the purchaser's view or in accordance with the purchaser's order.

TABLE 16.3. **FDA Type-Size Requirements for Net Quantity of Contents Package Declaration.**

Area of the PDP		Minimum Type Size
More Than	Less Than or Equal to	Letter Height Not Less than
	5 sq in	$1/16$ inch (1.6 mm)
5 sq in	25 sq in	$1/8$ inch (3.2 mm)
25 sq in	100 sq in	$3/16$ inch (4.8 mm)
100 sq in	400 sq in	$1/4$ inch (6.4 mm)
400 sq in		$1/2$ inch (12.7 mm)

Random-Weight Packages

Packages from one lot or shipment of the same commodity with no fixed weight pattern are random-weight packages (e.g., cheese cut from a bulk block). If the package label contains the net weight, price per unit weight, and total price, the requirements for type size, metric declaration, location, and copy-free area are waived. In addition, the weight may be declared in decimal fractions of a pound even if the weight is less than one pound.

Individual-Serving Packages Not Intended for Retail Sale

Small, unit-portion containers for use in restaurants, institutions, other food-service establishments, and passenger carriers that are not intended for retail sale are exempt from declaring the Net Quantity. These packages must be less than one-half ounce or one-half fluid ounce quantity.

U.S. DEPARTMENT OF AGRICULTURE DECLARATION OF NET QUANTITY OF CONTENTS

Although USDA products are exempt from incorporating metric labeling requirements into their Net Quantity of Contents declarations, the department encourages metric labeling.

UNITS AND TERMINOLOGY

Units employed in declaring the Net Quantity of USDA foods depend on the form of the food. Since most USDA foods are solid, semisolid, viscous, or a mixture of solid and liquid, a weight measure is generally used. A food sold in liquid form requires a fluid measure (e.g., chicken soup). Firmly established consumer usage or trade practice for declaring the contents of a liquid product by weight—or solid, semisolid, or viscous product by fluid measure—may be used, however.

When the Net Quantity is declared in a weight measure, the term "Net Weight" (or "Net Wt.") is used. If a fluid unit is used, either "Net Contents" or "Contents" should be used.

DUAL DECLARATION

On certain size packages, USDA requires that the Net Quantity be declared in ounces followed in parentheses by the quantity in pounds with any remainder expressed in ounces (similar requirements pertain to fluid measures). This format is referred to as "dual declaration." USDA continues to require dual declarations.

ABBREVIATIONS

USDA does not designate abbreviations that may be used in the Net Quantity declaration, and so meat processors and packagers may use the FDA-defined or other abbreviations.

SPECIAL LABELING PROVISIONS

USDA allows for modification of the Net Quantity declaration in specific situations. USDA requires a modified declaration on multiple primary packages. Some USDA requirements are relaxed to accommodate certain packaging situations and established trade practices.

Multiple (Secondary) Packaging

USDA-regulated packages that contain two or more individually primary-packaged, identical units that may also be sold individually have special labeling requirements. The Net Quantity declaration must be on the exterior surface of the secondary package if the individual primary package labeling is obscured by the exterior secondary packaging. The declaration must include the number of individual primary packages, the quantity of each individual unit, and, in parentheses, the total quantity of the multiple package in ounces or fluid ounces. The dual declaration requirement is waived.

Random-Weight Packages

USDA packages from one lot or shipment of the same commodity with no fixed-weight pattern are considered random-weight packages (e.g., intact meat cuts). The Net Quantity declaration must be applied to random-weight consumer packages before retail sale, but the declaration is exempt from type-size, dual declaration, and placement requirements, provided it appears conspicuously on the PDP.

Small Packages

If the shipping case of individually wrapped, small packages (less than one-half ounce) carries an accurate Net Quantity declaration, the USDA Net Quantity requirements are waived for the individual packages. When these small packages bear the net weight, price per pound, and total price, they are exempt from type-size, dual declaration, and placement requirements, provided the Net Quantity appears conspicuously on the PDP.

Margarine

Some margarine products contain animal fat and thus fall under USDA jurisdiction. Such margarine in one-pound rectangular packages (with the exception of whipped and soft margarine, or packages containing more than four sticks) is exempt from the requirements of placement in the bottom 30% of the PDP and the dual declaration. The Net Quantity declaration must appear as ''1 pound'' or ''One Pound'' in a conspicuous manner on the PDP.

Sliced, Shingle-Packed Bacon

USDA exempts certain size packages (i.e., 8-ounces, 1-pound, and 2-pound rectangular packages) of sliced, shingle-packed bacon from placement of the Net Quantity in the bottom 30% of the PDP and from dual declaration. These products are placed on boards that wrap around the top and serve as the PDP. Due to the space limitations of the PDP, these product packages have been granted more flexibility in incorporating the Net Quantity declaration. The declaration must appear conspicuously on the PDP.

In addition, sliced, shingle-packed bacon in any other size container (i.e., other than 8-ounce, 1-pound, and 2-pound) must show the Net Quantity declaration with the same prominence as the most conspicuous feature on the package.

CLAIMS-RELATED STATEMENTS

When a product package or other communication includes a claim, a number of elements must be incorporated into the package design. The regulations do not specify where a claim must be placed, but most claims appear on the PDP because it is the most conspicuous panel.

TYPES OF CLAIMS

The regulations authorize two categories of claims: nutrient content claims (sometimes referred to as ''descriptors'') and health claims. Nutrient content claims are statements about the level of a nutrient in a food. On the other hand, health claims link the nutrient profile of a food to a health or disease condition. Nutrient content claims are used widely by the food industry. Because the regulations governing health claims are more complicated and restrictive, health claims are not as common.

Nutrient Content Claims

Nutrient content claims characterize the level of a nutrient in a food. Only defined terms may be used on the label to describe the contained food's nutritional content. When these terms are used, a product must meet specific criteria. Nutrient content claims may be expressed or implied, or comparative or absolute. These distinctions are important in package graphics design because the type of claim dictates the information required.

Expressed Claim

Any direct statement about the level or range of a nutrient in a food is considered an expressed nutrient content claim. Specific terms are defined by regulation.

Implied Claim

An implied claim is any communication that might lead the target consumer to assume that a nutrient is absent or present in a certain quantity, or that the food may be useful in achieving dietary recommendations. Implied claims must follow the same requirements defined for expressed claims.

Comparative (or Relative) Claim

A claim comparing the level of a nutrient in one product to the level of that nutrient in another product or class of foods is considered a comparative or relative claim. For example, reduced, less, and more are comparative claims.

Absolute Claims

In contrast to comparative claims, absolute claims make a statement about the nutrient level in a food without stating or implying any comparison to another product. Free, low, very low, high, and source of are examples of absolute claims (see Table 16.4).

Health Claims

FDA has approved eleven health claims which may be used on food packages (Table 16.5), and has created very strict criteria for using these claims. In addition to these claims, processors may make health claims based on authoritative statements of certain scientific bodies. Written statements, third-party references, use of certain terminology in a brand name, symbols, and pictorials may be considered a health claim if the context in which they are presented

TABLE 16.4. **Nutrient Content Claims on Package Labels.**

Comparative	Absolute
• Light/Lite • Reduced • Less • More	• Free • Low • Very Low • High • Source of • Healthy • Lean • Extra Lean

either suggests or states a relationship between a nutrient and a disease. When a statement, symbol, pictorial, or other form of communication suggests a link between a nutrient and a disease, it is considered an implied health claim, and it is subject to all the requirements for health claims.

NUTRIENT CONTENT CLAIMS

When a product package bears a nutrient content claim (whether expressed or implied), certain information must be incorporated into the package graphics. Because claims usually appear on the PDP, this panel often is affected. When a claim appears elsewhere on a package, however, certain information must appear on the panel with the claim.

Required Statement

Depending on the type of claim and whether the product is governed by FDA or USDA, requirements vary for claims-related information. FDA

TABLE 16.5. **Health Claims Permitted by FDA: Nutrient and Disease/ Condition Influenced.**

- Calcium and osteoporosis
- Sodium and hypertension
- Dietary saturated fat and cholesterol, and risk of coronary heart disease
- Dietary fat and cancer
- Fiber-containing grain products, fruits, and vegetables and cancer
- Fruits, vegetables, and grain products that contain fiber, particularly soluble fiber, and risk of coronary heart disease
- Fruits and vegetables and cancer
- Folic acid and neural tube defects
- Sugar alcohols and dental caries
- Soluble fiber from oat bran and risk of coronary heart disease
- Soluble fiber from psyllium husk and risk of coronary heart disease

mandates more extensive information than USDA (i.e., the inclusion of a Disclosure statement). Both agencies require additional information on products bearing comparative claims.

Nutrient Content Claims on FDA-Regulated Products

When a product contains excessive levels of key nutrients associated with health risks, a Disclosure statement that indicates the nutrient(s) of concern and directs the consumer to the Nutrition Facts panel is required (Table 16.5).

Comparative Claims

Both FDA and USDA product package labels that bear a comparative claim must include a Nutrient Claim Clarification statement and Quantitative Information. On FDA-regulated products, this information is required in addition to a Disclosure statement.

The Nutrient Claim Clarification statement identifies the comparison food and states the percentage (or fractional) difference in the subject nutrient(s) between the product and its comparison food (e.g., "50% less fat than [comparison food], one-third fewer calories than [comparison food]").

The Quantitative Information provides the absolute amounts of the subject nutrient(s) in the product and in the comparison food.

Design Elements

USDA and FDA differ slightly in their package graphics design requirements for claims-related information. USDA does not require Disclosure statements; FDA applies the type-size standards defined for these statements to other claims-related statements.

Claims Statements

Claims may not have undue prominence because of typestyle in comparison to the Product Identity statement. A claim may be no larger than twice the size of the Product Identity.

Disclosure Statement

On FDA-regulated products that contain high levels of key nutrients, a Disclosure statement must be immediately adjacent to the claim. No intervening material may be placed between it and the claim, except for other claims-related statements and/or a standard name modified by the claim. It must appear on each panel where the claim is located, except for the panel that contains the Nutrition Facts (e.g., if the Nutrition Facts and the claim are both on the PDP, the statement is not required). If multiple claims appear on a panel, the Disclosure statement must be adjacent to the largest claim. It must be in easily legible, boldface type, in distinct contrast to other printed/graphic matter. The Disclosure may be no smaller than the Net Quantity declaration, unless the claim is less than twice the size of the Net Quantity declaration.

Nutrient Claim Clarification Statement

The Nutrient Claim Clarification statement must be placed in immediate proximity to the most prominent comparative claim. It must follow the same type-size requirements as the Disclosure statement. If different comparative claims appear on the same label, the Nutrient Claim Clarification statement must be placed in immediate proximity to the most prominent presentation of each claim.

Quantitative Information

Clear and concise Quantitative Information must appear adjacent to the most prominent claim or on the Information Panel.

HEALTH CLAIMS

Currently, only FDA products may carry health claims. USDA had proposed regulations for health claims; however, the agency withdrew the petition in 1998 and announced plans to issue a new proposal that will parallel FDA regulations.

When a product bears a health claim, very specific language must be used on the label. As stated above, the regulations governing health claims are complex, and so health claim situations should be handled on a case-by-case basis. The language associated with a health claim must conform to regulatory guidelines, which provide model health claim statements. Product packages bearing health claims must undergo careful review by legal and/or regulatory experts to be certain the language is accurate and any graphics are acceptable.

Model Statements

The regulations pertaining to the approval of health claims outline the assertions that may be made and any additional required statements. When a claim is implied through graphic representations, a complete claim statement must be included on the package surface.

Design Elements

All information required in a health claim must appear in one place, in the same type size, and without other intervening material.

Since the required statements can be quite lengthy, the complete claim may appear on a back or side panel. When this approach is taken, a reference statement may be placed on the PDP, flagging the claim and directing the consumer to the location of the claim (e.g., "See _____ for information about the relationship between _____ and _____). The first blank contains the location of the health claim on the package (e.g., back panel, attached pamphlet), the second blank declares the nutrient, and the third blank names the disease or health-related condition.

When any graphic material implying a health claim is communicated on the label or in accompanying labeling materials (e.g., pamphlet), the entire claim statement or a reference statement must appear in immediate proximity to the graphics.

ADDITIONAL USDA LABELING REQUIREMENTS

The USDA has additional labeling requirements: official inspection legend and "keep frozen/refrigerated" statements.

OFFICIAL INSPECTION LEGEND

The PDP of all USDA-regulated products must include an official Inspection Legend in accordance with established inspection procedures. The number of the official establishment must also be included in one of the following manners:

- Within the official inspection legend.
- Outside of the official Inspection Legend, but elsewhere on the label (e.g., lid of a can). When it is placed outside of the Inspection Legend, the prefix "EST" (establishment) must precede the establishment number and must be shown in a prominent and legible manner in a size to ensure visibility and recognition.
- Off the exterior of the container (e.g., on a metal clip used to close casings or bags) or on other package material in the container (e.g., on aluminum pans and trays within the container). When it is placed on the exterior of the container or on other packaging material, a statement of its location must be printed contiguous to the official inspection legend (e.g., Est. No. on metal clip, Est. No. on pan).

- On an insert label placed under a transparent covering, if it is clearly visible and legible and accompanied by the prefix ''EST.''

INFORMATION PANEL

A package may be designed to include an Information Panel on which some of the mandatory information may be placed (see Table 16.6). If the Ingredient List and Manufacturer Identity statement do not appear on the PDP, the package must bear an Information Panel that includes these statements. Whenever practical, the Nutrition Facts statement should appear on the Information Panel with these statements. Other label statements required on some products (e.g., a Percent Juice declaration, some warning statements, special handling instructions) also may appear on the Information Panel.

Statements intended for the Information Panel alternatively may appear on the PDP if space on the Information Panel is insufficient. Also, it is permissible to omit an Information Panel, incorporating all mandatory statements on the PDP. Exemptions from labeling requirements for nonretail packages apply to the components of the Information Panel, as well as to the PDP.

LOCATION OF INFORMATION PANEL ON PACKAGE

The Information Panel generally is considered the part of the package immediately contiguous to and to the right of the PDP as observed by an individual facing the PDP. For situations when this is not possible, alternative locations are specified.

NUTRITION INFORMATION

Only when the Nutrition Facts statement cannot be accommodated on the Information Panel or the PDP along with other mandatory information may it be placed elsewhere on the package.

TABLE 16.6.   Information Panel Information.

Mandatory Information	Additional Required Information
• Nutrition Facts • Ingredients List • Manufacturer Identity	• Percent Juice Declaration • USDA Safe Handling Instructions • Special Handling Instructions • Warning Statements

FEATURES OF THE NUTRITION LABEL

The title of the panel is "Nutrition Facts" (Figure 16.1).

Serving Size Information

Serving sizes are provided in common household measures (e.g., cups), followed by the metric equivalent in parentheses. The regulations define standardized amounts based on the servings people typically consume (e.g., 12-ounce can of soda is one serving, not two), and detail how to derive Serving Size declarations.

Nutrient List

Information on the following nutrients is mandatory: calories, total fat, saturated fat, cholesterol, sodium, total carbohydrate, dietary fiber, sugars, protein, and four vitamins and minerals. In addition, certain nonmandatory nutrients may be included. The actual amount of each nutrient present in a serving is listed beside the nutrient. In the right-hand column, the amount of the nutrient is expressed at "% Daily Value," which shows how a food fits into a 2,000-calorie reference diet.

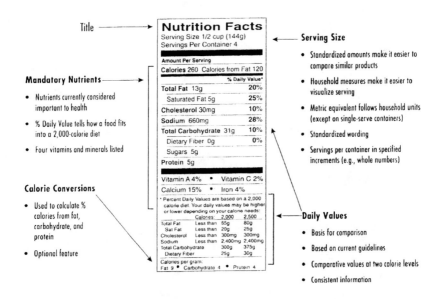

FIGURE 16.1 Nutritional label.

Daily Values Footnote

The Daily Values table at the bottom of the nutrition label presents the reference numbers used to calculate the % Daily Value. These numbers reflect current recommendations for health maintenance and disease prevention. They are provided at two calories levels and remain the same for all labels. On smaller packages and packages qualifying for the simplified format, this information may be omitted.

Calorie Conversion Footnote

Calorie conversions assist in calculating the percentage of calories from carbohydrate, protein, and fat. This feature is now optional.

MANDATORY DESIGN ELEMENTS

The regulations dictate a number of mandatory design elements that must be followed for all the Nutrition Facts layouts.

Hairline Box

The Nutrition Facts information must be set off from other printed material by a hairline box. The regulations are very precise on this point. No other information may be enclosed within the box. The hairline box must be in the same color ink used for the type.

Typestyle/Type Size

The type must be set in a single, easy-to-read typestyle. Helvetica is the recommended font, but any sans serif type is permissible. Upper- and lowercase letters must be used. Minimum type sizes are specified for each component of the label.

The nutrient list must be separated by 4 points of leading. All other text must be separated by at least 1 point of leading.

INGREDIENTS LIST

All products fabricated from two or more ingredients must include an Ingredient List on either the PDP or the Information Panel.

Ingredients are listed by their common or usual name in descending order of quantity by weight. Ingredients present in amounts of 2% or less may be listed at the end of the Ingredient List following an appropriate qualifying statement (e.g., Contains _____ percent or less of *[ingredient A]*, less than _____ percent of *[ingredient B]*).

The Ingredient List must be conspicuous and in a type size no smaller than $\frac{1}{16}$ inch.

SPECIAL PROVISIONS FOR BULK FOODS

Food sold in bulk containers at retail level may provide the Ingredient List by either displaying it on the bulk container or posting a counter card or sign with the required information.

MANUFACTURER IDENTITY

The Manufacturer Identity states the name and place of business of the manufacturer, packer, or distributor, typically on the Information Panel.

TERMINOLOGY

If the manufacturer is a corporation, the corporate name must be used and may be preceded or followed by the name of the particular division. If the manufacturer is an individual, a partnership, or an association, the name under which the business is conducted must be used. If the food is not manufactured by the entity whose name is on the label, the name must be qualified with "Manufactured for _____," "Distributed by _____," or other wording that expresses the relationship.

The statement must include the street address, city, state, and postal Zip Code. If the name is in a current telephone directory, the street address may be omitted. The principal place of business may replace the actual place where the food was manufactured, packed, or distributed, assuming this would not be misleading.

PERCENT JUICE DECLARATION

With certain exceptions, any beverage that purports or appears to contain a fruit or vegetable juice must declare the percent of juice in the product. Any of the following situations could trigger this requirement: using the name (or variation) of a fruit or vegetable in advertising, labels, or labeling; depicting a fruit or vegetable in a vignette or other pictorial representation; formulating a beverage to contain the color and flavor of a fruit or vegetable juice.

Percent juice labeling is generally required if a beverage has the appearance and flavor of containing a fruit or vegetable juice, even if the product in fact contains no juice. Certain exceptions are made for products containing minor amounts of juice (typically less than 2%) that are labeled with a fruit or vegetable name and a term such as "flavored."

TERMINOLOGY

The percentage of juice must be declared, and the type of juice may be declared. If a product does not contain juice but the labeling or color and flavoring suggest that it does, it must be declared as zero percent (or a similar phrase) (see Table 16.7).

USDA "SAFE HANDLING" INSTRUCTIONS

USDA has regulations requiring Safe Handling Instructions on all raw meat and poultry product packages not intended for further processing at another USDA-inspected establishment. These safety instructions are intended to heighten consumer awareness of proper food sanitation procedures and reduce the incidence and severity of foodborne illness.

The safe handling information must be presented on the label under the heading "Safe Handling Instructions" (see Table 16.8).

RATIONALE STATEMENT

"This product was prepared from inspected and passed meat and/or poultry. Some food products may contain bacteria that could cause illness if the product is mishandled or cooked improperly. For your protection, follow these safe handling instructions."

SAFE HANDLING STATEMENTS

"Keep refrigerated or frozen. Thaw in refrigerator or microwave." Any portion of this statement that is in conflict with the product's specific handling instructions may be omitted (e.g., instructions to cook without thawing).

TABLE 16.7. **Juice Content Information.**

If Beverage Contains Juice	If Beverage Does Not Contain Juice
• Contains—percent [or %] juice	• Contains 0% juice
• Contains—percent [or %] [type] juice	• Contains 0% [type] juice
• Contains—percent [or %] juice	• Does not contain [type] juice
• Contains—percent [or %] [type] juice	• Contains no [type] juice
	• Contains no fruit/vegetable juice

TABLE 16.8. **Safe Handling Instructions Model—USDA.**

Design Elements
• Set off by border
• Prominent location, visible at point of purchase
• Graphic illustrations next to safe handling statements
• Type no smaller than 1/16 inch, except heading, which must be larger
• One-color printing on single-color, contrasting background

"Keep raw meat and poultry separate from other foods. Wash working surfaces (including cutting boards), utensils, and hands after touching raw meat or poultry."

"Cook thoroughly."

"Keep hot foods hot. Refrigerate leftovers immediately or discard."

OTHER LABELING REQUIREMENTS

In certain situations, other labeling requirements apply. The requirements may be triggered by voluntary declaration of related information (e.g., stating number of servings, making a claim) or because of a key substance used in the product or packaging (e.g., aspartame, saccharin, self-pressurized containers).

QUANTITATIVE INFORMATION FOR COMPARATIVE CLAIMS

Products bearing a nutrient content claim that is considered a comparative claim must include a quantitative information either adjacent to the most prominent claim or on the Information Panel.

WARNING STATEMENTS

Certain products require warning statements that must be placed on the package Information Panel (or the PDP), also

- unpasteurized juice products
- irradiated foods
- foods containing psyllium husks
- shell eggs
- foods containing specific ingredients:
  —aspartame
  —sorbitol
  —mannitol
  —saccharin

—Olestra®
—products in self-pressurized containers
—protein products

BIBLIOGRAPHY

Altman, Tracy A. 1998. *FDA and USDA Nutrition Labeling Guide: Decision Diagrams, Checklists, and Regulations.* Lancaster, PA: Technomic Publishing Co., Inc.

Olsson, Frank and P. C. Weeda. 1998. *U.S. Food Labeling Guide.* Fair Lawn, NJ: The Food Institute.

Shapiro, Ralph. 1996. *Nutrition Labeling Handbook.* New York: Marcel Dekker, Inc.

Storlie, Jean. 1996. *Food Label Design: A Regulatory Resource Kit.* Herndon, VA: Institute of Packaging Professionals.

CHAPTER 17

Launching the New Product

JOHN B. LORD

Marketing a new food product is quite different from the marketing of an existing product. Members of the distribution channels and consumers are neither aware of nor familiar with the product. Everyone in the chain from retail buyers to retail store managers to consumers must become aware of the new product and try it to achieve the objective of repeat sales. Because so many new food products are introduced annually, retailers have established hurdles to control their limited space. Food marketers today must invest heavily in marketing to channel members as well as to consumers. Product launch also involves obtaining information from the marketplace to be used in the inevitable refining of the marketing process.

INTRODUCTION

A T this stage, development work on the product and package has yielded a product/package combination that delivers the benefits promised by our concept and has met requirements for safety, integrity, quality, and shelf life. At the same time, work on the introductory marketing program has proceeded, and individual elements of that program, including advertising copy, advertising media, package and label graphics, and price, have been evaluated. We are now ready to put the final marketing plan together, conduct final market testing, and, if all proceeds according to plan, launch our new product. In this chapter, we will cover three major topics: (1) the introductory marketing program, (2) market testing, and (3) the launch, including retail sell-in, execution of the launch program, and monitoring results during and after launch.

439

THE INTRODUCTORY MARKETING PROGRAM

The two most important decisions that any food manufacturer makes about the marketing program have to do with target market and positioning. Logically, all elements of the marketing program—product design, package design, pricing, distribution and merchandising, advertising, and promotion—all follow from the specification of our target market and positioning strategy.

These two decisions should be made very early in the development process. The target audience should be clearly specified during the opportunity analysis and ideation stages, and the positioning strategy should evolve and solidify during concept testing and subsequent product testing.

THE TARGET MARKET

The basic premise underlying both the opportunity analysis and idea generation stages of new product development is that the developing firm has identified a gap in the market. This gap is a combination of product benefits that a specified group of consumers demand but that are not currently being provided by existing products or competition. To the extent that we have followed this logic, the task of identifying our target audience is one of fine-tuning and elaboration. That is, we have a basic notion of our target customers, for example, mothers of school-aged children, and the task is one of expanding that definition to include appropriate demographic and lifestyle variables, and then overlaying media habits. Target marketing is only actionable when we have the ability to selectively and efficiently reach the target audience with marketing communications.

Target market information is available through our own sales data if the new product category is one in which we have prior experience. Results of early stage testing should yield some insight into segmentation patterns to the extent we have asked the right questions, particularly classification questions, and to the extent our sample is adequately diverse and our sample size large enough to find significant patterns in the data.

Several sources of syndicated marketing research can help us in this phase of new product development. Household panel data from a variety of sources, most notably Nielsen and IRI, will provide an overview of buying and consumption patterns along with demographic and geographic correlates, such as size of household, number and age of children, income level, region of the country, and so on. Data provided by Mediamark Research yield even richer information. These data show, by product category, purchasing indices according to demographic and geographic groupings, as well as the media habits, for both broadcast and print vehicles, of these customer segments. A population sub-group with a purchasing index of over 100 is a group that purchases more than the national average of a given commodity. For example, in the Spring

1994 report (Mediamark Research, 1994, p. 80), we find purchase indices for different classifications of women for ready-to-eat donuts. High school grads have an index of 111 while college grads have an index of 91. Therefore, we can conclude that women who graduated high school purchase 11% more ready-to-eat donuts than the average for all women while those who graduated from college purchase 9% fewer donuts. Women who are working parents have a purchase index of 125 while women who are single have a purchase index of 90. These indices clearly pinpoint the demographic groups with the greatest potential in a given category. Once the primary and secondary targets have been identified, we can use the same data to determine the media and vehicles that represent the most efficient choices to reach these groups.

POSITIONING

Positioning refers to the way consumers perceive our brand relative to others. Consumer behavior theory tells us that consumers develop beliefs about the characteristics and potential benefits of products through a combination of marketing communications (advertising, promotions, packaging, price, merchandising), word-of-mouth, and actual usage experience. We compare and contrast different products based upon these beliefs, which are filtered by our perception of the marketplace. Product characteristics and benefits define the mental framework we use to compare and evaluate brands. Thus, we can affect the way in which consumers fit a new product into that mental framework by associating with it, primarily through advertising but also through all aspects of product design and marketing strategy, specific characteristics and benefits. Note that the verb in the preceding sentence is "affect." Many phenomena impact the meaning a product ultimately has for the consumer, most of them uncontrollable from the point of view of the marketer. The major implication here is that we must find the right combination of benefits, and communicate them clearly, unambiguously and consistently to our target audience via all aspects of our product and marketing program.

The positioning statement succinctly notes the combination of sensory, rational, and emotional benefits offered by the brand. Sensory benefits refer to taste, appearance, texture, aroma, and so on. Rational benefits refer to what the product does for the consumer, such as being a "source of Vitamin C" or "can be prepared in five minutes." Emotional benefits refer to psychological benefits, such as those provided by eating "comfort foods" or the way in which we think others see us—image—because we consume a certain brand. Every positioning statement must cite one or more of these kinds of benefits.

Patrick (1997) notes that value-added positioning requires four ingredients. *Simplicity* means making our positioning understandable to the customer. *Specificity* means relating the brand to the specific needs of the target customer with specific benefits that will address these needs. *Durability* means developing a positioning strategy with staying power, meaning that we are addressing underlying and long-term consumer trends, not fads. *Advertisability* means that the strategy must lend itself to a number of different executions that can be communicated effectively via different media.

THE "A-T-R" MODEL

One of the most commonly accepted models in marketing is called the "hierarchy of effects model." This model provides a basis for planning, executing, and evaluating marketing strategies, particularly promotional strategies, and centers on the stages consumers go through in responding to marketing communications. Initially, consumers have no awareness of a brand, particularly a new brand. Through a combination of advertising, promotion, and word of mouth, consumers become aware of the brand, and then develop some beliefs about the brand, or what is termed "comprehension" of the brand's attributes and benefits. The first two stages are termed cognitive. The next stage is termed affective, meaning that the consumer forms some type of positive or negative associations, or degree of liking for the brand, normally based on some evaluation of what the brand offers versus what the consumer is seeking. If that liking is strong enough, the consumer may form a preference for the brand. Ultimately, if the brand fits within the consumer's lifestyle and budget, or even satisfies some need for novelty, the consumer may form an intention to purchase, meaning that he or she has made a judgment that the brand offers potentially want-satisfying benefits. In many instances, some type of promotional incentive, such as a coupon, may be the initiator because the incentive changes the value equation. The consumer may act on this purchase intention and buy the brand an initial time (trial) provided that the product is available and other situational variables do not interfere with the purchase decision. Finally, the consumer evaluates the brand, determines to what extent the new product has provided expected benefits, and, if satisfied, may decide to buy the brand again (repeat purchase). The latter stages of this process are called behavioral stages.

New product marketers have shortened this model to three stages: awareness, trial, and repeat purchase, or "A-T-R." These are commonly used both to set objectives for and measure the effectiveness of our new product launch program. We develop a marketing communication program to create a specified level of awareness. Aware consumers, if properly incented, may be induced to trial. Trial consumers, if satisfied with their initial purchase, and if the new product fits their lifestyle and consumption patterns, may be induced by buy the product again. We determine how much to spend, in what media, scheduled in a specific way, and we decide the types of consumer and trade promotions to use, all based on hitting the awareness, trial, and repeat levels judged to be necessary to create sales and share levels that meet targets.

THE PRODUCT NAME

Great product names come from a variety of sources, and require a great deal of research. A name should be simple, memorable, easy to pronounce, and indicative of product benefits. Brand names are legally protected, so we must use due diligence to ensure that the name we are planning to use is neither owned by someone else or will create some type of trademark infringement. Names must be tested on both an absolute basis and in a competitive setting. The name can make a big difference. Hershey's Nutrageous candy bar was originally named "Acclaim" but was changed after testing because while the candy tested very positively, consumers—kids—found no relevance in the name. Nutrageous is a play on "outrageous," at the time a commonly used kids' term to describe something they think is really different and "cool."

THE ADVERTISING PROGRAM

Major advertising decisions are listed in Table 17.1.

New product advertising must create awareness of the brand name and communicate product benefits clearly. Thus, our ads must capture the customer's interest up front and build involvement with the ad and the product. Our objectives must be stated clearly and quantitatively, such as, to "create awareness among 60% of our target customers within the first six weeks of the program."

Marketers have developed some principles of new product advertising, based on experience. Successful new product advertising copy must position the brand clearly in a specific product category, communicate product benefits, demonstrate how the product is different from competitors and how it will benefit consumers, and provide some type of evidence that the product will deliver the benefits promised. One leading consumer products advertiser maintains that new product advertising must have three specific qualities: relevance, originality, and impact. Advertisers face significant challenges today as viewership of network TV has fallen, new forms of advertising have emerged, and costs of media have skyrocketed. According to industry statistics, advertising

TABLE 17.1. **Major Advertising Decisions.**

- Objectives, typically stated in terms of awareness and trial levels
- What we will say about the brand, that is, our creative or copy strategy
- How much we plan to spend, that is, our budget
- Which media and vehicles we will use to reach our target customers, that is, our media plan
- How expenditures will be allocated to different geographic areas
- How the messages will be scheduled over the introductory campaign
- How to measure the effectiveness of our ads.

recall has fallen from 18% in 1965 to 4% in 1990, and things have only become worse in the 1990s. The average American is exposed to 500,000 advertising messages per year including 100,000 for new products. The challenge clearly is to create memorable advertising that is quickly recognized, and which serves to educate consumers about a product's important differences.

An interesting example of a new copy approach was reported in the *Wall Street Journal* (Parker-Pope, 1998). Although marketers of toilet tissue have traditionally used softness, communicated by images of puffy clouds or squeezable packages, or economy, communicated by comparing number of sheets per roll, Kimberly-Clark is using a novel approach. The copy strategies for a new toilet tissue involve communicating an important product difference, specifically, that a new ''rippled texture'' is ''designed to leave you clean and fresh.'' Results of advertising focus groups indicated that consumers responded negatively to words such as hygiene and cleansing, but not to the final benefit of being clean and fresh. This copy strategy is unique and even risky, but may be necessary in a competitively crowded and stagnant product category such as bathroom tissue.

Setting advertising budgets is, at best, an inexact science. The best approach is to test different expenditure levels in a controlled experiment. The next best approach is to use historical comparative data that show statistically reliable relationships between spending levels and awareness and trial levels. Certain principles guide the determination of introductory marketing expenditures:

- The first year budget should permit a heavy introductory schedule (''heavy up'') followed by ongoing expenditures equal to or greater than the second year budget.
- For a new product, share of voice (i.e., the brand's advertising expenditures divided by advertising expenditures for all brands in the category) should be approximately two times the market share objective. For example, if we are targeting a 10% share of market, we will need to spend to achieve about a 20% share of voice during the introductory period.

The short life cycle for most new food and allied products necessitates that trial be created through heavy advertising expenditures in the first few months. The most important period in the life of a brand is during and just after introduction, when sales build to an initial peak. The higher that peak share, the higher the probability the new brand will be a success.

We allocate consumer advertising expenditures in three ways: (1) across media and vehicles; (2) across regions, and (3) across time. Media planning is more complex today because of the proliferation of media choices and the fragmentation of audiences. The principle is simple. We must place our messages where our target audience will be exposed to them. However, the execution is difficult because there are so many choices. Media advertising is still the best way to build awareness quickly, and to build brand loyalty and equity through consistent messages focusing on product benefits. However, gross audience numbers are no longer relevant. The key is to use targeted communications to reach targeted audiences. These will necessarily include electronic (Internet) advertising as well as more traditional broadcast, print, outdoor, transit, and specialty media.

The most logical decision rule for geographical allocation of advertising dollars is straightforward: allocate according to market potential. Firms often resort to category development index (CDI) and brand development index (BDI) in allocating all types of marketing effort. A category development index (CDI) is an index number representing the sales of a product category in a specific geographic market relative to the average in all markets. As with all index numbers, the average is equal to 100. A brand development index (BDI) is analogous to a CDI except that it is used to measure the sales of a brand in a category rather than a category.

With respect to the timing of promotional expenditures, several points must be kept in mind. First, there is a threshold level of advertising below which the it will lack the intensity to break through the communications clutter all consumers face. Second, once advertising has had an impact in terms of creating awareness and comprehension, decay or forgetting will take place in the absence of reminder advertising. Third, rarely is it financially feasible to maintain high levels of advertising expenditure over an extended period of time. Fourth, advertising messages must be properly scheduled with respect to when the product is available in the store, when promotions are to be run, seasonal and/or special event sales patterns, and so on. Just as with expenditures, there is no magic or unique formula to guide our timing decisions; experience is the best teacher.

PACKAGING AND LABELING

Packaging strategy and design plus labeling have been covered in depth elsewhere in this book (see Chapters 8, 14, 15, and 16). As part of our introductory marketing program, packaging plays a vital role in accomplishing several objectives. The package must provide visual impact at point of purchase, clearly identify the brand, inform the consumer about the product and the package contents, provide adequate protection for the product, and provide functionality for the consumer in terms of opening and reclosing, dispensing, preparation, storage, and so on. The label must also provide ingredients, nutritional information as required by law (the Nutrition Labeling and Education Act), and storage, preparation, and serving instructions.

CONSUMER AND TRADE PROMOTIONS

Consumer promotions, such as coupons, cents-off deals, money back and refund offers, premiums, and so on, are commonly used, and in many cases necessary to build, awareness and especially trial to targeted levels. The most common type of consumer promotion is couponing. Coupons can be distributed through FSIs (free standing inserts in newspapers); in newspaper and magazine advertising; via direct mail; on packages, either "on-pack" or "cross-ruff" (cross-ruff coupons are coupons for one product which are contained on the package of another product); or in-store. In store coupons can be distributed via coupon kiosk, as a tear sheet on the shelf, or using Catalina Marketing coupons, which are targeted coupons delivered at checkout. Finally, coupons are now distributed on the Internet, and many companies post coupons on their web sites. Coupon redemption is typically very low (around 3% of all coupons distributed are redeemed) but FSI coupons remain an industry standard to support new product launches.

Another key consumer promotion is sampling, which typically allows firms to "force trial." Samples of packaged goods can be distributed with newspapers delivered to homes, via direct mail, or in-store. Sampling can also be accomplished through in-store demos. There are a large number of specialty merchandising companies that perform demos for client companies, and even a trade association—The National Association of Demonstration Companies. More and more, especially with prepared foods and HMR products, it is important to get the consumer to taste the product. This is accomplished through in-store demos that usually include coupon benefits derived, and the total outlay, including dollars, time, and effort, that the consumer must make to acquire and consume the product.

PRICE

The price for a new product does not have to be the lowest on the market, but it must be set with reference to perceived product quality and benefits so that consumers perceive higher value than they could obtain with the purchase of a competitive item. Therefore, competitive prices must be considered in this context as well. If consumers do not perceive meaningful differences among competing brands, then price becomes one of the key factors in determining consumer choice.

Second, the pricing structure must allow for adequate trade margins, without which the item will never reach the retail shelf. For each product category, the retailer has a profit objective that, along with sales velocity and cost factors related to stocking and displaying the product, drives the margins required for that category.

Third, the pricing structure must provide both adequate sales velocity and adequate margins for the manufacturer, so that objectives for profitability and return on investment can be met.

Fourth, psychological factors in responding to price must be considered. For instance, in certain product categories, there is a standard price level. In many instances, especially in cases where consumers have few objective cues with which to evaluate quality, consumers use price level as an indicator of product quality. And there is still a tendency on the part of marketers to price at $2.99 instead of $3.00 because of the belief that consumers will perceive the $3.00 price point to be significantly higher than the $2.99 price point, a practice known as psychological pricing.

FINANCIAL EVALUATION OF NEW PRODUCT PROJECTS

The bottom line for new product development is the "bottom line." Every new product project involves expending dollars for the purpose of making money. In publicly held companies, there is extreme pressure to meet short and intermediate term revenue and profit targets, as well as a need for projects to meet corporate targets for return on investment. While companies normally want line extensions to achieve bottom line profitability within the first year of introduction, really new products may take longer. For instance, Kraft Lunchables did not start making money until its third year on the market. Pepcid AC, a prescription pharmaceutical product converted (after FDA approval) to over-the-counter and launched through a partnership between Merck and Johnson & Johnson, is just starting to turn a profit after three years on the market. This is because the initial marketing expenditures to build awareness and trial, by plan, were extremely high. Companies with deep pockets and many successful brands at different stages of the life cycle can sustain new products through periods of heavy losses if there is the promise of long-term viability and profitability.

Except in the case of very straightforward line extensions, most new product development projects require investment capital, primarily for new or refitted plants and equipment, as well as other resources. As capital projects, product development projects must undergo financial evaluation for the primary purpose of determining whether it makes sense to commit resources to the project initially and at decision points throughout the process. Three major criteria are evaluated: projected demand, profitability, and return on investment.

Estimating demand is an exercise in sales forecasting. We have discussed projecting sales for new food products in Chapter 6. Profitability is always a primary consideration. Profits equal revenues less expenses. Estimates of revenue are derived from the sales forecast. Expenses include manufacturing costs (raw materials, supplies, labor, utilities, depreciation on plant and equipment, etc.), distribution costs (transportation, warehousing, inventory, order processing), marketing expenditures (measured media advertising, promotions, publicity, customer service) and general and administrative expenses. According to Boike and Staley (1996), several key financial measures must be estimated and validated against corporate standards and funds availability at the onset of the new product development project. These measures are shown in Table 17.2.

Measuring return on investment (ROI) involves determining the amount of net returns plus their timing and duration. Scarce capital resources must be

TABLE 17.2. **Financial Measures in NPD.**

- Development cost (e.g., hours, capital)
- Prototype and pilot plant costs
- Manufacturing costs: tooling and scaleup, in addition to ongoing manufacturing cost
- Related costs (e.g., advertising, packaging, promotion)
- Pricing
- Anticipated sales (e.g., units, revenue)
- Payback measures (e.g., ROI, profit contribution, anticipated margin)

put to use in projects that justify their deployment. In other words, new product development projects must be attractive projects. Attractiveness is evaluated by several methods. We will discuss four commonly used techniques, including the average rate of return method, the payback method, the internal rate of return method, and the present value method.

The average rate of return method is very simply to employ. We begin by calculating the ratio of average annual earnings after taxes to average investment in the project. For example, if average annual profits are $4 million and average annual investment is $48 million, the average rate of return is 8.3%. Note that this method does not consider the timing of either the investment or the profits. The projected return rate can be compared to some standard, such as an investment hurdle rate, to determine if we should fund the project.

The payback method uses as an investment decision criterion, the amount of time (number of years) we project that it will take to recover the initial investment in the project. The calculation employs as a numerator the initial capital investment, and as a denominator the annual cash flows. This computation yields the number of years to recovery of the initial investment. The computed payback period can be compared to a target to determine if the capital expenditures should be made. Like the average rate of return method, the payback method, in its simplest form, does not consider timing of the cash flows.

The internal rate of return method involves the discounted cash flow for the project. The internal rate of return is the interest rate that equates the initial investment and the discounted cash flow of the project according to the following formula:

$$A_0 = A_1/(1 + r) + A_2/(1 + r)^2 + \ldots + A_n/(1 = r)n$$

where r is the rate of interest, A_0 represents the initial investment occurring at time period 0 (at the beginning of the project), and A_1 through A_n represent cash flows accruing in periods 1 through n. This "r" must be greater than or equal to a hurdle rate set by management. The hurdle rate should be set in accordance with the riskiness of the project, so that riskier projects must surpass higher hurdle rates to be funded.

The present value method uses a criterion of net present value (NPV) for a project. NPV is computed by summing cash flows discounted by the rate of return required for a project by the company. The present value method uses a decision rule that if the NPV of the project is greater than zero, the project should be approved. In other words, if the present value of cash generated by the project is greater that the present value of cash expended, we should go ahead with the project.

Another type of analysis useful in financial evaluation of new product projects is break-even analysis. Break-even provides a very simple calculation to determine how much of a new product we need to sell to recover fixed costs of the project. The break-even formula is

B/E point = fixed costs/contribution margin ($SP - VC$)

where SP is selling price and VC represents variable costs. We can expand the analysis to include a desired profit, by adding desired profit to the fixed costs to be recovered. We can also do sensitivity analysis for different price levels to determine the impact of a price change on the break-even point.

The "financials" are detailed proforma profit and loss statements for the new product for year 1, year 2, year 3. They represent the financial plan for the project. The financial plan is important for the following reasons (Cooper, 1993):

(1) It serves as a budget for the new product—an itemized accounting of how much will be spent, and where.

(2) It is the critical input for the final go/no go decisions as the project moves closer to launch.

(3) It provides benchmarks critical to the control phase of the launch plan, making sure that the launch is on course.

We generally include both manufacturing and distribution costs in calculations of cost of goods sold or cost of sales, which is a key entry in the income statement. This is used to communicate sales, expense, and profit data. A simplified income statement can be formatted as follows:

	Period 1	Period 2	Period n
Income			
Net sales			
Other			
Less: cost of goods sold			
Equals: gross profit			
Less: general and administrative expense			
Less: marketing expense			
Equals: net profit before taxes			
Less: taxes			
Equals: net earnings			

Net sales are total sales revenue minus returns and allowances. General and administrative expenses typically include expenses not directly related to production output and include salaries, rent, insurance, utilities, supplies, and so on. Net profit or earnings before taxes measure performance of a firm's operations. The operating profit margin, which is calculated by dividing operating profit by net sales, shows the relationship between operating profit and net sales.

For most new food products, sales build slowly as distribution is achieved, dollars are granted to the trade via promotions, and consumers are incented to try the new product. Usually a brand will hit its peak share 6 to 12 months after introduction in a new area, and subsequently level off or even decline. Much of the initial sales volume for a new product is promotion driven. Even bad new products can achieve high initial sales, but only new products that deliver significant perceived benefits and substantial customer value will succeed after the initial flurry of sales activity.

MARKET TESTING

Market testing is all about beating the odds in new products. All food processing firms are looking to identify high potential opportunities, reduce the risk of costly new product failures, develop the strongest marketing plan possible, and execute effective launches. Market testing is designed to aid firms in achieving these objectives.

We pointed out earlier that the new product development process consists of a series of screens and evaluations—of the idea, the concept, versions of the prototype and finished product, package structure and graphics, plant and/or equipment (if new), and the marketing program. Having completed development, we are now ready for launch. Only one more preliminary step must be completed—market testing. Market testing refers to evaluation of the entire launch program through taking the finished product and marketing program to the final consumer, and testing consumer reaction to the new product. The firm must first decide whether to launch without a formal market test. For simple line extensions, such as new flavors of an existing product line, market tests may be unnecessary provided that early stage testing yielded necessary information on consumer reaction to the new item(s) as well as data to be used in forecasting sales and performing financial analysis. As pressure to get to market quickly has increased, companies have looked for ways to shorten the development cycle, and traditional test marketing has become a less-used option. Especially with new products that do not require significant investment in production technology, plant, and equipment, many firms opt to ''roll and fix.'' This means that the firm will initially launch in a limited geographical area, and roll the product into new markets while gathering consumer and trade feedback, making necessary adaptations to the program, and building up production capacity as demand increases.

As the new product takes on more dimensions of newness, such as a new brand name, new category, new technology, new plant, and new equipment, both capital investment and risk increase, and the need to perform some type of market testing increases as well. Let's examine market testing in some detail.

DATA REQUIREMENTS

Market testing uses research methods designed to provide several important types of data. Sales volume and market share projections provide important financial information plus direction for the sourcing, production, and logistical operations regarding expected demand. The test market allows evaluation of vendor and logistical and manufacturing performance. We test consumer reaction along several dimensions, including awareness, attitudes, trial, repeat purchase, satisfaction with the product, brand-switching behavior, and response to promotions such as coupons. We test trade reaction to our launch program in terms of adoption, distribution penetration, number of SKUs and facings, shelf position, extent of display and point of purchase activity, and feature advertising. By evaluating trade response, we also evaluate sales force or broker performance in executing the launch strategy. By examining consumer purchase dynamics and category sales figures, we can determine source of volume for the new brand. That is, to what extent does the new brand bring in new buyers to the category, raise usage among consumers in the category, take sales from competitive brands in the category, and take sales from our existing brands?

SOURCES OF DATA

Just as there are several important types of data to be gained from a market test, there are multiple sources of these data. These are listed in Table 17.3. Market test methodology will determine which and how many of these data sources are employed.

MARKET TEST ALTERNATIVES

Having made the decision to market test the launch program, the next step is to decide testing methodology. There are three major classes of market testing, with several variations of each: (1) simulated test markets, such as BASES, which has been previously discussed in Chapter 6, (2) controlled testing, and (3) traditional sell-in test marketing. These are not mutually exclusive categories of testing, and there are numerous instances of companies

TABLE 17.3. **Test Market Data Sources.**

- Point of sale transactions data, captured by retail scanners
- Store audits of displays and in-store promotions
- Household panel data showing purchase behavior by household
- Salesperson call reports
- Factory shipments and warehouse withdrawal data
- Tear sheets from retail feature advertising
- Results of surveys which measure awareness, attitudes, and response to advertising

employing more than one type of market testing methodology, such as conducting a "pretest market" early in the process, and a controlled test market or sell-in test market prior to launch.

SIMULATED TEST MARKETS (STMs)

Simulated test marketing is often conducted at the concept development and testing phase. STMs provide volume forecasts, assessment of consumer reaction to the concept, and diagnostics of different elements of the overall program. The primary advantage of a simulated test market is preservation of confidentiality for the product development firm in that we don't expose the product to our competition. Relative to traditional sell-in tests, STMs are less expensive, can be accomplished more quickly, and allow "what if" analysis in terms of testing different levels of price, advertising weight, and so on without compromising the launch. However, STMs are limited by their very nature. First, the purchase situation into which we place respondents is "simulated" or contrived. Although we have control over the variables, any laboratory experiment, by definition, removes the evaluation from the real world. In this case, the real world includes competition and the trade, and simulated test markets evaluate consumer response in a competitive vacuum. Further, we do so without assessing trade response to the program; instead the test establishes assumptions about trade penetration and distribution coverage. In addition, the assumptions about the effects of advertising and promotions on awareness may be unrealistic and therefore unreliable. Finally, the commitment to the launch, as measured by the level of spending, may not be sustainable. That is the bad news. The good news is that STMs represent possibly the best-validated tool of marketing research. Because several STM providers can claim highly accurate and projectable results, food and allied companies commonly use simulated test markets to generate sales volume estimates, especially when confidentiality is critical and some investment will be required.

SELL-IN TESTS

Regardless of whether the project incorporates a simulated test market, the firm may decide to conduct an in-market test. An in-market test provides the opportunity to assess consumer and competitive reaction to our launch, but to do so on a limited scale to avoid the investment, expense, and risk of a larger scale introduction. In-market tests come in two varieties: traditional sell-in tests and controlled tests. Traditional sell-in tests, which are conducted under normal market conditions and use the company's regular sales force and advertising and promotional programs, allow a much broader evaluation of the launch program than controlled tests. However, traditional sell-in tests have been falling out of favor during the 1990s because they have several potential disadvantages.

TEST MARKET METHODOLOGY

If the nature of the project dictates traditional sell-in test marketing, we must make several decisions about the design of the test. These include choice of test cities, length of the test, and specification of the data to be collected, and the sources of that data.

Test cities should possess certain characteristics. First, they should be representative of the broader geographic distribution target in terms of (1) population demographics, (2) category development and brand development, as measured by CDI and BDI, and (3) structure of the food retailing market. Second, the cities should be reasonably isolated in terms of media spill-in and spill-out. Media spill-in refers to a situation in which households in the test city have ready access to broadcast media from other cities. Spill-out refers to households in other cities having access to broadcast media from the test city. Limiting both spill-in and spill-out is necessary to accurately gauge the relationship between advertising weight and consumer response.

Choosing the length of the test always represents a series of trade-offs. For instance, the longer we are in test, the more reliable the data become because we "average out" seasonal sales trends as well as any short-term factors such as stock-outs. Longer tests also provide the opportunity to assess more repeat purchase cycles and give the firm a more in-depth read of both consumer and trade reaction. However, longer tests cost more, represent higher opportunity cost, and result in greater loss of lead-time.

Table 17.4, based on the work of Olson (1996), shows the benefits of test marketing. However, test markets have limitations and can cause problems. The limitations of test markets are shown in Table 17.5.

Sell-in test markets are most often utilized today in situations in which there is significant risk because the new product incorporates new technology and requires large capital investment in plant, equipment, and processing

TABLE 17.4. **Benefits of Test Marketing.**

- Reducing risk exposure, both in terms of avoiding potential damage to the company's image with the trade and consumers if the product does not succeed, and in terms of limiting investment to small-scale production
- Achieving success that can be used as a tool or showcase to gain trade acceptance when the new item is launched
- Gaining experience with fine-tune sourcing, manufacturing, logistical and sales operations via a pilot test
- Assessing consumer and trade reaction to specific elements of the marketing program for diagnostic purposes

technology. With significant technical and/or production barriers to entry, firms do not necessarily sacrifice the first-mover advantage by conducting a lengthy test market.

Dannon's new Yogurt Shake Drinkable Lowfat Yogurt was tested in about 15% of the country, including Chicago, Detroit, Baltimore, and Washington, DC, prior to a decision as to whether to launch nationally. Frito-Lay extensively tested olestra-based fat free salted snacks before rolling out. And, in a different situation, ConAgra tested Banquet Bakin' Easy, a seasoned dry baking mix designed to compete with Kraft's Shake 'n Bake, in Phoenix, Minneapolis, the Carolinas, and Georgia followed by rollout to the Midwest.

STM'S VERSUS SELL-IN TESTS

Packaged goods firms use different decision rules when deciding upon market tests. If STM results are positive, investment requirements are low to moderate, and there is a substantial risk of being beaten to market by competitors learning about the product in test market, we may skip formal test marketing. On the other hand, in very high-risk situations, or when a simulated test

TABLE 17.5. **Limitations of Test Markets.**

- Cost a lot of money, both in terms of the costs of the test, which include production and distribution of the product, marketing costs, and research costs, and in terms of opportunity cost of limiting distribution and sales to test areas instead of a larger market area.
- Expose new products and accompanying marketing programs to competition, a lack of confidentiality that may cause the firm to lose its first mover position by giving competitors an opportunity to develop their own versions of the item, and get to full-scale launch before the original firm can.
- Provide an opportunity for competition to interfere through such actions such as lowering price, engaging in heavy promotional activity, and even buying up large quantities of the test item, all of which lead to unreliable test data.
- May lead to unrealistic levels of sales force effort which cannot be sustained in a larger scale launch, creating unreliable data.

market uncovered potential problems, both the STM and the test market may be conducted. In situations where it is important to do in-market testing, but we want to conserve both time and expense, an alternative to sell-in testing is controlled testing.

CONTROLLED TESTING

Controlled testing is a term that incorporates different types of market tests that share the same basic methodology. The key difference between controlled testing and sell-in tests is that controlled tests employ outside companies hired by the manufacturer to handle the product's distribution, merchandising and, in some cases, consumer advertising and promotions. A research supplier administers the entire test, and the marketing variables, such as delivery of specific levels of media weight or advertising copy alternatives, are tightly controlled. Several popular and reliable controlled testing products are available to marketers. Here we profile two options: *RealTest* and *Behaviorscan*. RealTest can be classified as a controlled market test that "bridges the gap between concept and product testing and regional or national launch." Behaviorscan provides controlled in-market tests. [Author's note: the information presented below on RealTest and Behaviorscan has been taken from presentations made by Elrick & Lavidge and Information Resources Inc. respectively.]

BEHAVIORSCAN

Behaviorscan is a product of Information Resources Incorporated. Behaviorscan employs live, in-market testing methodology; however, IRI controls the retail environment including product distribution and placement as well as control over television advertising. The industry refers to this type of testing as "single source testing." Behaviorscan is available in eight minimarkets: Pittsfield, MA; Eau Claire, WI; Marion, IN; Rome, GA; Cedar Rapids, IA; Grand Junction, CO; Midland, TX; and Visalia, CA. Using its own production studio, IRI provides targetable television advertising in most Behaviorscan markets to individual households by cutting over regular advertising; all panelists are cable subscribers. In each minimarket, IRI has recruited a demographically balanced and representative household panel. Members of the panel use a shopper's card when they purchase food and allied products in supermarkets, drug stores, and mass merchandise outlets, enabling IRI to track household level purchasing. Partnerships between IRI and local retailers lead to broad outlet coverage: participating grocery stores cover 90%+ of market ACV; drug stores cover 60 to 75% of market ACV, and mass merchandisers cover 70 to 100% of market ACV.

The combination of targetable television advertising and household purchase data provides the opportunity to

- diagnose and respond to the test brand's performance (in terms of trial and repeat)
- quantify cannibilization of existing brands
- quantify volume taken from which competitor
- profile brand users to improve positioning/targeting
- gain most accurate new product national sales forecast

Significant advantages of the Behaviorscan methodology are listed in Table 17.6.

The use of Behaviorscan also involves some methodological limitations, which include limited choice of markets (since there are only eight B-Scan markets), the fact that the test does not provide for the assessment of either competitive or retailer response, and research costs, which can range as high as $500,000 for a test.

The major applications of Behaviorscan testing include: (1) the ability to evaluate the impact of different advertising expenditure levels, (2) the ability to evaluate different copy strategies, (3) the ability to compare two or more different marketing mix combinations, and (4) the ability to assess in-store variables such as shelf location, pricing, promotions, and so on.

REALTEST

RealTest is a product of Elrick & Lavidge. The RealTest methodology provides the client an opportunity to introduces a new product to a select group of consumers and monitors their normal (as opposed to simulated) purchase behavior. RealTest uses both a shopper and a user panel, specifically

TABLE 17.6. **Advantages of Behaviorscan Testing.**

- Tight control over marketplace conditions
- Quick turnaround
- Contained costs—lower media costs, less test product, no sales force involvement, no slotting allowances or failure fees
- Confidentiality of test results
- Ability to improve the marketing plan through use of simulation modeling techniques and manipulating marketing variables including media weight, copy, shelf location, and so on
- More efficient use of new product and marketing funds, because Behaviorscan results give marketers information that guides better launch decisions (validation: with no controlled market test, IRI data indicates that 72% of new products failed; of products launched after a Behaviorscan test, 84% were successful, for all Behaviorscan tests, 55% of the tested items were not rolled out, 38% were successful rollouts, and 7% were failures)

selected to replicate the client's target audience. Shopper panelists receive an announcement of the product with an advertising piece. For longer tests Real-Test usually provides additional awareness building announcements. Methodology includes measurement of awareness.

Test product is stocked in a limited number of stores, and panel members are provided a chance to purchase the test product in their normal stores, when they usually do it, within normal competitive activity and influence. The purchases of the shopper panel are monitored and trial is measured. The User Panel receives the product with an advertising piece. Whenever possible, a full size replica of the product is given to the User Panel, who are provided the opportunity to have a typical and full experience with the product. Purchase behavior is monitored by contacting household panelists during three to six purchase cycles. The purchase data are then used as data input into a volumetric model that been validated using individual household data from some new product launches. The shopper panel's purchase behavior becomes the source of trial and initial repeat inputs for the modeling while the user panel's purchase behavior becomes the source of initial and longer term repeat purchase.

The model provides forecasts of trial and repeat rates and an estimate of steady state volume, the equivalent of first year volume for many new items. It also provides diagnostics, including likes, dislikes, why not try, why not repeat, how used, what was replaced, and so on, so that improvements can be made before regular launch.

The major advantages of RealTest include time and cost savings relative to regular sell-in testing, due to control over the test variables including where the product is stocked, plus flexibility as the test approach can be adapted to specific needs of the client. RealTest represents a cross between a simulated test market and a controlled test market.

SELL-IN TO THE TRADE

The *Progressive Grocer* report (Mathews, 1997) cited a study conducted by Efficient Marketing Services (EMS) of a sample of new items launched in 1995. Results of the study showed that for the 300 new items examined, success was most negatively affected by four factors: slow distribution, incomplete distribution, lost distribution, and poor merchandising decisions. All of these problems are correctable, but only if the manufacturer has sufficient and timely knowledge of the situation and can take corrective action within a narrow window of time. Considering that a new product launch has about 12 weeks to take hold, the window is extremely tight. An industry study reported by Deloitte & Touche measured average lost sales per SKU per store due to slow, incomplete, and lost distribution at almost $90. With an average launch of 4.3 SKUs, the loss averages almost $400 per store per launch. All of the above point to the critical importance of the sell-in and monitoring stages of new product launch.

The retail environment presents significant challenges to the new product marketer. Retailers face increasing competitive pressures from alternative formats as well as from other supermarkets. Operators are driven to maximize their return on assets, at least at the headquarters level, and so are increasingly concerned about the high rate of new product failure. At the store level, the major emphasis is on increasing product turns and minimizing shrink.

Many retailers charge not only slotting fees to get initial distribution, but also failure fees if the product is delisted. For chain operators, the net effect of these fees is to help the distribution function and to increase the profits attributable to category managers. Shelf space is growing in value because the number of new products and the number of SKUs stocked by supermarkets has grown faster than store size and the amount of shelf space available. Increasing pressure on shelf space means that shelf space is available to manufacturers only at a premium. At the same time, there is a demand for special package sizes and variety packs, particularly by club stores, which causes increasing item proliferation and higher costs for both the processor and the trade channel.

Hill (1997) identified three consistent trends prevalent in the food business: (1) general agreement that new products are the lifeblood of the business, (2) increasing focus on managing manufacturer product lines for efficiency and cutting back on new item proliferation, and (3) increasingly stringent trade hurdle rates by retailers for new items. Suppliers who have taken ECR seriously have begun to emphasize category value as the key metric in determining the makeup of product lines and the decision to launch new products. Retailers are working with suppliers in category management partnerships to develop approaches to efficiently managing product mix, shelf space and location, promotion planning, and pricing. The key is to eliminate unproductive SKUs and emphasize product launches that build categories and create customer value.

New product hurdle rates are defined in terms of sales and profit contribution, sales per square foot, and return on inventory investment. Manufacturers attempting to launch new items must incorporate these hurdle rates into the process. Hill suggests that to be successful, manufacturers must carry out account-specific analysis to understand the existing category hurdle rates of retail customers and to prepare a proforma analysis for key customers to project the profit and productivity of each of our SKUs, including projections for the new items. Manufacturers must understand the specific impact new items will have on the total category; to do this manufacturers must understand consumer purchase dynamics in the category, including the interactions among brands and brand-switching behavior. They must be empathetic to the food retailers' perspective of efficient product assortment and efficient product introduction. Manufacturers should be able to answer for the retail customer questions that the retailer must ask such as: should we carry the new product? If so, where do we shelve it and how many facings do we give? What products lose facings? What will be the effect on the category, in terms of both sales and profits, if we cut in the new item? Carrying out this type of analysis successfully requires a commitment to true strategic partnering between manufacturer and retailer.

NEW PRODUCT SELECTION CRITERIA

Retail buyers and category managers specify a large number and wide range of criteria for evaluation new products, but the decision process should, if carried out objectively, boil down to basics. These include: projected sales volume of the new item, effect of the new item on category sales and profits, and the amount of money to be made on the sale of each unit and from stocking and advertising the item, and merchandising requirements. Individual decision criteria used by buyers are related to one or more of these basic factors. Unfortunately, both retail and manufacture executives will tell you that the one-on-one dynamic between the manufacturer sales rep and the retail buyer/merchandise manager can lead to some very subjective decisions on the part of the retailer.

Projected sales volume is a function of the several different variables and factors. These are shown in Table 17.7. Retail margins and profitability are both dependent on the following:

- gross margin for the item, both normal and during deal periods
- introductory trade incentives: off-invoice allowances, bill-back allowances, display allowances, advertising allowances, and so on
- slotting or performance-based monies available
- pallet or shipper programs offered to build in-store displays
- ability of the retailer to feature the item in a circular as a "hot deal"
- special terms available, for instance if normal terms are 2/10 net 30, offer by manufacturer of 2/40 net 41 (for example) during the introductory period

Merchandise managers and buyers want several questions answered regarding the merchandising plan for a new item. These include

- How does the item fit into the retailer's variety strategy?
- Where (in which category, where in the store) should the product be merchandised?
- If a multi-item launch, how many SKUs?
- What type of shelf placement is preferred?

TABLE 17.7. **Factors Affecting Sales Volume for a New Product.**

- What consumer need is being addressed?
- Does the new product build on an emerging and significant consumption trend?
- For a chain, in which particular stores in the trading area will this product fit best (for instance, supermarkets in college towns need to stock more snacks, health foods, and vegetarian items, while stores in areas with a heavy concentration of certain ethnicities must stock more ethnic foods)?
- Performance of the category
- Marketing research information, such as the results of concept tests, product tests, and market tests that suggest strong consumer purchase intent
- Manufacturer's track record with recent new product launches
- Projected source of new product volume and impact on the category, both volume and profitability
- Uniqueness of the item relative to items currently stocked; are there similar items already available?
- The amount of consumer value delivered by the item
- The strength of the manufacturer's advertising program to create awareness
- The strength of the manufacturer's consumer promotion program—coupons, deals, sampling, demos, etc.—to induce trial
- Whether the manufacturer will provide shippers and/or display units, which make it easy for the store to put a display on the floor
- Whether the manufacturer will sponsor some type of local event

- Will an existing item or items be deleted to make room on the shelf?
- How many facings are needed?

THE NEW PRODUCT PRESENTATION

New item presentations by manufacturers to the trade do not follow a standard format. Every retailer, every retail buyer or merchandise manager, and every category is somewhat unique. There are, however, some general guidelines for planning a sales presentation. The first issue involves which information the sales representative will provide to the buyer. Table 17.8 shows what needs to be covered in a new item sales presentation to the trade.

Many manufacturers proclaim (or at least would like to proclaim) "We don't pay slotting." The fact remains that most food retailers demand some type of financial incentive to take on a new product. This becomes the basis for negotiation of performance funds, which many manufacturers have adopted. Firms may provide a fixed percentage per case for performance funds for the retailer to put toward advertising, demos, and so on. For example, Helene Curtis set aside a significant amount of money to promote an extension of its Degree line of antiperspirants to retailers. The company was flexible with its guidelines for using the money to allow for customization of the retailers' needs. A budgeted amount was set for each retailer, followed by a

TABLE 17.8. **Recommended Content of New Item Sales Presentations.**

- Discuss the category in terms of size, growth, and how the new item will help the retailer to sell more product in the category.
- Review the manufacturer's position in the category, overall category consumption and any relevant information about subcategories, such as chocolate versus nonchocolate in the candy segment.
- Provide the consumer rationale for new item in terms of basic demographics, lifestyle, and consumption trends, and the type of opportunity that exists for the new item.
- Identify who is doing the shopping in this category, and specify their needs, wants, and unsolved problems.
- Discuss the track record of new product launches by the manufacturer, and the proven winners from that company.
- Provide results of marketing research, such as results of concept tests, packaging and product tests that justify the contention that the product will move off of the shelf.
- Use "fact-based selling" providing a thorough review of category data, especially that specific to the retail customer. Companies practicing category management must share information, helping the retailer and manufacturer to work as partners to grow the category. Provide a plan-o-gram with a recommended shelf set, and project the impact on category movement and margin for the retail customer.
- Present the consumer program, which includes advertising and promotions, provide the specifics of the advertising program: budget, weight, media mix, and schedule. Explain consumer promotion tactics including FSI coupons and other promotions, including the schedule. Discuss any retail-specific programs including in-store demo program.
- Present the trade program including pricing, terms, and margin structure, plus all monies and incentives available: discounts, allowances, free goods, and slotting fees or performance funds. When presenting introductory deals, there is a need to be specific, for example: "We are offering an off-invoice allowance of $3.40 per case during the introductory buy-in period, along with a bill-back of $1.50 per case, providing a gross margin of 26% on the intro at $2.49 suggested retail."
- Discuss possible advertising and display allowances and programs.
- Include an overview, typically in the form of a calendar with months in the columns and consumer and trade events along the rows, of all of the key dates in the launch program: start ship, advertising flights, FSIs, etc.
- Discuss product-specifics—packaging, case pack and size, shippers, display modules available.
- Provide a product fact sheet that includes information about the package, including weight, dimensions, and the UPC code; the case, including pack and size, case weight and dimensions, cube configuration, and truckload quantity in cases; and the pallet, including weight, pallet pattern—number of layers, cases per layer, cases per pallet, and pallet dimensions—length, width, height, cube, plus truckload quantity in pallets.
- Discuss merchandising recommendations for the customer—distribution, display, plan-o-gram, and a recommendation for removal of slow moving items in the category. This should include a suggested initial quantity for stores, based on anticipated movement. Look for cross-merchandising promotional opportunities in the store, for example, the manufacturer's brand merchandised with a complementary private label item.

joint decision on how the investment would be used. Typically, funds were used for distribution allowances, promotion allowances, and feature allowances. Smaller companies may not be able to duplicate large company performance funds. However, if the retail believes its customers will buy the product, most retailers will ''sacrifice'' smaller dollars to take on the product.

One final note about the sell-in. Food manufacturers sell products through various classes of trade, including supermarkets, club stores, mass merchandisers, drug stores, and convenience stores. Manufacturers must be aware of the need to tailor pack sizes and merchandising recommendations based on the class of trade. In addition, supermarkets follow different pricing and merchandising strategies, with ''high-low'' operators acting differently with regard to passing along promotional dollars in the form of lower consumer prices than ''everyday low price'' or EDLP operators. The use of performance funds allows manufacturers to tailor their promotions to the specific needs of the customer. The risk lies in retailers ''knowing'' that someone else is getting a better deal.

THE LAUNCH

The launch involves executing all of the activities involved in manufacturing, distributing and selling the new product to both the trade and the final consumer. Each functional area—sourcing, production, distribution, sales force activity, and marketing—must effectively and efficiently perform its designated activities; just as importantly, these activities must be properly timed and coordinated. The manufacturer must make sure that product is in the stores in the intended time frame and that advertising and promotional events are scheduled accordingly. This means that retail sell-in and distribution of the product will have to be scheduled with adequate lead-time to ensure product availability in the store. For example, Hershey tries to start new item sales presentations as much as 6 months prior to the rollout, and may wait two months or more after initial retail distribution to break advertising on a national basis. Having scheduled when product must be in the store, the start ship date can be determined, then production runs and sourcing of ingredients and supplies must be scheduled. Considerations such as any seasonal patterns of product sales, the geographic extent of the launch (crash versus rollout), and possible supply and production problems must be taken into account. We also must plan for fine-tuning and follow-up work with the trade after the initial sales presentations are completed.

TRACKING THE LAUNCH

A quick read of the new product's performance, using objective scan data from a provider such as Nielsen or IRI, is necessary soon after launch. The trade-off between early information and accurate information is important, as early buyers, especially those buying only because of novelty considerations or because of large promotional incentives, may not be fully representative of buyers who adopt the product for regular use. The tracking program must be designed to provide clear and actionable indications about both areas of success and specific problems.

Both the types and sources of data we use to track the launch are similar to those we use in test markets. We need to track consumer data such as sales volume, share and share trends, reasons for purchase, likes and dislikes, awareness, trial and repeat purchase levels, consumer attitudes, and source of volume, including the effect of the new item on the category and brand shifting patterns. We need to track wholesale and retail data such as wholesale inventory levels and sell-through, retail distribution penetration, shelf location, number of facings, and display activity. We should disaggregate statistics by market area and customer segments to determine any patterns that may exist. We should attempt to determine the portion of our initial sales that is accompanied by a coupon redemption or other promotional incentive. And if we are executing a multiple SKU launch, we need to gather this data for each individual item.

Initial consumer sales will reflect primarily early consumer trial of the new item, which is normally inflated by consumer promotion events such as FSI coupons and samples. As a result, sales velocity inevitably peaks in the first couple of months and then falls off as trial flattens. Understanding the month-by-month sales growth and decline curves is essential to projecting sales accurately (Olson, 1996). It is also critical to ascertain which sales are trial sales and which sales are due to repeat purchase. Heavy trial with low re-purchase means that sales may decline rapidly after we have tapped trial households. Sources of data include scan data, such as InfoScan from IRI; household panel data; and results of consumer surveys.

Finally, the firm needs to have some type of contingency and postlaunch plans in place for responding to the results of our launch. Accurately diagnosing the launch data is a critical step. High trial and poor repeat might indicate a strong concept and strong trial-inducing promotions but a weak product or package. Lower than expected sales may be attributable not to poor product quality but instead to poor distribution penetration. Successful companies anticipate problems that may occur, gather necessary data and analyze the source of problems that do occur, and have remedial programs in place to respond. Successful companies also have what might be termed "year 2" strategies as a follow-up to a successful launch. These might involve introduction of new flavors, rollout into new geographic territories, or taking a product from retail to foodservice distribution.

SUMMARY

This chapter covers the final stage of the new product process. We discussed the introductory marketing program, including positioning, advertising, and promotional strategy. We discussed options for market testing, which is the final type of dress rehearsal prior to launch. The sell-in process was covered, including specific criteria buyers use to evaluate new product programs and the elements of a productive sales presentation. Finally, the need to properly time, coordinate, and track the launch was discussed.

BIBLIOGRAPHY

Ailloni-Charas, D. 1997. "In-Store Sampling Accelerates the Establishment of New Products," *New Product News*, August 11, p. 15.

Boike, D. G. and J. L. Staley. 1996. "Developing a Strategy and Plan for a New Product," in *The PDMA Handbook of New Product Development*, Rosenau, M. D., A. Griffin, G. A. Castellion, and N. F. Anschuetz, eds. New York: John Wiley & Sons, pp. 139–152.

Cooper, R. G. 1993. *Winning at New Products*, 2nd Ed. Boston: Addison-Wesley.

Cooper, R. G. 1996. "New Products: What Separates the Winners from the Losers," in *The PDMA Handbook of New Product Development*, Rosenau, M. D., A. Griffin, G. A. Castellion, and N. F. Anschuetz, eds. New York: John Wiley & Sons, pp. 3–18.

Cooper, R. G. and E. J. Kleinschmidt. 1990. *New Products: The Key Factors in Success*. Chicago: American Marketing Association.

Crawford, C. M. 1997. *New Products Management*. Boston, MA: Irwin McGraw-Hill.

Findley, J. 1998. "New Product Pacesetters 98." Presented at the *Annual Convention of the Food Marketing Institute*. Information Resources, Inc.

Fuller, G. W. 1994. *New Food Product Development: From Concept to Marketplace*. Boca Raton, FL: CRC Press.

Fusaro, D. 1995. "Kraft Foods: Mainstream Muscle," *Prepared Foods*, 164(5):8–12.

Hill, J. 1997. "The New Product Hurdle That May Cost You the Race," *New Product News*, 33(8):13.

Hoban, T. J. 1998. "Improving the Success of New Product Development," *Food Technology*, 52(1):46–49.

Mathews, R. 1997. "Efficient New Product Introduction." July 1997 Supplement to *Progressive Grocer*.

Mediamark Research Inc. 1994. *Mediamark Research Bake Goods, Snacks & Desserts Report Spring 1994*. New York: Mediamark Research Inc., p. 80.

Olson, D. W. 1996. "Postlaunch Evaluation for Consumer Goods," in *The PDMA Handbook of New Product Development*, Rosenau, M. D., A. Griffin, G. A. Castellion, and N. F. Anschuetz, eds. New York: John Wiley & Sons, pp. 395–410.

Ottum, B. D. 1996. "Launching a New Consumer Product," in *The PDMA Handbook of New Product Development*, Rosenau, M. D., A. Griffin, G. A. Castellion, and N. F. Anschuetz, eds. New York: John Wiley & Sons, pp. 381–392.

Parker-Pope, T. 1998. "The Tricky Business of Rolling Out a New Toilet Paper," *Wall Street Journal*, January 12, pp. B1, B8.

Patrick, J. 1997. *How to Develop Successful New Products*. Lincolnwood, IL: NTC Business Books.

Thompson, S. 1997. "Private Label Marketers Getting Savvier to Consumer Trends," *Brandweek*, 38(44):9.

Thompson, S. 1998. "Minute Maid Aims Tangerine Blend with Calcium at OJ Resistant Kids," *Brandweek*, 38(30):4.

CHAPTER 18

Public Policy Issues

ERIC F. GREENBERG

Laws and regulations cover far more than labeling on package surfaces. Various agencies have authority to ensure food safety, an increasingly important facet of delivered product; intentional or unintentional adulteration; misbranding; misrepresentation; and good manufacturing practices in food processing operations. Hazard analysis critical control points (HACCP) programs are now mandatory in most meat plants to control food safety, and are being considered for the entire food chain.

INTRODUCTION

A study of the legal requirements, or more broadly, the public policy issues, applicable to food products can be viewed as an inquiry into (1) requirements relating to food as such and (2) requirements relating to food packaging. The universe of packaging-related requirements is broader but, as explained below, less well defined than the universe of substantive food laws and regulations. Nevertheless, because a packaged consumer food product contends with more than just food laws and regulations, some understanding of that broader realm is crucial.

First, therefore, below is an overview of the key elements of U.S. food law and regulation, and second, a discussion of the wider realm of packaging law of which food law is a component, and an overview of its other elements.

FOOD LAWS AND REGULATIONS

FDA/USDA

When it comes to summarizing food legal requirements, the primary focus will be the requirements of the U.S. Food and Drug Administration for most food products, and those of the United States Department of Agriculture for meat and poultry products. Other agencies, such as the U.S. Environmental Protection Agency, which regulates food-related pesticides, also have a hand in food regulation. Other examples are the United States Customs Service, which works closely with FDA to inspect imported foods, and the Bureau of Alcohol, Tobacco and Firearms, which regulates alcoholic beverages. In addition, many states have counterpart agencies to these federal bodies, who often cover similar ground.

Some observers consider this FDA/USDA division of labor to be an odd historical accident, and suggest that it be remedied by creating a single food regulatory body. The somewhat anomalous circumstance of having two important agencies regulating the nation's food supply, operating under different statutes and in different ways, can be mitigated when the agencies work closely together and coordinate their requirements as much as possible. Recent areas in which these two agencies have worked closely include food labeling and food safety.

Conceptually, it is best to think of those marketing food in the U.S. as operating in a heavily regulated realm. Through the prohibitions against adulteration and misbranding, federal law and regulations impose strict sanitation standards and complex labeling specifications. There are no federal requirements that food makers be registered or get pre-approval of foods, as there are for makers of drugs and medical devices. Still, there are plenty of other forms of regulatory control. For some categories of food, such as seafood and low-acid canned foods, complex quality control programs for manufacturing are mandatory.

These systems are the wave of the future, or, for some industry segments, are the main story of the present. The modern regulatory approach to controlling food safety is a framework referred to as Hazard Analysis and Critical Control Points. A HACCP system requires manufacturers to identify the likely hazards, like contamination from outside sources, or deterioration due to failure to maintain refrigeration. These "critical control points" are spots in the process where a failure could lead to a hazard, most notably, unsafe levels of dangerous microorganisms. Manufacturers establish critical limits, then, if a manufactured product falls outside the limits, the maker knows something needs correction. HACCP systems require extensive record keeping, so the history of manufacturing processes can be easily reviewed and monitored.

Adulteration

Food is one of the primary product categories under the jurisdiction of the U.S. Food and Drug Administration. It is defined as including (1) articles used for food or drink by man or other animals, (2) chewing gum, and (3) components of any of these [21 U.S.C. §321(F)].

The conceptual framework regulating foods is probably best summarized as a series of prohibitions against adulteration or misbranding. Adulteration can mean a number of different things, generally summarized as something being wrong with the product. For example, adulterated food may contain a poisonous or deleterious substance that renders it injurious to health, or it may be filthy or otherwise unsanitary, or it may not meet the standards it purports to meet.

Once a food product is in production, compliance with current Good Manufacturing Practices ("GMPs") is a central concern. GMPs are a body of requirements, set forth in general terms in regulations, that are designed to assure that food is manufactured in a consistently sanitary manner. Building such steps into the manufacturing process is considered a preferable technique to merely testing some end products for compliance with contamination or sanitation standards. The GMP concept builds preventative measures into production of every unit, instead of relying on periodic testing at the end of the line.

It is useful to think of food as a combination of food components, that is, ingredients traditionally consumed as food or drink, together with food additives. "Food additive" is a very important term of art. The concept of "food additives" is quite broad in scope. Food additives must be used in compliance with a fairly complex web of requirements, or else the food in which they are found will be considered adulterated, and subject to enforcement action.

Food additives are defined as substances added to food, or, when used as intended, are reasonably expected to become components of the foods, and that are not Generally Recognized As Safe (GRAS) by qualified scientists [21 U.S.C. §321(S)]. Color additives, which are intended to impart color to food, are not included in the definition of food additive. Food additives must be the subject of an existing FDA regulation permitting their use, or be otherwise exempted by virtue of a prior sanction.

The "reasonably expected" clause encompasses, at least theoretically, virtually all components of food packaging, since components of packaging that contact food have the potential of migrating into the food. Because the components of packaging are not directly added to food, they are referred to as "indirect food additives." Food packaging manufacturers and users must be alert to the fact that components of their packaging are candidates for treatment as food additives, and will be considered food additives unless one of the relevant exceptions applies, such as an existing approval, general recognition of safety, or a prior sanction issued by the government.

When a use of an unapproved food additive or a new use of an approved additive is contemplated, submission to FDA of a food additive petition (FAP) may be required before its use is permitted. An FAP must contain a variety of types of information about the product and its use, including its chemical identity and composition; its physical, chemical, and biological properties; specifications for use, with identification and limits on reaction by-products;

the amount proposed for use, and directions, recommendations, and suggestions for use, and labeling specimens; data establishing that the additive will have the intended effect; practical methods to determine the amount of the food additive in food; reports of investigations of the safety of the use; and a proposed tolerance (limit) for the additive, if needed, to assure safety, among other details. FDA provides scientific guidance documents to assist applicants with petition preparation, for example regarding the relevant chemistry considerations. The crux of the FAP process is always the safety of the proposed use of a material.

FDA reviews the petition, and, if approves it, regulation is added to the Code of Federal Regulations describing the new approved use. Other manufacturers can also use the material as described in the regulation. FDA has 90 days to review the FAP, plus another 90 if needed. Frequently (that is, almost always), FDA fails to meet this deadline.

In late 1997, Congress created a new, alternative food additive approval procedure for indirect food additives, now referred to as "food contact substances," such as packaging components. The FDA Modernization Act of 1997 created this new "notice" system, in which marketers submit notice of their intended new use of a food contact substance. The submission details information similar to that in a FAP. The major difference under the notice system is that the proposed use is permitted unless, within 120 days, FDA gives notice to the petitioner of questions it may have about the safety of the proposed use. If the 120 days pass without action by FDA, the use of the substance is automatically approved. This new notice system should add considerable predictability to the timing of the petition review process by FDA, which might be a boon for new food packaging research and development. (The new law also made implementation of the notice system dependent on certain FDA funding conditions in future years; these conditions are expected to be met.)

In recent years, FDA has added another ground for exemption, its Threshold of Regulation rule, which is a cutoff line for very small exposures to certain substances (21 CFR §170.39).

The new rule simplifies the approval process for food-contact articles such as packaging that migrate in very small amounts. Because food-contact articles that migrate into foods are often indirect food additives, they must be subject to an approved food additive petition, unless they are Generally Recognized as Safe or have another exemption. Under the Threshold of Regulation procedure, packagers submit a petition to the agency and get its approval, but the information that must be submitted can be much simpler, and the agency review time could be much shorter, more a matter of months than years.

One important feature of the rules is that substances being considered for exemption may not have been the subject of any toxicological testing at all. Their safety comes in the low level of their migration, which is so low that the agency is comfortable with it despite the lack of data.

Another important point is that once a substance has been exempted, other manufacturers may use the substance in accordance with the exemption. As long as the conditions of use (e.g., use levels, temperature, type of food contacted, etc.) are equivalent to those for which the exemption was issued. In this sense, the exemptions are the same as food additive approvals.

Finally, in many instances, companies will find it useful to examine the question of whether a use of a substance meeting the Threshold of Regulation criteria also qualifies for a self-determination as Generally Recognized as Safe because it is by definition an exposure FDA considers safe.

FDA makes available a list of the exemptions it grants at its Dockets Management Branch, and updated lists are available from FDA's Office of Premarket Approval, HFS-200, 200 C-Street, SW, Washington, DC 20204. FDA does not publish a description of each exemption granted in the *Federal Register.*

Misbranding

Misbranding can also mean a variety of things, best summarized as something being wrong with the product's labeling. A product is misbranded, for example, if its labeling is false or misleading in some detail or misrepresents the quantity of its contents, or fails to reveal material information, or contains some other defect.

The prohibitions against misbranding say that a label that fails in any detail to comply with FDA's labeling requirements is misbranded, and subject to enforcement action. Certain information is required to appear on most food labels: a statement of identity of the product; its net quantity of contents; the name of the manufacturer, packer, or distributor; a list of ingredients, in descending order of predominance; and nutrition labeling. Other requirements apply if the label makes specified claims about the attributes of the food.

Anything incorrect on a label makes the product misbranded; false labeling is prohibited, but so is merely misleading labeling. Failure to present information on a label that is required to be there makes the product misbranded. Misstated or omitted ingredients, overstated product attributes, false claims of 100% purity, or false claims of geographic origin ("Imported cheese") are just some of the types of label problems that would render a food misbranded.

The most significant recent development in this area was the Nutrition Labeling and Education Act of 1990, and the regulations it spawned. These are responsible most notably for the appearance of the familiar Nutrition Facts box on the labels of most packaged foods. Other highlights of the NLEA and its regulations included standardized definitions for commonly used terms such as "lite," "high," "low," and others, newly standardized food servings sizes, and new rules for the making of claims about the nutrient content of food, and so-called "health claims" making a connection between a food and a disease-related condition.

The NLEA was inspired by the recognition by both Congress and FDA of the increasingly strong evidence of the connection between diet and health. This made it more important than ever that consumers be provided with accurate and complete nutrition information on their food labels so they could use them to control their intake of certain nutrients or coordinate their overall dietary patterns.

The outlawing of misbranding can be seen as achieving a variety of public policies. It prevents fraud and deception of the public. This is really a type of theft by deception, so its prevention is obviously desirable. Standardization of food labeling also reduces confusion about the differences between different products, and makes side-by-side comparisons of foods easier. Additionally, now that the connection between diet and health is better understood, the current, more informative food labels are designed to allow consumers to make more informed food choices with specific health goals in mind, such as reducing their intake of fat, sodium, or calories, or increasing their intake of vitamins, minerals, or dietary fiber.

FDA Enforcement Powers

FDA's primary powers for enforcing its adulteration and misbranding prohibitions, including its labeling requirements, are its powers of seizure, injunction, and criminal prosecution. It initiates its enforcement mechanisms most commonly through a plant inspection. FDA has the right to inspect food plants, without advance warning, and examine only (1) pertinent equipment, (2) finished and unfinished materials, (3) containers, and (4) labeling in the plant. FDA has broader inspection powers with respect to inspections of plants where seafood is packed, or where infant formula or low acid canned foods are manufactured.

It is a violation of the law to refuse to permit an FDA inspection. Congress determined that this relatively free reign for FDA was necessary to allow it to effectively protect the public against adulterated or misbranded foods.

If an FDA inspector finds what he or she thinks are violations, FDA commonly follows up with a Warning Letter giving the company notice that violations were found. If more serious violations are discovered, product seizure, or injunction against manufacture and distribution, or criminal prosecution seeking fines against the company and fines and imprisonment against involved individuals can be sought.

Those seeking to understand the legal significance of FDA's requirements should not fail to recognize that the penalties built into the Federal Food, Drug and Cosmetic Act include court actions or criminal penalties of fines or imprisonment. More importantly, these penalties can be imposed on companies or even individuals regardless of whether the violations were committed intentionally.

The underpinning for this state of affairs is the Congressional recognition of the importance of a safe and wholesome food supply to the American public. Given the lack of control that the average consumer has over the safety of products that are packaged and distributed over long distances, Congress determined that it would place the highest possible duty of vigilance on those responsible for food in interstate commerce. If an individual shares some of the responsibility for a violative product's appearance in interstate commerce, they can be held criminally liable for the violation. Only if the person with responsibility can prove that it was objectively impossible for him or her to have done anything to prevent the violation, will there be a sufficient legal defense to the charges.

How USDA Regulates Meat and Poultry

USDA regulates meat and poultry products. The Department's Food Safety and Inspection Service carries out the legally required, mandatory inspection of meat and poultry production. Labeling and packaging of meat and poultry are also governed by USDA. Recent developments include promulgation of "safe handling" labeling requirements for retail packages of raw meat and poultry (1993), and an overhaul of the food safety standards to require implementation of Hazard Analysis and Critical Control Points ("HACCP") systems. Under these systems, hazards that could affect products are identified and evaluated, control systems to avoid those hazards are put into place, and then compliance with the control systems is carefully documented and monitored.

USDA borrows from FDA requirements with respect to the approval status of food additives, including components of packaging for meat and poultry.

USDA had long been known for its preapproval of meat and poultry labels. Effective July 1, 1996, USDA changed its long-standing practice of preapproving labels for many meat and poultry products. In fiscal 1991, USDA's Food Labeling Division processed 167,500 labels. Individual on-site USDA inspectors, called Inspectors In Charge (IICs), who visit or work full-time at plants, also have the authority to approve certain types of labeling. IICs approved another 43,000 labels in fiscal 1991.

The USDA always had a class of labels that were passed without preapproval, called generic labels. USDA has now made most of the system generic. Instead of submitting labels to a USDA headquarters office for review, or letting the individual IIC approve labels, no prior submission is needed. The individual IIC at each registered official establishment will have the responsibility to make decisions on label compliance, along with product and process compliance.

Under the old system, IICs could approve labels for single ingredient products for which no claims were made. Labels for products that contain a single ingredient, such as beef steak or chicken legs, no longer require approval unless the manufacturer makes specific claims about quality, nutrient content, etc.

Even the system for labels that must get prior approval received an overhaul. Manufacturers would only need to send USDA one set of sketches of printer's proof quality or equivalent, instead of first sending sketches and later sending final labels. Also under the new system, if a manufacturer has several locations that make the same product, it may submit only one label instead of one label for each location.

USDA tries to minimize variation in interpretation through guidance to inspectors, but some in industry worry that product will be unnecessarily detained over label issues.

FOOD SAFETY: A NEW EMPHASIS

In 1997, U.S. government emphasis on food safety was amplified. In an era when many consumers are fearful of minuscule amounts of what are, to them, mysterious and unidentified chemical residues, such as pesticides, in their foods, government has instead focused much of its attention, with a good deal of justification, on more traditional foodborne hazards. Food processing requirements and controls aimed at reducing or eliminating hazards like *Salmonella, Listeria,* botulism and *E. coli* have grown in recent years. Both USDA and FDA have expanded the applicability of the HACCP principle, requiring it for more and more categories of food. In 1997, seafood processors were, for the first time, required to impose HACCP systems.

Other aspects of the new emphasis on food safety, heralded under the heading National Food Safety Initiative, included a broad array of measures, such as these: expanding the Federal food safety surveillance system; increased inspection; and consumer education and actively involves a large number of federal agencies.

From an enforcement perspective, having foods produced under a system that requires extensive recordkeeping has the potential to ease government inspection.

Although under current law, FDA does not have the power to inspect production documents at food plants, future changes in the law requiring HACCP and expanding FDA's powers could make food plant inspections resemble drug plant inspections, where stacks of production "Batch records" are reviewed, instead of whatever equipment, product, or labeling happens to be on hand at the time of the inspection. Government controls over food manufacturing processes are increasing, and the requirements can be expected to become more detailed and stringent in the future, all aimed at increasing the safety of the U.S. food supply.

PACKAGING LAW

SUMMARIZING PACKAGING LAW

The other major category of laws applicable to food are those applicable to food packaging as such.

There are two primary obstacles facing those who seek to study the legal requirements applicable to packaging. First, the body of federal, state, and local laws and regulations applicable to packaging has not historically been referred to as "packaging law." Only recently, perhaps as an outgrowth of the increasing self-identity of the packaging industries, has demand grown for a systematic, orderly study of the legal and regulatory pressures on packaging generally.

The second obstacle is that, as a practical matter, many packages and products face private standards just as onerous, complex, and important as any government-derived legal requirement. Sophisticated operations of a variety of types are controlled by private standards, or demanded by customers or industry pressures relating to quality or other parameters.

In short, not only is the law applicable to packaging, and food packaging in particular, difficult to summarize because it hasn't been summarized very often in the past, it is also not the whole story, since private standards and requirements also affect virtually all modern operations.

Summarizing packaging law is, then, a pioneering activity. It is logical to begin with a list of the subject matter areas within packaging law. Not included are areas of law that affect packaging and its businesses, but not specifically because they are packaging; so worker safety regulations, corporate law (the rules about stock issuance, corporate organization, and so on), and even product liability to some degree are certainly substantive areas of law, but are not "packaging law" as such. These substantive areas are part of packaging law:

(1) Regulatory law, including requirements of the U.S. Food and Drug Administration, United States Department of Agriculture, Federal Trade Commission, Consumer Product Safety Commission, and others, concerning primarily labeling, but also food additives and other topics

(2) Environmental law, encompassing specific requirements of the Environmental Protection Agency and states regarding solid waste handling and disposal, laws controlling hazardous wastes, laws limiting heavy metals, and other topics

(3) Intellectual property law, the umbrella term referring to patents, trademarks, copyrights, the increasingly important doctrine of trade dress, and unfair competition.

Intuition has probably told you that this list of subject areas is incomplete as an attempted expression of the universe of packaging law. It does not account for the extraordinary breadth and depth of the packaging industries, nor of the legal requirements that fall upon them. Some of the other factors that contribute variations to specific products or classes of products are

(1) The material used to make the package
(2) The contents of the package

The list still is incomplete, for an accurate list of the elements that contribute to packaging law must include recognition of forces such as:

(1) Politics, as the vagaries of modern politics is as responsible as anything else for why specific requirements and exceptions read as they do
(2) Scientific advances, which are the mothers of new packaging law invention
(3) Federalism, the philosophy that calls for state law and policy to control unless there is a good reason for federal, nationwide action; unique, or at least varied, state legal requirements that do not match federal requirements or those of other states, are a crucial component of the packaging law landscape

TOWARD A PRACTICAL APPROACH

The question of how to think about food law and regulation in a total quality environment can be summarized in one word: early. Consistent with the total quality philosophy of minimizing defects and maximizing consistency and predictability, consideration of legal burdens should be among the earliest activities in the development of a food product. Questions such as these should be made part of the product development strategy for a new product almost before the inventor's shouts of "Eureka!" fade away:

(1) Is any aspect of the new product able to be patented? Is the packaging of a unique design such that it might be able to be patented?
(2) Are any of the ingredients in the food product potential "food additives" under the federal Food, Drug and Cosmetic Act? If so, do any approvals for its use already exist? Will they be used in accordance with existing approvals for those additives? Alternatively, are the additives Generally Recognized As Safe; are they candidates for treatment under FDA's Threshold of Regulation; or are they subject to a prior sanction permitting their use?
(3) Which FDA or USDA requirements will control the manufacturing processes employed to manufacture this product? Are they subject to HACCP?
(4) Are any of the components controlled by other, more general regulatory requirements, such as control on handling of hazardous materials, worker-protection right-to-know laws or OSHA requirements?

(5) Can the product be labeled as desired? Can the marketers' intended label claims for the virtues of the product pass muster under government requirements? If not, should changes be made to the label, or the product?

When questions like those listed above are handled as afterthoughts, companies too often find themselves quite far along in the product development process before first discovering a significant hitch in their plans. Production and marketing schedules have to be pushed back to accommodate the time it takes to examine questions of these types, and, in some cases, to obtain necessary government approvals or make necessary changes in product, labeling, or packaging to achieve compliance.

BIBLIOGRAPHY

FDA *Code of Federal Regulations*, Title 21 Food and Drugs.

FDA. *Federal Food, Drug And Cosmetic Act*, 21 U.S.C. §301 et seq.

Greenberg, Eric F. 1996. *Guide to Packaging Law, a Primer for Packaging Professionals.* Herndon, VA: Institute of Packaging Professionals.

USDA. *Federal Meat Inspection Act*, 21 U.S.C. §601, et seq.

USDA. *Poultry Products Inspection Act*, 21 U.S.C. §451, et seq.

Index

477